Phylogeny, Ecology, a

D1383723

Erratum for Brooks and McLennan PHYLOGENY, ECOLOGY, AND BEHAVIOR

When this book was printed, the last line on page 103 was inadvertently omitted. That line should read,

occur either while the hosts and associates remain in sympatry (*sympatric*

PHYLOGENY, ECOLOGY, AND BEHAVIOR

A Research Program in Comparative Biology

Daniel R. Brooks and
Deborah A. McLennan

The University of Chicago Press
Chicago and London

Daniel R. Brooks, associate professor of zoology at the University of Toronto, is the coauthor of *Evolution as Entropy: Toward a Unified Theory of Biology,* also published by the University of Chicago Press (2d ed., 1988). Deborah A. McLennan is a research associate in zoology at the University of Toronto.

The University of Chicago Press, Chicago 60637
The University of Chicago Press, Ltd., London
© 1991 by The University of Chicago
All rights reserved. Published 1991
Printed in the United States of America

00 99 98 97 96 95 94 93 92 91   5 4 3 2 1

Library of Congress Cataloging-in-Publication Data
Brooks, D. R. (Daniel R.), 1951–
    Phylogeny, ecology, and behavior : a research program in comparative biology / Daniel R. Brooks and Deborah A. McLennan.
        p.   cm.
    Includes bibliographical references and index.
    ISBN 0-226-07571-0.—ISBN 0-226-07572-9 (pbk.)
    1. Phylogeny.   2. Evolution.   3. Ecology.   4. Animal behavior.
I. McLennan, Deborah A.   II. Title.
QH367.5.B76   1991
574.5—dc20   90-11051
                                                                                            CIP

⊗The paper used in this publication meets the minimum requirements of the American National Standard for Information Sciences—Permanence of Paper for Printed Library Materials, ANSI Z39.48-1984.

Contents

Preface

Evolution has been the unifying concept of biology for more than 130 years. Productive research programs in ecology, ethology, and phylogenetics—nurtured and strengthened by a cross-fertilization of ideas—originated and developed around this concept. Paradoxically, as the data base and methodological sophistication of these programs grew, interdisciplinary communication diminished, making it increasingly more difficult to "reconstruct the (evolutionary) elephant." It is our hope that this book will demonstrate the value of reestablishing the channels of communication and the means by which these channels can be reopened. We recognize that, like everyone, our views are shaped by the systems we study, the questions that engage our interest, the methods of analysis that we use, and the biases of our teachers. We are grateful to the reviewers, official and unofficial, who so enthusiastically highlighted such unintentionally myopic portions of the original manuscript. Because of their input, coming from a variety of different levels and interests, this is a better book.

Throughout this book, we will attempt to convince you that the integration of phylogeny with ecology and behavior produces exciting new research. Our fascination with this evolutionary perspective, which we will call "historical ecology" for short, does not in any way diminish our enthusiasm for other areas of research. Because of this, we do not believe that historical ecological approaches to studying ecology and behavior are the only, most relevant, or most interesting way to address evolutionary questions. They are simply a different way. But this of itself is exciting because, to paraphrase an adaptationist argument, when different research programs converge on the same answer, we have strong evidence for the robustness of that answer. Nevertheless, we do think that there are some important aspects of evolution that must be addressed within a phylogenetic framework, and it is to those aspects that we direct our attention in this book. In some cases you will discover that the historical ecological approach produces answers that agree with our standard expectations, while in other cases, the results are unexpected. When one method of analysis draws conclusions that are at odds with an accepted theory, there are two explanations. The first is that the method of analysis is flawed. The second is that the accepted theory is incomplete. This is a book about possibilities. It is not a book about the theoretical and philosophical underpinnings of phylogenetics. So, for the examples demonstrating new twists to old questions, we will explore the second explanation. This means that we adopt, a priori, the assumption that phylogenetic systematics is the best method currently available for reconstructing phylogenetic relationships.

In the movie *How the Americans Live,* Cary Grant, playing a worldly and weary novelist, is advised by his publisher that some people will read his book "to find out what they missed; others will read it to find out what you missed." In the best of all possible worlds, we would not have missed anything; nature and science would have provided us with a profusion of unambiguous examples liberally sprinkled across all phyla. Fortunately for current and future biologists, there are gaps in the data base, so our choice of examples was constrained by the quantity and the quality of our collective knowledge. This knowledge, in turn, has been rigorously developed within a phylogenetic framework for only a few groups of organisms, most notably the parasitic flatworms, North American freshwater fishes, and North and South American reptilomorphs and amphibians. Although numerically small within the vast realm of the biological sciences, studies of these organisms have formed a strong foundation for the development of new evolutionary perspectives. We intend to build upon the strength and clarity of these examples to demonstrate the logic of a historical ecological approach and the rewards of collaborative efforts to reunify phylogeny, ecology, and behavior.

The development of this type of research, and the subsequent writing of this book, represent one of the first attempts at such a collaboration, the first author being a systematist and the second author a behavioral ecologist. Having discovered that the process was not too painful, we now offer historical ecology as a research program that permits productive communication among a variety of biologists. We hope that the ideas presented herein can help to tear down the walls between ecologists, ethologists, systematists, and evolutionary biologists and help us develop a more robust, and truly unified, theory of evolution.

We dedicate this book to the memory of five biologists who consistently advocated an integration of phylogeny, ecology, and behavior: Charles Darwin, Willi Hennig, Konrad Lorenz, Herbert Ross, and Niko Tinbergen.

DANIEL R. BROOKS
DEBORAH A. MCLENNAN

Acknowledgments

The end product of every creative enterprise represents the work, ideas, and support of more than just the authors. This book is no exception. For their dynamic discussions of the ideas presented herein, we would particularly like to thank Susan Bandoni, University of New Mexico; George Benz, Storrs, Connecticut; Arthur Boucot, Oregon State University; James Brown, University of New Mexico; Nancy Butler, University of British Columbia; Joseph Carney, University of Toronto; Joel Cracraft, University of Illinois Medical Center; Gregory Deets, Institute of Parasitology, California State University–Long Beach; Virginia Ferris, Purdue University; William Fink, University of Michigan; Darrel Frost, University of Kansas; Vicki Funk, Smithsonian Institution; Robert Gregory, University of British Columbia; Robert Geesink, Rijksherbarium, Leiden; John Gittleman, University of Tennessee; David Glen, Victoria, British Columbia; Eduardo Gudynas, Montevideo, Uruguay; Gordon Haas, University of British Columbia; Eric Hoberg, University of Prince Edward Island; Arnold Kluge, University of Michigan; George Lauder, University of California–Irvine; John Lynch, University of Nebraska–Lincoln; Cheryl Macdonald, University of British Columbia; Doyle McKey, University of Miami; J. D. McPhail, University of British Columbia; Brian Maurer, Brigham Young University; Richard Mayden, University of Alabama; Richard O'Grady, Smithsonian Institution; William Presch, California State University–Fullerton; Bruce Rannala, University of Toronto; Douglas Siegel-Causey, University of Kansas; James Smith, California State University–Fullerton; Hans-Erik Wanntorp, University of Stockholm; Peter Watts, University of British Columbia; E. O. Wiley, University of Kansas; and Rino Zandee, Institute for Theoretical Biology, University of Leiden. We would also like to thank our families, with special thanks to Ann and Rudy Koenig, Donald (F_0, F_1, and F_2), Josephine, Della, Janice, David, and Ian McLennan, Donald and Ethel Heller, Bernice Good, and Vera, Robert, and Mae Hamlyn, in Vancouver and Nanaimo, British Columbia, for their constant, if often bemused, support throughout this entire project.

Many of the empirical studies reported in this book were made possible through a continuing operating grant (no. A7696) from the Natural Sciences and Engineering Research Council (NSERC) of Canada to DRB, and grants to study neotropical stingrays from the National Geographic Society (to Thomas B. Thorson, University of Nebraska). Our thanks for institutional support, in the form of visiting professorships to DRB, from the University of Michigan (1983), University of Canterbury, Christchurch, New Zealand (1986), Smithsonian Institution (1986), Brigham Young University (1987), and the University of Miami (1988). We also deeply appreciate the opportunity to present these ideas before groups of enthusiastic students and faculty at three workshops: a Nordic Council of Ecology graduate short-course, Tovetorp, Sweden (1987); a workshop on phylogenetic systematics sponsored by the CNPq and Brazilian Academy of Sciences, Rio de Janeiro (1989); and a workshop in historical ecology held at the First Latin American Congress of Ecology, Montevideo, Uruguay (1989).

On the home front, we extend special thanks to Betty Roots, chairman, and Sherwin Desser, associate chairman, Department of Zoology, University of Toronto, to Richard Winterbottom and Robert Murphy, Royal Ontario Museum, and to the phylogeneticists, ecologists, and ethologists, students and faculty alike, at the University of Toronto and the Royal Ontario Museum who provided an atmosphere of collegiality, patience, and interest that contributed to both the quality of this project and the sanity of its authors.

Finally, our heartfelt thanks to Brent Mishler, Duke University; Peter Wainwright, University of California–Irvine; and Joseph Travis, Florida State University, the three formal reviewers of this book, for extensive, thoughtful, and constructive criticism that greatly improved the final manuscript.

PART ONE
The Basic Issues

1 Setting the Stage

Carvings and paintings on cliffs and in caves throughout the world attest to a long-standing human fascination with the relationship between organisms and their environments. This fascination has accompanied us down many cultural pathways, from the bestiaries of medieval Catholicism, through the surrealistic paintings of Rousseau, to whale songs. It has also resulted in centuries of natural history observations, the outcome of which was the accumulation of a vast, but only loosely connected, biological data base by the early nineteenth century. Darwin forged the connections with his proposal that evolution was the unifying principle in biology. His original conceptual framework included two components. First, all organisms are connected by common genealogy (Darwin 1872:346).

> The characters which naturalists consider as showing true affinity between any two or more species, are those which have been inherited from a common parent, all true classification being genealogical.

And, second, the form and function of organisms is closely tied to the environments in which they live (Darwin 1872:59).

> Slight modifications, which in any way favoured the individuals of any species, by better adapting them to their altered conditions, would tend to be preserved; and natural selection would have free scope for the work of improvement.

Over the past century, many specialized research programs have emerged from these two postulates. Every one of these budding disciplines initially incorporated both genealogical (phylogenetic) and environmental (adaptational) factors into their explanations of evolutionary change. However, the role of phylogeny has been progressively diminishing in some fields, most notably in ecology, ethology, and the physiological sciences, while in other fields, most notably systematics, the role of the environment has been virtually eliminated from evolutionary explanations. This, in turn, has led to the emergence of markedly different worldviews even within evolutionary biology. Gareth Nelson summarized these perceptual differences in a discussion at a biogeography conference at the American Museum of Natural History in 1979. He told an apocryphal story of two biologists, one an ecologist and the other a systematist, who stepped into a large room together. Suspended from the ceiling by a variety of supports were thousands of balls of many different colors and sizes. All at once the supports were cut, and all the balls dropped from the ceiling, hit the floor, and began bouncing around the room. The ecologist exclaimed, "Look at the diversity!" whereupon the systematist said, "Hmm, thirty-two feet per second per second!"

The Darwinian revolution was founded on the concept that biological diversity evolved through a combination of genealogical and environmental processes. Although in theory the majority of biologists still adhere to this proposition, in practice phylogenetic and ecological studies are often conducted quite independently. Is this a problem? In order to answer this question, let us consider the following thought experiments. Suppose we were to pick, at random, any organism from a designated tide pool and a crab from anywhere in the world. If we then asked for a list of morphological, behavioral, and ecological characteristics of the unknown organism from a given environment and of the known organism (a crab) from an undetermined habitat, we would expect that more of the predictions would be correct for the crab than for the unknown tide-pool organism. In this system, we can make better predictions by reference to genealogy than to current environments. Now, consider an alternative example. Suppose we chose, at random, a finch from a Galapagos island and attempted to predict its beak morphology based upon knowledge of its feeding ecology and its phylogenetic position. In this case, we would expect ecology to be more informative than genealogy. So it appears that Darwin's original intuition was correct: evolutionary explanations require reference both to phylogeny and to local environmental conditions. The answer to the preceding question is thus, Yes, the dissociation of phylogenetic and ecological studies is an important problem because the exclusion of either perspective will weaken our overall evolutionary explanations.

This answer leads us to two new questions: Given the conceptual framework proposed by Darwin, how did this dissociation come to be? How can communication between ecology and systematics and between behavior and systematics be reestablished? In order to answer the first question we must examine the history of the two disciplines. This in itself has formed the central theme for numerous papers, books, and book chapters (see, e.g., Kingsland 1985; McIntosh 1985, 1987; Lauder 1986; Hull 1988; McLennan, Brooks, and McPhail 1988; Burghardt and Gittleman 1990), so we will present only a brief summary of the subject in this chapter. Answering the second question requires the development of a research program that will allow us to integrate ecological, behavioral, and historical information to produce a more robust picture of evolution. We are calling this integration historical ecology, and we will dedicate the remainder of this book to delineating the conceptual, methodological, and empirical foundations of this research program.

The "Eclipse of History" in Ethology

Ethology, as a science, was founded upon a tradition of investigating behavior within an explicitly phylogenetic framework. Darwin started the ball rolling when he compared, among other things, the behavior of two species of ants within the genus *Formica* in an attempt to trace the evolution of slave making in ants. Following this example, the "founding fathers" of ethology, Oskar Heinroth and Charles O. Whitman, proposed that there were discrete behavioral patterns which, like morphological features, could be used as indicators of common ancestry. Whitman's (1899) views mirrored Darwin's: "instincts and organs are to be studied from the common viewpoint of phyletic descent." This perspective served as the focal point for a plethora of studies in the early twentieth century. Behavioral data were examined with an eye to their phylogenetic significance for birds, including anatids (ducks and their relatives: Heinroth 1911; Herrick 1911), weaver birds (Chapin 1917), cowbirds (Friedmann 1929), and birds of paradise (Stonor 1936); and for insects and spiders, including wasps of the family Vespidae (Ducke 1913), bumblebees (Plath 1934), caddisfly larvae (Milne and Milne 1939), termites (Emerson 1938), social insects in general (Wheeler 1919), and spiders (Petrunkevitch 1926). Wheeler (1928:20) reiterated Darwin's and Whitman's perspective and reaffirmed the basis of ethological studies at the time.

> Of late there has been considerable discussion . . . as to the precise relation of biology to history . . . and what most of us older investigators have long known seems now to be acceded, namely that biology in the broad sense and including anthropology and psychology is peculiar in being both a natural science and a department of history (phylogeny).

Comparative behavioral studies flourished under the direction of Konrad Lorenz and Niko Tinbergen during the 1940s and 1950s. Both of these ethologists repeatedly emphasized two distinct but related points: behavioral patterns are as useful as morphology in assessing phylogenetic relationship, and behavior does not evolve independently of phylogeny. Lorenz (1941) stated that "all forms of life are, in a way, phylogenetic attainments whose special objects would have to remain completely obscure without the knowledge of their phylogenetic development," and (1958), "every time a biologist seeks to know why an organism looks and acts as it does, he must resort to the comparative method." Tinbergen (1964) outlined the comparative method:

> The naturalist . . . must resort to other methods. His main source of inspiration is comparison. Through comparison he notices both similarities between species and differences between them. Either of these can be due to one of two sources. *Similarity* can be due to affinity, to common descent; or it can be due to convergent evolution. It is the convergences which call his attention to functional problems. . . . The *differences* between species can be due to lack of affinity, or they can be found in closely related species. The student of survival value concentrates on the latter differences, because they must be due to recent adaptive radiation.

In other words, the phylogenetic relationships among species provide the **pattern** from which explanations of **processes** responsible for behavioral evolution within species must be derived.

Although the comparative approach to studying behavioral evolution flourished during the 1950s and 1960s, skepticism mounted about Lorenz's assertation that species-specific behavioral characters were valuable systematic characters. By the centenary of the publication of Darwin's book, two widely divergent viewpoints were held.

> To assume evolutionary relationships on the basis of behavior patterns is not justifiable when such findings clearly contradict morphological considerations. The methods of morphology will therefore remain the basis for the natural system [of classification]. (Starck 1959, cited in Eibl-Eibesfeldt 1975:223)

> If there is a conflict between the evidence provided by morphological characters and that of behavior, the taxonomist is increasingly inclined to give greater weight to the ethological evidence. (Mayr 1958)

This difference in opinion was founded, in part, upon continuing unresolved debates among ethologists. Two questions recurred; first, how well can sequences of ancestral and derived traits be determined for attributes that left no fossil record, and, second, how well can similarities due to common ancestry (homology) be distinguished from similarities due to convergent evolution (analogy) (Boyden 1947; Lorenz 1950; Tinbergen 1951; Schneirla 1952; Michener 1953)? The question of homology was problematical because homologous characters were defined by their common origin and, at the same time, were used to reconstruct phylogenetic relationships. The inherent circularity in such a method bothered many biologists. Remane (1956) proposed a set of criteria for testing hypotheses of common origin (homology) without a priori reference to phylogeny. These were (1) similarity of position in an organ system, (2) special quality (e.g., commonalities in fine structure or development), and (3) continuity through intermediate forms. Although authors did not agree about the universal applicability of Remane's criteria to behavior, the majority accepted that the criterion of special quality, studied at the level of muscle contractions (fixed action patterns), was the fundamental tool for establishing behavioral homologies (Baerends 1958; Remane 1961; Wickler 1961; Albrecht 1966). Initial attempts to homologize behavior in this way were admittedly vague and simplistic when compared to the more quantitative methodology of comparative morphology, but this reflected more the youth of the discipline than a fundamental flaw in the behavioral traits themselves. Time and again, phylogenies reconstructed using behavioral characters mirrored those based solely on morphology. However, in a scathing review of the ethologists' research program, Atz (1970) made only a cursory reference to these successes when he concluded,

> The number of instances in which behavior has provided valuable clues to systematic relationships has continued to grow but it should be made clear that the establishment of detailed homologies was seldom, if ever, necessary to accomplish this. . . . Functional, and especially behavioral, characters usually do not involve demonstrable homologies, but depend instead on resemblances that may be detailed and specific but nevertheless cannot be traced, except in a general way, to a common ancestor. . . . Until the time that behavior, like more and more physiological functions, can be critically associated with structure, the application of the idea of homology to behavior is operationally unsound and fraught with danger, since the history of the study of animal behavior shows that to think of behavior *as* structure has led to the most pernicious kind of oversimplification.

This review marked the end of attempts to homologize behavior, and the beginning of the "eclipse of history" in ethology.

Lorenz (1941) cautioned, "The similarity of a series of forms even if the series structure arises ever so clearly from a separation according to characters, must not be considered as establishing a series of developmental stages." In his opinion, without reference to phylogenetic relationships, the criterion of similarity was, of itself, a dangerously misleading evolutionary marker. Unfortunately, the Gordian knot of behavioral homology drove ethologists towards a new methodology based, in direct contrast to Lorenz's warning, upon arranging behavioral characters as a "plausible series of adaptational changes that could easily follow one after the other" (Alcock 1984:432). Although intuitively pleasing, this method relies heavily on subjective, a priori assumptions concerning the temporal sequence of ethological modifications and dissociates character evolution from underlying phylogenetic relationships. This dissociation of history from behavioral evolution has had an important impact on both the nature and direction of ethological research.

The "Eclipse of History" in Ecology

Ecology is founded upon the search for an understanding of the interactions between an individual and its environment. This simple aim masks a Herculean challenge, for the term "individual" encompasses practically all biological levels, from the organism through the species to the ecosystem. The complexity of this search prompted Moore (1920), in the opening paper of the first number of *Ecology,* to call for an integration of ecology with other sciences.

> There have been three stages in the development of the biological sciences: first, a period of general work, when Darwin, Agassiz and others amassed and gave their knowledge of such natural phenomena as could be studied with the limited methods at hand; next, men specialized in different branches and gradually built up the biological sciences which we know today; and now has begun the third or synthetic stage. Since the biological field has been reconnoitred and divided into its logical parts, it becomes possible to see the interrelations and to bring these related parts more closely together. Many sciences have developed to the point where contact and cooperation with related sciences are essential to full development. Ecology is in this third stage.

Over the next thirty years, the call for integration and cooperation was answered by disciplines such as forestry and geology. Communication with systematists developed more slowly, however, and this period saw only a handful of studies exploring ecological questions within a historical framework (see, e.g., Baker 1927; Rau 1929, 1931; Parker 1930; Talbot 1934, 1945, 1948; Park 1945; Park and Frank 1948; Smith and Bragg 1949). Although numerically small, this research foreshadowed the emergence of a phylogenetically based perspective in ecology at the same time that this theme was being developed in ethology. On one side of the Atlantic, Lorenz (1941), drawing on his observations of anatid ducks, was emphasizing the importance of phylogeny to studies of behavioral evolution. On the other side of the ocean, Bragg and his co-workers were reaching a similar conclusion from their extensive studies of the ecology and natural history of toads (Bragg and Smith 1943).

> Since variations in ecological conditions (physical or biotic) markedly affect the lives of individual organisms, and through this, of species, it follows that there is a broader line between the usual

ecological emphasis upon succession of communities to the climatic or edaphic climax of a given region, on the one hand, and the taxonomic and geographic distributional emphasis of taxonomists and biogeographers on the other. The study of habits of animals, interpreted in the light of both ecology and taxonomy is, thus, an aid— indeed an absolute essential—to a complete understanding by either group of workers of the peculiar problems of either.

The next twenty-five years were characterized by two significant changes: the appearance of papers by systematists in ecological journals, echoing this sentiment of cooperation (e.g., Sabrosky 1950; Davidson 1952; Constance 1953; McMillan 1954), and a burst in the number of comparative studies (see, e.g., Pavan, Dobzhansky, and Burla 1950; Hairston 1951; Dobzhansky and da Cunha 1955; Carpenter 1956; MacArthur 1958; Kohn 1959; Cade 1963; Rand 1964; Schoener 1965, 1968; Shoener and Gorman 1968; Brown 1971; Preston 1973; Laerm 1974; Roughgarden 1974; McClure and Price 1975). The ascension of the comparative approach coincided with the appearance of the "new" evolutionary ecological perspective developed by Hutchinson and MacArthur. This research program was primarily concerned with attempting to answer the general question, Why are there so many species? and its corollary, How do these species manage to coexist? Answers to these questions had traditionally been sought within a comparative framework, an approach that was reinforced by MacArthur's (1958) statement that "ecological investigations of closely-related species then are looked upon as enumerations of the diverse ways in which the resources of a community can be partitioned." The importance of searching for evidence of competitive exclusion within a closely related group of organisms was emphasized by King (1964) in his critique of MacArthur's broken-stick model of species abundance.

As realized by Darwin (1859), the principle of competitive exclusion is most applicable to closely related sympatric species (that is, to species of high taxonomic affinity) having similar but not identical niches. This may be related to the MacArthur model since when competitive exclusion has taken place, the species of high taxonomic affinity that remain may be expected to have niches which are nonoverlapping but contiguous. Hairston (in Slobodkin, 1962) suggests that tests of these species should display better fits to the MacArthur model than do tests of all species occurring in the habitat. That these predictions are valid was first indicated by the striking fits obtained by Kohn (1959, 1960) when only members of the genus *Conus* were examined. Subsequent investigations of fresh-water fishes . . . reveal that in one collection from a single locality members of the class do not fit well, but when members of the same family are considered the fit is much better.

MacArthur set the tone for ecological studies of species coexistence and the search for correlations between changes in a species' ecology and changes in the environment. However, although evolutionary ecologists were examining experimental data within a comparative framework, few researchers were incorporating phylogenetic information into their evolutionary explanations. The difference between asking a question within a historical context and incorporating historical information into the answer is a critical and, at first, counterintuitive one. Consider the following simple example. Suppose you are interested in the question of species coexistence. As MacArthur noted, the best place to look for the factors involved in species coexistence is among sympatric populations of congeners. The assumption behind this recommendation is a historical one: members of the same genus should theoretically share a number of ecological, morphological, and behavioral characters in common because they are all descended from a common ancestor. The recognition that the genealogical relationships among species may influence the outcome of an experimental investigation is the first step in any evolutionary ecological study. Having discovered an appropriate group of sympatric congeners, you set about collecting a wealth of data concerning feeding behavior, habitat preference, and breeding cycles, in order to identify the way(s) in which the species are partitioning their environment. This second step in your study is primarily nonhistorical because it requires that you make assumptions about the evolutionary *past* of species' interactions, based upon characters and interactions observed in the *present* environment. What is missing here is information about the evolutionary origin and elaboration of the characters and of the associations themselves. So, when we talk about "incorporating phylogenetic data into an evolutionary explanation," we are referring to the combination of both *the history of the species* and *the history of the traits that characterize interactions among those species.*

By the early 1970s the recognition that a collaboration between ecologists and systematists would be mutually beneficial had progressed so far that E. O. Wilson (1971) submitted a paper about the "plight of taxonomy" as a research program to an ecological journal.

> In the fashion rankings of academic biology, substantive taxonomy long ago settled to the bottom. This must not be permitted to continue. Ecologists, now beginning to savor the windfall of popularity and growing financial support, should recognize their dependence on substantive taxonomy and special responsibility to it. Most of the central problems of ecology can be solved only by reference to the details of organic diversity. Even the most cursory ecosystem analyses have to be based on sound taxonomy. . . . It is to be hoped that ecologists, in their newly acquired influence, will accept that . . . aid to their intellectual kindred, the taxonomists, is both part of their

larger responsibility to science and in their own immediate self interest.

Nevertheless, the number of historically based studies began to decrease within the rapidly burgeoning field of ecology at about the same time that the comparative method was waning in ethology (but see, e.g., Fraser 1976; Huey and Webster 1976; Huey and Pianka 1977; May 1977; Pitelka 1977; Hubbell and Johnson 1978). This trend continued through the 1980s (but see, e.g., Hixon 1980; Hairston 1981; Keen 1982; Horton and Wise 1983; Kingsolver 1983; Davidson and Morton 1984; Schroder 1987; Armbruster 1988) and, paradoxically, paralleled an increase in the number of studies concerned with examining ecology within a specifically evolutionary context. We cannot offer any particular explanation for this observation. Part of the answer may stem from the perception that historical effects would confound ecological predictions. Although "historical effects" were often considered to be a within-species, genealogical phenomenon, there was a tendency to extrapolate from the genetic to the phylogenetic level. As a consequence, evolutionary ecologists were advised to adopt a "Goldilocks" approach (Wiens 1984) in which the scope of their studies would not be too large for interesting patterns to be found and not so small that similar patterns would be due strictly to historical effects. Part of the answer may simply be that the theoretical foundations for ecology were well developed by the 1970s, so more ecologists turned their attention towards a rigorous examination of the assumptions underlying those theories. Although painstaking, there is no other way to test assumptions than by a careful species-by-species examination. And still another part of the answer may lie in an observation by Stenseth (1984) that ecology was once the "handmaiden" of taxonomy, but became a science on its own in the 1960s. If many ecologists felt that they had been under the yoke of taxonomy, perhaps the break had more to do with desires for individual identities. If so, it would be unfortunate, because many systematists have felt the same way about the subordination of their discipline within ecology. All this really tells us is that perception of subordination has been based on mutual misapprehensions.

Whatever the reason, Ricklefs (1987) suggested that this "eclipse of history" had a profound and adverse effect on the field of community ecology. He argued that community ecology has relied mostly on local-process theories for explanations of patterns that are strongly influenced by regional processes. Local explanations rely on the action of competition, predation, and disease to explain patterns of species diversity in small areas, from hectares to square kilometers. According to this perspective, the community is maintained at a saturated equilibrium by biotic interactions. However, independent lines of evidence from different communities suggest that regional diversity plays a strong role in structuring local communities. For example, the observations that (1) there are four to five times more mangrove species in Malaysia than Costa Rica and four times more chaparral plant species in Israel than California, (2) the number of cynipine wasps on a species of California oak is strongly related to the total number of cynipines recorded from the whole range of the oak species, and (3) local species richness in Caribbean birds is strongly related to total regional bird diversity, cannot be explained solely by the assumption of local, saturated equilibria—otherwise similar states would be attained in systems exposed to similar environmental conditions.

Ricklefs concluded that "the responsiveness of the equilibrium diversity of a locality to regional processes and historical circumstances argues that co-evolved interrelations among component species do not buffer community structure against externally imposed change. Accordingly, the function of a system, including its stability, does not strongly depend on its diversity." He then pointed out just how high the stakes are in our attempt to understand the origin and maintenance of biological diversity: "The threat of habitat destruction and pollution derives primarily from direct impacts rather than from loss of system stability after depauperization. However, to the extent that local communities depend on regional processes, reduction and fragmentation of habitat area will initiate a decline in both regional and local diversity to a local equilibrum, from which there can be no recovery." Ricklefs recognized the need for alternative explanations in community ecology. He also recognized the potential benefits of historical approaches, explicitly including the perspective on phylogenetic history provided by new approaches in systematics, to provide those alternatives.

Brown and Maurer (1989) reinforced this conclusion with their suggestion that general statistical regularities in ecological associations occur on much larger spatial scales than previously considered. They proposed a research field, called macroecology, in which the emphasis is on large-, rather than small-, scale studies. Like Ricklefs, they recognized that enlarging the spatial scale of evolutionary ecological studies would increase the amount of phylogenetic influence in the systems under investigation. Given the existence of these effects, then, Brown and Maurer called for ways to incorporate them into the explanatory framework of macroecology. Fortunately, this call coincides with the development of systematic methodologies that will allow us to investigate changes in behavioral and ecological characters within a phylogenetic framework.

A Revolution in Systematics

While evolutionary ecology and ethology were experiencing a surge of interest in the comparative approach, the attention of systematists was being focussed in the opposite direction. The "new systematics," prompted by the successes of the neo-Darwinian program, emphasized studies of population variation and downplayed phylogenies. The reasons for this shift in perspective were straightforward: systematists shared the general concern that phylogenies could not be reconstructed in a noncircular manner, and evolutionary biology in general was heavily influenced by the quantum leaps occurring in population genetics. Under the influence of theoreticians such as Fisher, Haldane, Wright, and Dobzhansky, researchers sought the golden fleece of evolution in a new arena: the changes in gene frequencies within and among populations under different environmental conditions.

The "new systematics" began its reign in the 1940s. By the late 1950s and early 1960s, systematic biology experienced another revolutionary change, triggered as a reaction against a perceived lack of repeatable methodology and quantitative rigor in the discipline. Some theorists thought that these problems were inherent in any attempt to reconstruct phylogeny, and suggested evolution-free systematics (Sokal and Sneath 1963). Other believed that there could be more rigor in the evolutionary approach to systematics. These researchers, however, were faced with solving three long-standing and thorny problems: homology, levels of generalities in similarities, and characterizing useful traits. As previously discussed, Remane proposed a set of criteria for testing hypotheses of common origin (homology) without a priori reference to phylogeny. These criteria work well for establishing that some traits that appear to be "the same" are, or are not, "the same." However, certain traits that are homologous under Remane's criteria could conceivably be nonhomologous evolutionarily. This would occur, for example, if two species showing the same ancestral polymorphism experienced similar selection pressures leading to fixation of the same trait. Because the fixed trait arose more than once evolutionarily, its various manifestations among different species are not evolutionary homologues. What was needed, then, was a homology criterion that would allow workers to recognize evolutionary sequences of ancestral-to-derived traits (levels of generality) that would not be circular.

The "evolutionary homology criterion" (see Wiley 1981) that has emerged in systematics is based on the assumption that (evolutionarily) homologous traits all covary with phylogeny (since they are products of a single evolutionary history), whereas nonhomologous traits do not covary with phylogeny. To implement this criterion, systematists needed a method for reconstructing phylogeny independent of assumptions of phylogenetic history. Taxa could not be grouped according to overall similarity, because similarity embodies three different phenomena. First, there is similarity in **general homologous traits** (e.g., humans, gorillas, and elephants all have vertebrae, and vertebrae appear to have evolved only once, but the presence of vertebrae does not help determine that humans and gorillas are more closely related to each other than either is to the elephant). Second, there is similarity due to convergent and parallel evolution (jointly termed **homoplasy**), which conflicts with phylogenetic relationships. And third, there is similarity in **special homologous traits** (e.g., birds and crocodilians have submandibular fenestrae, a trait found in no other vertebrates), which is evidence of phylogenetic relationships. Given this, two problems must be solved: distinguishing general from special traits and distinguishing homology from homoplasy. A solution to these problems was provided by the German entomologist Willi Hennig (1950, 1966).

Hennig suggested that homology should be assumed whenever possible by applying criteria such as Remane's. General homology could then be distinguished from special homology by using what is now called the "outgroup criterion" (Wiley 1981). Briefly, the outgroup criterion states that any trait found in one or more members of a study group that is also found in species outside the study group is a general trait. Hence, the presence of vertebrae in mammals is a general trait because there are nonmammals that also have vertebrae. Those traits occurring only within the study group are special similarities. The members of the study group are then clustered according to their special shared traits. If there are conflicting groupings, it means that some traits assumed to be evolutionary homologies on the basis of nonphylogenetic criteria are actually homoplasies. Because all evolutionary homologies covary, and homoplasies do not covary, the pattern of relationships supported by the largest subset of special similarities is adopted as the working hypothesis of phylogenetic relationships. As more and more traits are sampled, there will be progressively more support for a single phylogenetic pattern. Traits that are inconsistent with this pattern are interpreted, post hoc, as homoplasies. Thus, the phylogenetic systematic method works in the following way: (1) assume homology, a priori, whenever possible; (2) use outgroup comparisons to distinguish general from special homologous traits; (3) group according to shared special homologous traits; (4) in the event of conflicting evidence, choose the phylogenetic relationships supported by the largest number of traits; (5) interpret inconsistent results, post hoc, as homoplasies. So homologies, which indicate phylogenetic relationships, are determined *without* reference to a phylogeny, while homoplasies, which are inconsistent with phylogeny, are determined as such *by* reference to the phylogeny.

The advent of phylogenetic systematics marked a return to the position advocated by Darwin in 1872: "community of descent is the hidden bond which naturalists have been unconsciously seeking, and not some unknown plan of creation, or the enunciation of general propositions and the mere putting together and separating of objects more or less alike" (346). Armed with a noncircular method for use in formulating, testing, and refining explicit hypotheses of phylogenetic relationships by the late 1960s and early 1970s, systematists were in a position to begin contributing detailed information about phylogenetic effects on evolving systems of many kinds. And indeed, a variety of applications, derived from phylogenetic analyses using morphological data (including the micromorphological data represented by biological molecules) have been suggested (e.g., Michener 1967; Ashlock 1974; Rosen 1975; Eldredge and Cracraft 1980; Brooks 1981; Lauder 1981; Nelson and Platnick 1981; Wiley 1981; Fink 1982). By this time, however, systematists had virtually abandoned ecological and behavioral data as primary indicators of phylogenetic relationships. Their apprehensions stemmed, in part, from legitimate concerns about the dynamic nature of functional, as opposed to structural, traits. After all, verbs are intuitively more labile than nouns. These apprehensions have persisted, and a vast data base of ecological and behavioral characters remains virtually unexplored by systematists. In fact, the current state of affairs is still best summarized in a paper presented by R.D. Alexander (1962) during a symposium on the usefulness of nonmorphological data in systematic studies.

> Anyone with more than a passing curiosity about the study of animal behavior soon acquires the feeling that it has been neglected too frequently in many aspects of zoology, but especially among the systematists, who have almost a priority on the comparative attitude. . . . Behavioral attributes are . . . too often at the core of diverse problems in animal evolution to allow us to get by with the vague feeling that structure and physiology can be compared but behavior cannot—that a structural description is important information but that a behavioral description is a useless anecdote.

And so today we stand at a branching point between evolutionary ecology/ ethology and systematics. The first has increasingly turned its gaze towards patterns of variation within species (microevolution), whereas the second has become preoccupied with among-species patterns (macroevolution). Strangely, this dichotomy has returned us, more than a century later, to Darwin's two theories of evolution, one emphasizing genealogy, and the other adaptation. What is strange is not that the disciplines have separated along these lines, but that they separated at all. Darwin's greatest contribution lay in his attempt to consolidate his two theories within a unified framework of "evolution," in which genealogy, or common history, explained the similarities that bound all living organisms together, and natural selection, or adaptation, explained the differences. A reunification of ecology and systematics will return us to this multidimensional view of evolution. And, as many biologists are beginning to realize, this reunification is long overdue.

The Reemergence of Macroevolution as an Evolutionary Phenomenon

The distinction between microevolution and macroevolution was first made when Goldschmidt (1940) proposed that there were two separate evolutionary mechanisms at work. He believed that microevolution encompassed processes such as natural selection and genetic mutations. These processes operated at the population level to produce differences within species. Macroevolution, on the other hand, involved the production of new species from chromosomal mutations regardless of the effects of natural selection. Goldschmidt's ideas were opposed by the major founders of the new synthesis, or neo-Darwinism. Dobzhansky (1937), Mayr (1942), and Simpson (1944) advocated the "extrapolationist view" (Eldredge 1985) in which macroevolution was seen simply as microevolution "writ large" (see also Lande 1980b; Charlesworth, Lande, and Slatkin 1982). Mayr (1942:291) summarized this position succinctly.

> There is only a difference in degree, not one of kind, between the two phenomena. They gradually merge into each other and it is only for practical reasons that they are kept separate.

In other words, although macroevolutionary patterns may exist, their existence is due to the effects of microevolutionary processes.

The ascension of phylogenetic systematics provided biologists with an additional rigorous methodology to study both the patterns and mechanisms of evolution (Eldredge and Cracraft 1980). As a consequence, a new perspective is now emerging, based on the concept that evolution results from a variety of interacting processes, termed "forces" or "constraints," operating on different temporal and spatial scales. Evolutionary processes that occur at rates fast enough to be manifested as change within a single species lineage (**within-species** patterns) are included within the domain of **microevolution.** By contrast, processes that occur at slower rates, so that their effects are manifested in **among-species** patterns, are consigned to the realm of **macroevolution.** Microevolution and macroevolution are thus considered to be parts of a more inclusive whole represented by the hierarchical nature of biological systems (e.g., Gould 1981; Salthe 1985; Eldredge 1985; Futuyma 1986; Brooks and Wiley 1988). Since macroevolution is neither autonomous from nor reducible to microevolution, robust evolutionary explanations require data from both sources. For example, because macroevolutionary processes operate so slowly, they help define the boundaries within which microevolution takes place. That is, they can affect the ways in which and the extent to which local populations respond to selection pressures. In a complementary vein, microevolutionary processes strongly affect the building blocks upon which macroevolutionary processes work. The emphasis in this new view is on holistic explanations in which the relative contributions of a number of processes can be assessed.

Despite this clarification of issues, Cracraft (1985b) noted that the term macroevolution is currently used in two different ways by evolutionary biologists. The **transformational** view emphasizes large-scale rules governing the origin of form, and large-scale adaptive changes. Researchers in this macroevolutionary realm search for major phenotypic differences among members of a group of species and its close relatives (see, e.g., Bock 1979; Stebbins and Ayala 1981; Alberch 1982; Ayala 1982a; Charlesworth, Lande, and Slatkin 1982; Gould 1982a,b; Maderson 1982; Levinton 1983; Stearns 1983). By contrast, the **taxic** view stresses changes in species richness within and among evolutionary groups of species. Researchers in this area are concerned with detecting influences on the rate of speciation and of extinction (see, e.g., Eldredge 1979; Eldredge and Cracraft 1980; Vrba 1980, 1984a,b; Cracraft 1982a,b; Fisher 1982; Padian 1982). These influences, in turn, may include adaptive processes (see, e.g., Mayr 1963; Jackson 1974; Vrba 1980, 1983; Jablonski 1982; Hansen 1983; Valentine and Jablonski 1983), so the distinction between the transformational and taxic approaches to macroevolution tends to blur upon close examination (Eldredge 1989). Nonetheless, Cracraft's description of these different perspectives on the evolution of biological diversity is an excellent starting point for understanding applications of phylogenetic systematics in evolutionary ecology. Over the next eight chapters, we hope to demonstrate that the "transformational" and "taxic" views of macroevolution are quite complementary.

Both the theoretical and empirical aspects of microevolution have been well developed over the last fifty years. This has allowed ecologists to make extensive use of microevolutionary principles in developing evolutionary ecology (see Collins 1986 for a review). By contrast, detailed investigations of macroevolutionary patterns were delayed until the advent of noncircular methods for reconstructing phylogenies. Prior to such investigations, evolutionary ecologists assumed that processes that could be detected *within* species, such as local demic equilibria and intraspecific competition, were also responsible for structuring (organizing) *among*-species interactions. However, the revolution in systematics was followed by a revolution in evolutionary theory, as some biologists began to believe that macroevolutionary patterns were not always microevolutionary patterns "writ large" (see Eldredge 1985 for a discussion). Gould (1981:170) summarized this new perspective.

We maintain that nature is organized hierarchically and that no smooth continuum leads across levels. We may attain a unified theory of process, but the processes work differently at different levels and we cannot extrapolate from one level to encompass all events at the next. I believe, in fact, that . . . speciation by splitting . . . guarantees that macroevolution must be studied at its own level. . . . If macroevolution is, as I believe, mainly a story of the differential success of certain kinds of species and, if most species change little in the phyletic mode during the course of their existence . . . then microevolutionary change within populations is not the stuff (by extrapolation) of major transitions.

As macroevolutionary theory developed, evolutionary ecologists began to search for other processes to explain observed patterns of biological diversity, and the call was raised for the "return of history" to ecological studies. Fortunately this came at a time when some ecologists and ethologists (Ridley 1983; Clutton-Brock and Harvey 1984; Dobson 1985; Lauder 1986; McLennan, Brooks, and McPhail 1988; Pagel and Harvey 1988; Wcislo 1989) and some systematists (Brooks 1980a, 1985; Wanntorp 1983; Coddington 1988; Donoghue 1989) had already begun to bridge the gap between the disciplines. The movement to reestablish the channels of communication was based, in part, upon the (re)discovery that phylogenetic, or genealogical, constraints have played an important role in shaping the patterns of biological diversity on this planet (see Lauder 1982; Dobson 1985; Ricklefs 1987; McLennan, Brooks, and McPhail 1988; Pagel and Harvey 1988; Brown and Maurer 1989; Ferris and Ferris 1989; Gittleman and Kot 1990; Wanntorp et al. 1990). These biologists believe that persistent ancestral traits constrain the scope of the adaptively possible at every point in evolution (Brooks and Wiley 1988). Some innovations that occurred in the past have been fixed, integrated into the phenotype (the "Bauplan" of Gould and Lewontin 1979), and function as constraints on the evolution of other characters today (the "phylogenetic inertia" of Cheverud, Dow, and Leutenegger 1985; the "phylogenetic constraints" of Brooks and Wiley 1988). As a consequence, evolutionary biologists are beginning to reexamine and refine the concept of organization through constraints, one of which may be the "pull of history" (see Brooks and Wiley 1986, 1988; Endler and McLellan 1988; Gould 1989 for discussions of the various types of constraints operating in evolution).

The Emergence of Historical Ecology

By the early 1970s some researchers had begun to focus their attention on macroevolutionary patterns of diversity. Ross (1972a,b) was particularly interested in explaining these patterns for a variety of groups within the most diverse taxonomic class on this planet, the insects. Based upon his discovery that approximately only one out of every thirty speciation events in these groups was correlated with some form of ecological diversification,

Ross suggested that ecological change was consistent with, but much less frequent than, phylogenetic diversification. Furthermore, since he could not uncover any predictable patterns to explain the shifts that did occur, he proposed that ecological change constituted a biological "uncertainty principle" in evolution. Ross's interpretation of the relationship between ecological/functional and phylogenetic diversification was certainly at odds with the traditional perspective of ecologically driven evolutionary change. Within a few years, Ross's insights were corroborated by other studies. Boucot (1975a,b, 1981, 1982, 1983) reported that the majority of ecological changes leaving some trace in the fossil record occurred out of time phase with periods of phylogenetic diversification. Like Ross, he concluded that ecological change lagged behind morphological and phylogenetic diversification, or "evolution takes place in an ecological vacuum" (Boucot 1983).

Brooks (1985) consolidated the research of authors such as Ross and Boucot, as well as the results from his own studies with parasitic organisms, into a discipline that he called **historical ecology.** Subsequent to this, we discovered a paper by Rymer (1979) in which the same term had been used to describe a research program that, today, is generally called paleoecology. This overlap highlights an important and often confusing aspect of the term "historical": it has been applied in at least two contexts. "Historical" *sensu* Rymer refers to the reconstruction of past environments, whereas "historical" *sensu* Brooks refers to the reconstruction of phylogenetic relationships. Although different, both are components of historical ecology, because phylogenetic hypotheses about changes in ancestral species are strengthened by information about those ancestors' environments.

Initially, historical ecology *sensu* Brooks (1985) was concerned with studying macroevolutionary components of ecological associations, such as host-parasite or herbivore-plant systems, or communities and biotas. In this book, we will expand the boundaries of historical ecology to include two general evolutionary processes, speciation and adaptation. We will explore the macroevolutionary effects of these processes in the production of both evolutionary groups of organisms and multispecies ecological associations.

We will ask two kinds of macroevolutionary questions about groups of organisms. First, how did a given species arise? In order to answer this, we must explore a variety of ways in which descendant species are produced from an ancestral species (speciation). Second, how did a given species acquire its repertoire of behavioral/ecological characters? This question moves us into the more familiar realm of the relationships between an organism and its environment (adaptation). In this case, however, these relationships will be examined within the context of phylogeny.

Answering both these questions will provide a data base for investigating the macroevolutionary components of biological diversity. According to the traditional evolutionary scenario, diversity is influenced by speciation, which provides the raw materials, and adaptation, which shapes these materials to fit the environment. However, the studies by Ross (1972a,b) and Boucot (1983) indicate that ecological change may be evolutionarily conservative. If this is true, then we must ask, What is organizing biological diversity, if not adaptation? Part of the answer to this question lies in the cohesive influences of persistent ancestral traits. For example, some adaptations that originated in the past may become fixed and inherited relatively unchanged for long periods of time. These slowly evolving traits come to characterize genealogical groups of species, or clades, and function to constrain the scope of the adaptively possible at every point in the evolution of those clades (Brundin 1972; Riedl 1978; Lauder 1982; Brooks and Wiley 1988). Since the existence of persistent ancestral traits is a reflection of common ancestry, and common ancestry is a component of the speciation process, it appears that speciation may have been a more important evolutionary influence than previously thought. If this is true, then biological diversity should be structured primarily by shared ancestral characteristics (phylogenetic patterns), and secondarily by adaptive changes (functional fit to a current environment).

Having investigated speciation and adaptation processes within a group of organisms, we will turn our attention to the effects of these processes on interactions between groups. We will ask three basic questions within the context of multispecies ecological associations: How did a set of co-occurring species come to be in the same geographic area? How did co-occurring species exhibiting strong ecological interactions come to be associated? How did the traits characterizing those interactions come to be? Answering these questions will provide a data base for investigating the macroevolutionary component of biological interactions.

So, let us apply the "theory" of historical ecology to some hypothetical, biological observations. This example will center on a large, white predator (species X) living in the arctic, and the question, Why is this species white? A traditional approach to answering this question might be: (1) observation: "white" individuals are cryptic in the arctic environment; (2) hypothesis: white is selectively advantageous (i.e., at some time in the past individuals bearing white coats gained a large enough selective advantage over nonwhite individuals to promote the spread of white throughout the population/species); (3) prediction: white individuals will be "better" at acquiring either food or mates than their nonwhite counterparts; (4) research: test the predictions with a series of studies on mate choice and foraging behavior. The success of these studies requires that "white" is sufficiently variable to be quantified and compared among individuals. If coat color is a fixed trait, however, there is no way to test the hypothesis. Additionally, even if coat color does prove to be variable, studies concentrating on only one species are investigating only one aspect of the evolution of white coats, that is, its maintenance in that species. This kind of analysis does not address the mechanisms by which a shift from the ancestral coloration to white occurred, nor does it address the question of the environmental conditions under which the trait arose and was fixed in the first place.

A historical ecological analysis would approach the problem by investigating the questions presented in table 1.1. Suppose that our investigations of the first two questions revealed (1) the ancestor of species X lived in an arctic habitat, and the speciation event producing species X and its closest relative occurred in that environment, and (2) white coat color originated in the ancestor of X. The first step in the evolution of white coat color has now been traced: white originated in an arctic environment, and its presence in species X is an ancestral legacy. Flushed with success, we turn our attention to the next problem, the potential mechanisms promoting the shift from ancestral coat color (say, brown) to white in the ancestor of X. This phase of the analysis requires that we search for correlations between white coat color and other ecological traits throughout the evolutionary history of the group of species to which X belongs. For example, one hypothesis might be that coat color in general is correlated with hunting behavior, and that white in particular is associated with hunting seals. Examining the distribution of feeding behaviors and coat color on a phylogenetic tree for this group of organisms will reveal information about the relationships between these characters

Table 1.1 Investigation of the evolutionary ecological question, Why is species X white? from a historical ecological perspective.

1. How, when, and in what environment did species X originate?
   Process: **Speciation**
   Minimum requirements for study: A phylogeny for the group including species X; geographical data.

2. When and in what environment did white arise? (i.e., is it a novel trait in species X or a legacy from its ancestor?)
   Process: **Adaptation**
   Minimum requirements for study: A phylogeny for the group including species X; paleoecological data would be useful but are not necessary for a preliminary investigation (reconstruction of ancestor's habitat).

3. Is the appearance of white correlated with a change in the interactions between the appropriate ancestor and other community members?
   Processes: **Cospeciation** and **coadaptation**
   Minimum requirements for study: Phylogenies for the group including species X and for other groups in the community; geographical, ecological, and behavioral data; paleoecological data would be useful but are not necessary for a preliminary investigation (reconstruction of ancestor's interactions).

through evolutionary time. Let us focus our attention on the ancestor of species X. There are three possible macroevolutionary patterns of the relationships between coat color and diet in this ancestor: (1) The shift to seal hunting arose before the appearance of white coloration. This indicates that brown, noncryptic individuals were capable of surviving by feeding on seals. (2) The shift to seal hunting and white coat color appeared at the same time (in the ancestor of species X). This relationship provides strong evidence that there is a causal link between color and hunting, supporting a hypothesis that white coat color conferred an adaptive advantage to its owners. This explanation could be strengthened if changes in coat color and feeding preferences covary throughout the evolutionary history of the entire group. (3) The shift to seal hunting arose after the appearance of white coloration. In this case, we are faced with the possibility that white individuals could not compete successfully with other members of the community and were forced to change their feeding habits in order to survive. So far, we have sought an explanation for the ancestral shift from brown to white coat color by examining character correlations within a clade. How can the possible influence of other community members on the evolution of this trait be investigated? We can address this problem by examining the distributions of color and feeding preferences on phylogenetic trees for other members of the community with which the ancestor of species X potentially interacted (answer to question 3). Basically, we are attempting to uncover the interactions occurring in the community at the time of the feeding shift and coat color change in that ancestor. For example, we may discover that the change in feeding preference was associated with both a change in coat color and the appearance of a new competitor in the community. This additional piece of macroevolutionary information further strengthens our adaptive hypothesis; white coat color conferred an adaptive advantage to its owners in terms of hunting ability (predator-prey interaction) and competition with other organisms (predator-predator interactions, both intra- and interspecific).

This discussion about the evolution of white coat color in the mysterious, predatory species X illustrates some of the new perspectives that will emerge from a phylogenetic (historical) comparison. Such an analysis allows us to investigate questions concerning (1) the evolutionary origin of a character and (2) the mechanisms promoting the spread of that character once it appeared in the population. This second area is of particular interest to evolutionary ecologists, because information about the associations between traits through evolutionary time, both within one group of species and among species interacting in a community, is relevant to adaptive hypotheses. Historical ecology also offers researchers a way to investigate the evolution of traits that are currently fixed in a species; in fact, the methods of phylogenetic systematics work best with such characters.

Historical ecology is primarily interested in incorporating the origins of diversity and the historical constraints on that diversity into causal evolutionary explanations in ecology and behavior. Patterns of biological diversity, in turn, are influenced by the interactions of the evolutionary processes speciation/cospeciation and adaptation/coadaptation. In the following chapters we will investigate these processes in a variety of examples ranging through plant, invertebrate, and vertebrate systems. Our choice of examples is constrained by only two considerations, the availability of explicit phylogenetic hypotheses for the study organisms and the clarity of the examples. Since historical ecology is a young discipline, such examples are limited, but every study contributes a new piece of information to the ecological puzzle. And because of its youth, historical ecology is not committed to any particular theoretical perspective, beyond the belief that evolution unifies all living organisms. So, this is truly the most exciting time in the development of any scientific discipline: the stage of discovery. This journey of discovery will lead us down many pathways and open many previously inaccessible doors, leading to a richer explanatory framework for that most important of biological processes, evolution. As Moore (1920) concluded,

> Will we be content to remain zoologists, botanists and foresters, with little understanding of one another's problems, or will we endeavor to become ecologists in the broad sense of the term? The part we will play in science depends upon our reply. . . . the future is in our own hands.

2 Tools of the Trade

In this book we will focus on the use of the comparative method to unveil patterns of biological diversity in ecological and ethological settings and to help study the mechanisms underlying these patterns. Although comparison is one of the cornerstones of biological investigation, the "comparative method" has been approached from several directions (e.g., Ridley 1983; Clutton-Brock and Harvey 1984; Bell 1989). For one group of researchers this method consists of comparing distantly related species living in a common environment and explaining any similarities as convergent adaptations to that environment. For other researchers, the comparative method embodies reconstructing the phylogenetic relationships within a group of species, then using those relationships as a template for explanations about the evolutionary origin and diversification of other characters (see also Coddington 1988). The attraction of the first approach lies in the assumption that the evolutionary history of the test groups is unimportant. However, because the decision that two groups are "distantly related" is often highly subjective, this approach may fail to distinguish similarity due to common environments from similarity due to common inheritance. The second approach will distinguish various kinds of similarity but requires a direct estimate of phylogeny. Since a robust methodology for estimating phylogeny is available in the postulates of phylogenetic systematics (Hennig 1950, 1966; see also Eldredge and Cracraft 1980; Nelson and Platnick 1981; Wiley 1981, 1986a,b,c,d,e,f), we advocate using the second form of the comparative method because of its greater explanatory scope.

Comparative biologists often use "similarity" as an indicator of "relationship." Hennig, noting that there are different kinds of similarities and relationships, emphasized that the primary goal of systematics should be the delineation of a special type of similarity (homology) that, when used to reconstruct relationships, would provide a general reference system for comparative biology. Hennig reasoned that this system should be based on reconstructing phylogenetic relationships from shared homologous traits because **all homologies covary with each other and with phylogeny.** Other types of similarity (homoplasy), although evolutionarily interesting, are not phylogenetically informative because **homoplasies need not covary with anything.** Evolution, with its underlying assumption of the preeminence of genealogical ties, has been the unifying biological principle since the emergence of Darwinian ideas, so Hennig's perspective would seem conceptually unobjectionable. However, response to this new perspective was reserved, originating, in part, from a long-standing problem with the relationship between homology and phylogeny. Specifically, if homology is both defined by phylogeny and required to reconstruct phylogeny, then a researcher needs to adopt an Orwellian doublethink strategy in order to "know the phylogeny, obtain the homologies, and build the phylogeny." This relationship makes phylogenetic reconstruction irreducibly circular.

Hennig's solution to this problem stemmed from his belief that genealogical influences in evolution are so pronounced that homologous characters will outnumber covarying homoplasious characters within any given group. He suggested that researchers begin with the assumption that all characters that conform to *nonphylogenetic criteria for homology,* such as those proposed by Remane (1956), are, in fact, evolutionarily homologous. In some cases this will lead to the incorrect, and initially undetectable, identification of homoplasious traits as homologues. When a phylogeny is reconstructed by grouping taxa according to their shared homologies, these misidentifications will be revealed because the homoplasious characters will not covary with the majority of the other characters. These traits can then be recognized, using the phylogenetic hypothesis, as homoplasies. The distinction between the non-Hennigian and Hennigian approaches is a subtle but vital one. Consider the following example. Suppose, while describing the behavior of four different avian taxa, you notice that all of the males perform the same type of mock-preen display during courtship. A non-Hennigian systematist would say, "Since this display looks the same in these four taxa *and* is performed by different members of a closely related group ('birds'), it is a homologous trait. We can use this trait to assess the phylogenetic relationships among these birds." A Hennigian systematist would say, "Since this display looks the same in these four taxa, it is the same (is homologous). We can use this trait to assess the phylogenetic relationships among these organisms." In the first case, homology is assumed because of similarity among characters, *coupled with* presumed relatedness among the taxa bearing the characters. Hence, there is an underlying assumption of prior knowledge of phylogeny. In the second case, homology is assumed solely on the basis of similarity among characters (the Wiley criterion "if it looks like a duck and quacks, it's a duck"). Hence, the approach advocated by Hennig is not circular because **homologies, which indicate phylogenetic relationships, are determined** *without* a priori reference to a phylogeny, while homoplasies, which are inconsistent with phylogeny, are determined as such *by* reference to the phylogeny.

Hennig recognized that there are three types of homologous characters: shared general characters, which identify a collection of taxa as a group; shared special characters, which indicate relationships among taxa within the group; and unique characters, which identify particular taxa within the group. In addition to these three categories of homology there is a separate category, homoplasy or "false homology," which tells us nothing about relationships among taxa. Since only shared special homologies denote particular phylogenetic relationships within a study group, characters must be assigned to one of the three homology categories before their usefulness in a phylogenetic study can be determined. Once again, the risk of circularity is high: "if two taxa are related, then a character that they share in common is a shared special homology; therefore, this character can be used to determine if the taxa are related." Hennig suggested that this determination be made by comparing the state of each character in the study group to the state of the same characters in one or more species outside the study group (outgroups). In this way, **each character is independently assigned a particular homology status (general, special, or unique) depending upon properties of species for which the phylogenetic relationships are not being assessed (the outgroups).** This "outgroup comparison" distinguishes among traits that are shared between the outgroups and at least some members of the study group (shared general traits), traits that are restricted to some members of the study group (shared special traits), and traits unique to single members of the study group (unique traits).

In the remainder of this chapter we will present a detailed discussion of the basic methods involved in phylogenetic systematics adapted largely from Wiley et al. (1990). More-advanced applications of phylogenetic systematic methodology pertinent to historical ecology will be introduced in chapters 5 and 7.

Terminology

There is a perception that researchers are required to learn an inordinate number of new and specialized words before their initiation into phylogenetics can be completed. This apprehension is based, in part, on the incorporation of numerous old terms such as monophyly, ancestor, homology, and homoplasy into the field of phylogenetic systematics. Since most evolutionary biologists are familiar with these terms, this is not really so daunting after all. There are of course some new words to learn, apomorphy and plesiomorphy being the most important and, for many researchers, the most

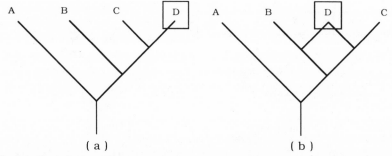

Fig. 2.1. The endpoints of speciation. *Letters* = species. (*a*) Cladogenesis: species D is produced by the division of an ancestral species into two new daughter species C and D. (*b*) Reticulate speciation: species B and C hybridize, forming species D.

perplexing. We hope that, by the end of this chapter and definitely by the end of this book, these words will no longer be fraught with such mystical connotations. There are two reasons for adopting the terminology, both the old and the new, that we will use in this book. First, the terms used in any empirical field should be as unambiguous as possible because hypotheses and explanations are framed within this terminology. Second, because evolution is a genealogical process, it is critical that systematic data be incorporated explicitly into evolutionary explanations.

Groups of Organisms

A **taxon** is a group of organisms that is given a name. The relative position (or rank) of a taxon in the Linnaean hierarchical system of classification is indicated by the use of categories (e.g., "family," "genus"). *You should not confuse the rank of a taxon with its reality as a group.* For example, the taxon Aves includes exactly the same organisms whether it is ranked as a class, an order, or a family. A **natural taxon** is a group of organisms that exists as a result of evolutionary processes. There are two kinds of natural taxa: species and monophyletic groups. A **species** is a lineage, a collection of organisms that share a unique evolutionary history and are held together by the cohesive forces of reproduction and development. Every species originates from a single **ancestral taxon** through either **cladogenesis,** the division of one ancestral species into new daughter species (fig. 2.1a), or **reticulate speciation,** the formation of a new species through the hybridization of two ancestral species (fig. 2.1b).

A **monophyletic group,** or **clade,** is a group of taxa encompassing an ancestral species and all of its descendants (fig. 2.2). Members of a monophyletic group are bound together by common ancestry relationships that they

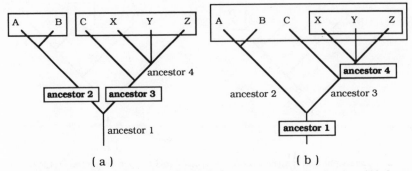

Fig. 2.2. Monophyletic groups on a phylogenetic tree. *Letters* = species. Species within boxes are part of the following monophyletic groups: (*a*) ancestor 2 and all its descendants (ancestor 2 + species A + B); ancestor 3 and all its descendants (ancestor 3 + species C + ancestor 4 + species X + Y + Z); (*b*) ancestor 1 and all its descendants (ancestors 1 + 2 + 3 + 4 + species A + B + C + X + Y + Z); ancestor 4 and all its descendants (ancestor 4 + species X + Y + Z).

do not share with any other taxa. Each monophyletic group begins as a single species, the ancestor of all subsequent members of the clade. Because of the nature of speciation, groups of species cannot give rise to other groups or to a single species. In brief, the various processes involved in speciation allow one species to give rise to another (or two species to produce a new taxon of hybrid origin), but there are no documented processes that can produce a genus from a genus ("geniation") or a family from a family ("familiation"). This occurs because species are the largest units of taxic evolution; they are real, evolutionary entities (we will discuss this in more detail in chapter 3). Higher-level categories, on the other hand, are artifacts of our propensity to classify our surroundings. As figments of our collective imaginations, supraspecific taxa have no evolutionary substance, whereas species, and the array of speciation processes that form them, lie at the very heart of "descent with modification" or evolution.

An **artificial taxon** represents an incomplete or invalid evolutionary unit. **Paraphyletic groups** are artificial because one or more descendants of an ancestor are excluded from the group, making such groupings incomplete units (fig. 2.3). For example, most researchers think that *Homo sapiens* shares a common ancestor with the African great apes (chimpanzees and gorillas). A group within a classification comprising the African great apes plus orangutans and gibbons (the Asian great apes) while excluding humans would be paraphyletic.

Polyphyletic groups are artificial because taxa that are separated from each other by more than two ancestors are placed together without the inclusion of all the descendants of that common ancestor (fig. 2.4). Since the relationship between the two taxa is so distant, this type of grouping misrepresents the evolutionary relationships that have arisen from speciation events following the divergence of the shared common ancestor, making the grouping an invalid evolutionary unit. A classic example of a polyphyletic group would be a classification that placed bats and birds in the same taxon.

The **ingroup** is any group of theoretically closely related organisms of interest to an investigator (see fig. 2.5). *Choice of the ingroup is constrained only by the rule that it must contain more than two species* because it is

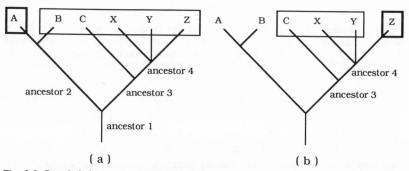

(a)   (b)

Fig. 2.3. Paraphyletic groups on a phylogenetic tree. *Letters* = species. (*a*) Two groups have been distinguished: group 1 includes ancestors 1 + 2 + 3 + 4 + species B + C + X + Y + Z; group 2 contains species A. Species A should be included in group 1 because it shares ancestors 1 and 2 with that group. (*b*) Two groups have been distinguished: Group 1 includes ancestors 3 + 4 + species C + X + Y and group 2 contains species Z. Once again, Z should be included in group 1 because it shares both ancestors with that group.

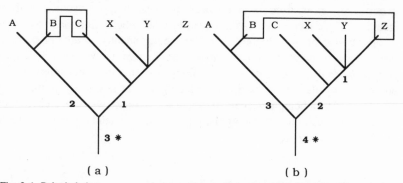

(a)   (b)

Fig. 2.4. Polyphyletic groups on a phylogenetic tree. *Letters* = species. (*a*) Species B and C are grouped together because they "look the same" even though they do not share a recent common ancestor (you have to count back through two ancestors before arriving at an ancestor, marked with an asterisk, that the taxa share). (*b*) Species B and Z are placed together; you have to count back through three ancestors before arriving at a common link between the two taxa.

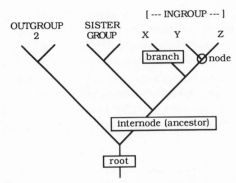

Fig. 2.5. Components of a phylogenetic tree. On this diagram, the internode is the common ancestor of the sister group and the ingroup, the node represents the speciation event that gave rise to species Y + Z, and the branch refers to species X. In all, there are six nodes, five internodes, and seven branches on this particular tree.

impossible to determine phylogenetic relationships for only two taxa. For example, an investigator studying the genealogical relationships between tigers and gorillas can only say that, by virtue of their status as biological entities, they are related. However, add lions to the picture, and we increase our degrees of freedom from none to three, because there are four possible hypotheses of relationships: they are all equally related to one another, gorillas are more closely related to tigers, gorillas are more closely related to lions, or lions are more closely related to tigers. A **sister group** is the taxon that is most closely related genealogically to the ingroup (see fig. 2.5). For example, Old World monkeys are the sister group of the great apes. The ancestor of the ingroup cannot be its sister because it is a member of the ingroup. An **outgroup** is any group used for comparative purposes in a phylogenetic analysis (see fig. 2.5). Since outgroups are used to assess the evolutionary sequence of appearance of homologous characters independently, *choice of the outgroup is constrained by the rule that it cannot contain any members that are part of the study group.* Because genealogy is so important in evolution, it is not surprising that the most important outgroup in any study is the sister group to the taxa being investigated. So, if we were interested in studying the phylogenetic relationships within the great apes, we would use the Old World monkeys as our outgroup.

Relationships of Taxa

In phylogenetic systematics the term **relationship** refers strictly to connections based on genealogy. In other systems "relationship" may be equated with "similarity" without evolutionary implications or with the implication that taxa that are more similar to each other are more closely related evolutionarily. The latter system, based on the a priori assumption that things that "look the same are the same," can lead to the recognition of polyphyletic groups because it lacks a rigorous methodology to test the validity of the assumption. *Degree of similarity is never equated with degree of relatedness in the phylogenetic system.* **Genealogical descent** at the taxic, rather than the individual, level is based on the proposition that ancestral species give rise to daughter species through speciation. A **phylogenetic tree** is a branching diagram depicting the sequence of speciation events within a group. As a graphical representation of genealogy, *a phylogenetic tree is a hypothesis of the genealogical relationships among taxa.* Since phylogenetic trees are hypotheses and not "facts," they are dependent upon both the quality and quantity of data that support them. A tree is composed of several parts (fig. 2.5): a **branch point,** or **node,** sometimes highlighted with a circle, representing an individual speciation event; a **branch,** the line connecting a branch point to a terminal taxon, representing the terminal taxon; and an **internode,** the line connecting two speciation events, representing an ancestral species. The internode at the bottom of the tree is given the special term **root.**

Classifications

A **natural classification** contains only monophyletic groups and is thus consistent with the phylogenetic (evolutionary) relationships of the organisms. In other words, the genealogical relationships depicted on the phylogenetic tree can be reconstructed from the classification scheme. An **artificial classification** contains one or more paraphyletic or polyphyletic groups, rendering it inconsistent with the phylogeny of the organisms. In such cases the phylogenetic tree cannot be wholly reconstructed from the classification scheme. An **arrangement** is a classification of a group whose phylogenetic relationships have not yet been delineated, so it can be either a natural or an artificial classification. The overwhelming majority of current classifications are arrangements, serving as necessary but interim vehicles for classifying organisms until their phylogenetic relationships have been determined. Neither artificial classifications nor arrangements have been constructed via a rigorous, phylogenetic methodology. *It is therefore inappropriate to convert such classification schemes into phylogenetic trees, because you cannot assume a priori that taxonomic relationships are consistent with phylogenetic relationships.*

The **category** of a taxon indicates its relative place (or rank) in the hierarchy of the classification. The Linnaean hierarchy is the most common taxonomic classification scheme. Within this scheme, the formation of category names occupying specific places in the hierarchy is governed by rules contained in various codes of nomenclature. It is important to remember that the rank of a taxon does not affect its status in the phylogenetic system. *All monophyletic taxa are equally important and all paraphyletic and polyphyletic taxa are equally misleading to the phylogeneticist.*

Features of Organisms

A **character** is any observable part, or attribute, of an organism. Characters have two evolutionary options: they can either remain the same and be passed on genetically from ancestor to descendant unaltered, or they can change in one species and be transmitted in the new form to its descendants. If a trait changes, it is transformed from its existing (ancestral) condition into an **evolutionary novelty.** Two characters found in different taxa can thus be assigned homologous status because they are either the same character that is found in the common ancestor or they are different characters that are genealogically linked by passing through the transformation from an ancestral condition to a novel condition. The ancestral character is termed the **plesiomorphic character** (*plesio* = close to the stem; *morpho* = shape), while the descendant character is termed the **apomorphic character** (*apo* = away from the stem; *morpho* = shape). A **homoplasy** is a character shared among taxa that is similar but does not meet either of the two preceding criteria of homology.

A **transformation series** is a collection of homologous characters: two homologous characters produce a binary transformation series, while three or more homologous characters create a multicharacter or multistate transformation series. An **ordered transformation series** is a hypothesis of the particular pathway a character travelled during its evolutionary modification(s); however, without further information we cannot tell which direction the character moved along the pathway. If a transformation series is **unordered,** there are several possible routes open to explain the character changes. Information about the evolutionary direction of character change is provided by polarization. For **polarized transformation series,** the relative apomorphic and plesiomorphic status of characters has been determined, so we have a hypothesis of which character state represents the ancestral condition and which the derived condition. And finally in **unpolarized transformation series,** the direction of character evolution remains unspecified. There are thus four possible types of character transformation series based on the amount of information available concerning the pathway and direction of evolutionary change: ordered, unpolarized (fig. 2.6a); unordered, unpolarized (fig. 2.6b); ordered, polarized (fig. 2.6c); and unordered, polarized (fig. 2.6d). Not surprisingly, ordering and polarization of multicharacter transformation series

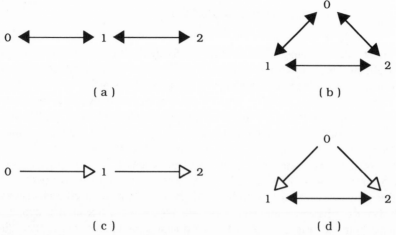

Fig. 2.6. Types of character transformation series. (*a*) Ordered, unpolarized: information about pathway; no information about direction. Character modification may be either 0 to 1 to 2, or 2 to 1 to 0. (*b*) Unordered, unpolarized: no information about either direction or pathway. (*c*) Ordered, polarized: information about both pathway and direction. (*d*) Unordered, polarized: information about direction; no information about pathway. Zero is the plesiomorphic state, but we do not know whether character modification moved from 0 to 1 to 2, or 0 to 2 to 1, or whether 1 and 2 arose independently from 0.

can become very complicated. Binary characters are much simpler because they are all automatically ordered (but not necessarily polarized!).

Character argumentation is the logical process of determining which characters in a transformation series are plesiomorphic and which are apomorphic based on a priori deductive arguments using outgroup comparison. Frequently termed "polarizing the characters," this is the pivotal process in phylogenetic systematics. **Polarity** refers to the plesiomorphic or apomorphic status of each character. **Character optimization** consists of a posteriori arguments concerning how particular characters should be polarized given a particular tree topology.

Phylogenetic systematists are quickly converting to computer-assisted analysis of their data. Such analyses require the production of a **data matrix** composed of transformation series and taxa. Each character in the matrix is assigned a numerical **code.** By convention, the code 0 is usually assigned to the plesiomorphic character, while 1 is reserved for the apomorphic character of a transformation series if the polarity of that series has been determined (hypothesized) by outgroup comparisons. If a transformation series consists of more than two characters, the situation becomes more complex, as we will discuss later in this chapter.

Hennig Argumentation: Building Trees

Phylogenetic systematists assume that all organisms, both living and extinct, occupy a unique position on one phylogenetic tree rooted at the origin of life on this planet. Since characters are features of organisms, they should have a place on this tree corresponding to the point at which they arose during evolutionary history. So ultimately we are seeking to reconstruct phylogenetic trees in which the taxa are placed in correct genealogical order and the characters are placed where they arose. For example, consider the tree for some major land-plant groups (fig. 2.7). This diagram provides us with a hypothesis of the genealogical ties between plant groups (i.e., tracheophytes and mosses are more closely related to each other than either is to hornworts). It also provides us with a hypothesis of the evolution of specific characters. Notice that characters are depicted on the phylogenetic tree at their hypothesized point of origin. This is the shorthand notation for stating that "the ancestor in which character x arose, and all of its descendants, display character x, unless it is modified again later in the evolutionary history of the group." For example, the phylogenetic hypothesis states that xylem and phloem originated in the common ancestor of mosses and tracheophytes. In other words, these characters arose in an ancestral species between the time of origin of the hornworts and the speciation event that produced the mosses and tracheophytes. Since xylem and phloem are postulated to be homologous in all plants bearing these tissues, each trait appears only once, at the level on the tree where it is thought to have arisen as an evolutionary novelty.

Now, even without a phylogenetic tree we might suspect that all plants bearing xylem and phloem shared a common, unique ancestor because both characters are morphologically and developmentally similar in these plants.

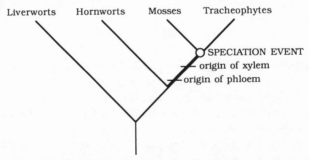

Fig. 2.7. The phylogenetic relationships of four groups of plants. Xylem and phloem originated in the ancestor of the mosses and tracheophytes. (Redrawn from Bremer 1985; cited in Wiley et al. 1990.)

In the phylogenetic system such detailed similarity is always considered as a priori evidence that the characters are homologous. This concept is so important that it has been termed

Hennig's Auxiliary Principle: *Never assume convergent or parallel evolution; always assume homology in the absence of contrary evidence.*

The principle is a powerful one. Without it we could assert that all characteristics "probably" arose multiple times by convergent evolution. Of course, just because we use Hennig's auxiliary principle we do not necessarily believe that convergences are rare or nonexistent. Convergences are facts of nature and rather common in some groups. But, in order to pinpoint convergence without invoking ad hoc assumptions, you must first have a tree, and without Hennig's auxiliary principle you will never get one. So back to the xylem and phloem: with Hennig's auxiliary principle you can deduce that plants that have xylem and phloem shared a common ancestor not shared with other plants. Of course, you don't make such a deduction in a vacuum. You "know" that "more primitive" plants lack these characters, and thus it is a good guess that the development of xylem and phloem is the derived state. This deduction is a primitive sort of outgroup comparison (we will discuss the ins and outs of outgroup comparisons in more detail later).

This principle represents the first step in Hennig argumentation. Basically, Hennig proposed that the phylogenetic puzzle should be solved by investigating individual characters and then combining the information from each character according to a set of rigorous rules.

Grouping Rule: *Only synapomorphies (shared special homologies) provide evidence of common ancestry relationships. Symplesiomorphies (shared general homologies) and convergences and parallelisms (homoplasies) are useless in this quest.*

Convergent and parallel characters (both termed homoplasies) are useless indicators of common ancestry relationships because they evolved independently in each taxon that displays them (fig. 2.8a). The futility of using plesiomorphies in an attempt to reconstruct a particular phylogeny is more problematical. After all, plesiomorphies are homologies, so why can't they be used to seek common ancestry relationships? In fact, the answer is quite straightforward: depending upon the level of your analysis, these characters can be used because, *since evolution is an ongoing process, the plesiomorphic or apomorphic status of a character is a relative condition.* All plesiomorphies begin as evolutionary novelties (autapomorphies). So, a symplesio-

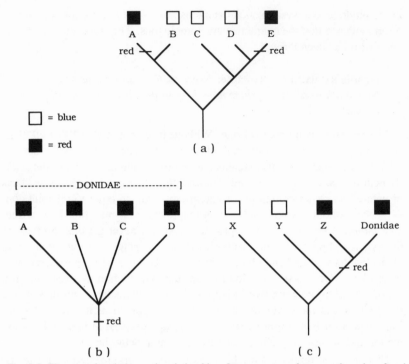

Fig. 2.8. The quest for phylogenetic relationships. *Letters* = taxa; *boxes* = the color of each taxon below. (*a*) Taxa A and E are both red, and other members of the group are blue; however, clustering A with E would be incorrect because, according to the phylogenetic tree based on numerous other characters, red arose independently in both taxa. (*b*) Since all members of the Donidae are red, that character does not tell us anything about individual relations among taxa A, B, C, and D (symplesiomorphy). (*c*) On a larger scale the Donidae share the character red with group Z (the tree is supported by many other characters); therefore red is useful in determining sister-group relationships at this level (synapomorphy).

morphy (character possessed by all members of the ingroup) cannot show common ancestry relationships within the group you are studying because it originated earlier than any of the taxa in your study group (fig. 2.8b). If you increase the temporal scale of your investigation, say by examining relationships among genera within one family instead of among species within one genus, this character will eventually prove to be useful (fig. 2.8c).

Finally, we have to consider how to combine the information from different transformation series into hypotheses of genealogical relationships. There are several ways to accomplish this. For now, we will use an old-fashioned (and perfectly valid) grouping rule that returns us to the roots of phylogenetic methods.

Inclusion/Exclusion Rule: *The information from two transformation series can be combined into a single hypothesis of relationship if that information allows for the complete inclusion or the complete exclusion of groups that were formed by the separate transformation series. Overlap of groupings leads to the generation of two or more hypotheses of relationship, since the information cannot be directly combined into a single hypothesis.*

The inclusion/exclusion rule is directly related to the concept of logical consistency. Trees that conform to the rule are logically consistent with each other, while trees that do not are logically inconsistent with each other. You can get an idea of how this rule works by studying the examples in figure 2.9. In figure 2.9a we have four characters and three potential trees. The first tree contains no character information, and since it provides no resolution of the phylogenetic relationships within the group, it is logically consistent by default with any tree that has character information. The second tree states that B, C, and D form a monophyletic group based on characters from two transformation series (1 and 2). The third tree states that C and D form a monophyletic group based on two additional characters (3 and 4). Note that the group C + D is completely included within the group B + C + D. Based on this distribution of characters, we hypothesize that the tree for these taxa includes C + D as a monophyletic group enclosed within a second monophyletic group B + C + D. Figure 2.9b shows the result of the inclusion of two monophyletic groups (A + B and C + D) within a larger monophyletic group (A–D). In this case, the groupings are consistent because A + B and C + D completely exclude each other. Finally, the example presented in figure 2.9c violates the inclusion/exclusion rule. Although both C + B and C + D can be included within the group B + C + D, the transformation series for characters 2 and 3 groups C + B and excludes taxon D, while the transformation series for characters 4 and 5 groups C + D and excludes taxon B. Thus, the phylogenetic information gleaned from these characters conflicts and the groups overlap (C is included in two different groups), producing two equally parsimonious trees that are locally inconsistent with each other.

The relationships of ABCidae

1. Transformation series (TS) 1 is composed of characters in the first column of the data matrix (table 2.1). Recall that plesiomorphies are coded as zero, synapomorphies are coded as one, and relationships among taxa are reconstructed based on shared derived traits or synapomorphies (the grouping rule). Given this, we can draw a tree with the groupings implied by the syn-

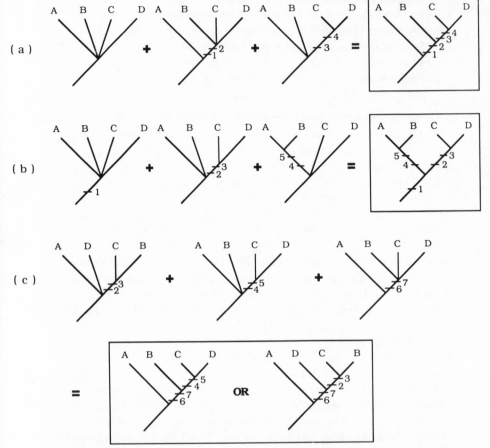

Fig. 2.9. The use of the inclusion/exclusion rule for combining information from different character transformation series into trees. (*a*) The group BCD completely includes the group CD. (*b*) The group ABCD completely includes the groups AB and CD, and, within this, the group CD completely excludes the group AB. (*c*) The group BCD completely includes the groups CD and CB, but groups CD and CB overlap because they both contain C. This results in the production of two different trees.

apomorphy found in the first transformation series (fig. 2.10a). We can then repeat the process for TS 2 (fig. 2.10b). Both trees, based on the distributions of two different characters, imply the same groupings; therefore, we can say that the trees are topologically identical, or isomorphic. Combining trees a and b according to the inclusion/exclusion rule produces the tree depicted in figure 2.10c (i.e., both characters support the group ABC to the complete exclusion of X). We can calculate a tree length for tree c by simply adding

Table 2.1   Data matrix for determining the relationships among taxa A, B, and C.

	Character Transformation Series						
Taxon	1	2	3	4	5	6	7
X (outgroup)	0	0	0	0	0	0	0
A	**1**	**1**	0	0	0	0	0
B	**1**	**1**	**1**	**1**	0	0	0
C	**1**	**1**	**1**	**1**	1	1	1

Notes: The matrix is composed of seven character transformation series and four taxa, the outgroup X, and the ingroup A + B + C. Synapomorphies are in bold type.

the number of synapomorphies that occur on it. In this case, the tree length is two steps.

2. Now, repeat this procedure for TS 3 and 4. Inspection of the data matrix reveals that the synapomorphies for these characters have identical distributions, implying that B and C form a monophyletic group (fig. 2.11a and b). If we combine the information from both characters, the results should look like the tree in figure 2.11c. This tree is also two steps long.

3. Only taxon C has the apomorphies listed in TS 5, 6, and 7. Apomorphic characters that are unique to one taxon are termed **autapomorphies.** Although they can tell us nothing about relationships among different taxa (fig. 2.12), such characters are useful diagnostic traits for identifying a particular taxon. For example, if we were to collect individuals displaying the autapomorphic state for characters 5, 6, and 7 (denoted by a one in the data matrix), we would assign those individuals to taxon C. On the other hand, collecting organisms bearing the synapomorphic condition for character 4 (also denoted by a one in the data matrix) only tells us that they are members of either taxon B or taxon C. Autapomorphies also count when figuring tree length, so the length of this tree is three steps.

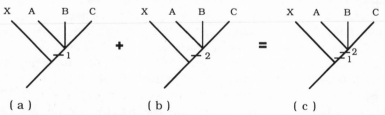

(a)   (b)   (c)

Fig. 2.10. Trees for the ABCidae, based on characters 1 and 2. (a) Tree produced by applying the grouping rule to character transformation series 1. (b) Tree produced by applying the grouping rule to character transformation series 2. (c) Tree produced by applying the inclusion/exclusion rule to the information provided by both characters.

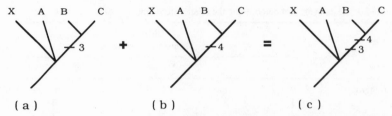

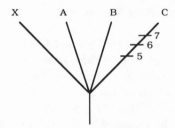

Fig. 2.11. Trees for the ABCidae, based on characters 3 and 4. (*a*) Tree produced by applying the grouping rule to character transformation series 3. (*b*) Tree produced by applying the grouping rule to character transformation series 4. (*c*) Tree produced by applying the inclusion/exclusion rule to the information provided by both characters.

Fig. 2.12. Tree for the ABCidae, based on characters 5, 6, and 7. Since autapomorphies are not useful for grouping, this tree shows no resolution of relationships among the taxa.

4. We now have three different tree topologies (figs. 2.10c, 2.11c, and 2.12). If we examine these trees more closely, we discover that, although they are topologically different, they do not contain any conflicting information. For example, since TS 5–7 only imply that C is different from the other three taxa, this tree (fig. 2.12) does not conflict with the other two trees. Further, the distributions of TS 1 and 2 do not conflict with the distributions of TS 3 and 4 because TS 1 and 2 imply that A, B, and C form a monophyletic group, while TS 3 and 4 imply that B and C form a monophyletic group without saying anything about the relationships of A or the outgroup, X. Trees that contain different but mutually agreeable groupings are **logically compatible** or **fully congruent.** They can be combined without changing any hypothesis of homology, and when combined the length of the resulting tree is the sum of the lengths of each subtree. For example, all of the information in the data matrix can be combined to produce one tree (fig. 2.13) with a length of seven steps, exactly the number of subtree steps (2 + 2 + 3). This example produces a pattern similar to the one depicted in figure 2.9a: the group ABC completely includes the group BC and completely excludes the taxon X.

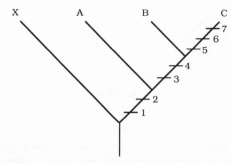

Fig. 2.13. Combining all the information in the data matrix (table 2.1) produces one hypothesis (tree) of the phylogenetic relationships within the ABCidae. This is the best estimate of the relationships, based on the available data. The tree proposes that taxa B and C are sister groups bound together by the possession of synapomorphies for characters 3 and 4, and that ABC is a monophyletic group based on the common shared characters 1 and 2.

The Relationships within the RSTidae

1. The data matrix for TS 1 and TS 2 (table 2.2) implies that R, S, and T form a monophyletic group (fig. 2.14).

2. TS 3 and TS 4 imply that S and T form a monophyletic group (fig. 2.15).

3. TS 5, 6, and 7 imply that S and R form a monophyletic group (fig. 2.16).

4. At this point you should suspect that something has gone wrong. TS 3 and 4 imply a monophyletic group that includes S and T but excludes R, while TS 5–7 imply a monophyletic group that includes R and S but excludes T. There must be a mistake, since we have violated the inclusion/exclusion rule. In such a situation we invoke the first principle of phylogenetic analysis: **there is only one true phylogeny.** Thus, one or more of our groupings must

Table 2.2 Data matrix for determining the relationships among taxa R, S, and T.

	Character Transformation Series						
Taxon	1	2	3	4	5	6	7
X (outgroup)	0	0	0	0	0	0	0
R	**1**	**1**	0	0	**1**	**1**	**1**
S	**1**	**1**	**1**	**1**	**1**	**1**	**1**
T	**1**	**1**	**1**	**1**	0	0	0

Notes: The matrix is composed of seven character transformation series and four taxa, the outgroup X, and the ingroup R + S + T. Synapomorphies are in bold type.

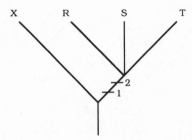

Fig. 2.14. Tree for the RSTidae, based on characters 1 and 2. This tree was produced by applying the grouping rule to character transformation series 1 and 2, then combining this information via the inclusion/exclusion rule.

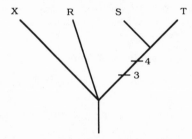

Fig. 2.15. Tree for the RSTidae, based on characters 3 and 4. This tree was produced by applying the grouping rule to character transformation series 3 and 4, then combining this information via the inclusion/exclusion rule.

be wrong. Fortunately the auxiliary principle keeps us going until we can *demonstrate* which of the groupings is incorrect. We are now faced with the problem of trying to differentiate between two logically incompatible trees (fig. 2.17). Note that there is some congruence between the trees based on their possession of the apomorphies from the first two TS (characters 1 and 2).

5. You have probably guessed by now that each of the trees in figure 2.17 is incomplete. Tree a lacks the transformation series 3 and 4, while tree b lacks the transformation series 5, 6, and 7. **Leaving characters out of an analysis is not acceptable.** In fact, eliminating characters that do not "fit" your hypothesis ranks among the top three heinous "crimes against phylogenetics" (the other two being grouping by symplesiomorphies and equating taxonomy with phylogeny). Adding the missing characters into both trees requires that we postulate that some of the evolutionary changes within this group are due to homoplasy (fig. 2.18). There are basically two types of homoplasy: a character may have risen independently more than one time (convergent or parallel character evolution), or there might be a reversal to

the "plesiomorphic" condition. We must consider both types of homoplasy in this example. Recall that our first tree (fig. 2.17a) neglected to include characters 3 and 4. There are potentially two ways to portray the distribution of these characters on the tree: either 3 and 4 arose independently in taxa S and T (fig. 2.18a), or 3 and 4 arose in the common ancestor of the group RST and were subsequently "lost" in taxon R (reversal to the ancestral [plesiomorphic] character conditions, fig. 2.18b). Examination of the distributions of characters 5, 6, and 7, which are missing on the second tree (fig. 2.17b), produces a similar pattern of homoplasy: either 5, 6, and 7 arose independently in taxa R and S (fig. 2.18c), or 5, 6, and 7 arose in the common ancestor of the group RST and taxon T subsequently reverted to the ancestral (plesiomorphic) character conditions (fig. 2.18d).

6. The question now becomes, Which of these trees should we accept?

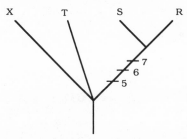

Fig. 2.16. Tree for the RSTidae, based on characters 5, 6, and 7. This tree was produced by applying the grouping rule to character transformation series 5, 6, and 7, then combining this information via the inclusion/exclusion rule.

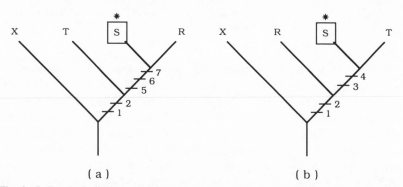

Fig. 2.17. Two logically incompatible trees produced from the information in the data matrix (table 2.2). Taxon S (marked with an asterisk) is the problem: characters 5, 6, and 7 place it with R, while characters 3 and 4 group it with T. Both trees cluster RST together based on possession of the apomorphic form of characters 1 and 2.

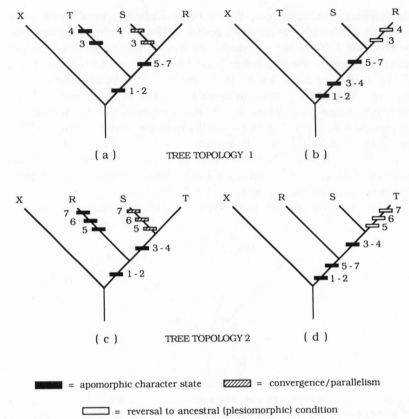

Fig. 2.18. Alternate hypotheses for the relationships of R, S, and T due to homoplasious characters. (*a*) Convergent evolution of 3 and 4 in taxa S and T. (*b*) Characters 3 and 4 revert to the plesiomorphic condition in taxon R. (*c*) Convergent evolution of 5, 6, and 7 in taxa S and R. (*d*) Characters 5, 6, and 7 revert to the plesiomorphic condition in taxon T.

That turns out to be a rather complicated question. If we adhere to the auxiliary principle, we should strive for a tree that includes the greatest number of homologies and the fewest number of homoplasies. Although these qualities are usually consistent with each other (i.e., the tree with the greatest number of synapomorphies is also the tree with the fewest number of homoplasies), you can find exceptions. Fortunately, numbers of homologies and homoplasies are related to tree length. Before we begin counting steps, notice that trees a and b (fig. 2.18) have the same topology; in fact, *these trees are the same hypothesis (tree) of phylogenetic relationships among the RSTid taxa even though they are based on different hypotheses of character change. This occurs because the topology of a tree is determined by synapomorphic relationships, not by the distributions of homoplasies.* The same is true for trees c and d. So, as previously shown (fig. 2.17), we really have only two phylogenetic trees for this group. When you count the number of steps on each tree you discover that tree type 1 (fig. 2.18a and b) has nine steps while tree type 2 (fig. 2.18c and d) has ten steps. We accept tree 1 as the best estimate of phylogeny because it has a shorter length and thus incorporates the greatest number of homology statements and the fewest number of homoplasy statements for the data set. *Note that such statements are relative only to trees derived from the same data set.* The auxiliary principle coupled with the principle that there is only one phylogeny of life carried us to this point. Methodologically, we have employed the **principle of parsimony.** In the phylogenetic system the principle of parsimony is nearly synonymous with the auxiliary principle. At the moment, we cannot choose between the different sequences of character evolution postulated by trees a and b because both trees are equally parsimonious (the same length). This requires the use of criteria other than tree length. The examples we present in subsequent chapters of this book will not require such assessments. For those who are interested in these matters, we refer you to Wiley et al. (1990).

7. Finally, we can evaluate the performance of each character originally coded as a synapomorphy by calculating a **consistency index** (CI) for it. The CI of a character is simply the reciprocal of the number of times that character appears on the tree; therefore, true homologues (real synapomorphies) have a CI of 1.0. The CI (usually reported as a percentage) is a favorite summary "statistic" in computer programs such as PAUP and MacClade, so it is worthwhile practicing some hand calculations to remove some of the computer mystique! For example, in our most parsimonious tree (topology 1 in fig. 2.18), the apomorphy coded one in TS 3 appears twice on the tree, so CI = 1/2 = 50%. Based on this CI, we can see that this character is not really a synapomorphy. Of course, our best estimate of a true homologue is an a posteriori test because we can only calculate the CI after we have determined our best estimate of common ancestry relationships. There is no way to know in advance that a particular derived similarity will turn up with a CI less than 1. The interesting point about calculating consistency indices for individual characters is that, if you have the tree before you, it is completely unnecessary. Examination of the tree will tell you everything you need to know about the number of times a particular character has appeared during the evolution of your group. This calculation is simply a numerical convenience. More important to considerations of choices among multiple trees is the CI for the entire tree (Kluge and Farris 1969). Basically, this is a goodness-of-fit measure designed to indicate the degree of support for a particular tree by a

particular data set. The CI ranges from 0 to 100%, with a value of 100% indicating that all characters support the tree (no homoplasy). Once again, the calculation is very simple.

$$CI = \frac{\text{number of characters in the data matrix}}{\text{total number of characters on the tree}} \times 100$$

So, for the trees shown in figure 2.18, the consistency indices would be 7/9 = 77.8% (fig. 2.18a and b) and 7/10 = 70% (fig. 2.18c and d). For a given data set, trees that have the same length have the same CI. Tree length gives an indication of the quantity of character support, whereas the CI gives an indication of the quality of character support for a given tree.

Hennig (1966) and Brundin (1966) characterized the essence of phylogenetic analysis as the "search for the sister group." They recognized that if you could find the closest relative or close relatives of the group you are working on, you have the basic tools for deciding which characters are apomorphic and which are plesiomorphic in a transformation series. The argument goes something like this: You discover that members of a group have two different but homologous courtship characters, "zigzag dance" and "pummel dance." As a phylogeneticist you realize that one of these characters (the apomorphic one) might provide information about relationships within your study group, but that both cannot be equally informative because *you cannot group taxa based on plesiomorphic characters* (recall fig. 2.8b). If you find zigzagging in the sister group or in closely related groups (outgroups) of the taxon you are studying, then it is fairly clear that this dance type is older than pummeling; so zigzag dance must be the plesiomorphic character in the transformation series. The characteristics of members of related groups are thus vital components to decisions regarding the polarity of characters within the study group. The simplest rule for determining polarity is the

Relative Apomorphy Rule (outgroup comparison): *Homologous characters found in the members of a monophyletic group and in the sister group are plesiomorphic, while homologous characters found only in the ingroup are apomorphic.*

Actual polarity decisions can be more complicated than our simple example. For example, what if (1) we don't know the exact sister group but have only an array of possible sister groups, (2) the sister group also has both characters, (3) either the ingroup or the sister group is not monophyletic, or (4) zigzagging evolved in the sister group independently? Answers to these questions depend on our ability to argue character polarities using some formal rules. We think the best discussion of these rules was published by Maddison, Donoghue, and Maddison (1984), although the issues have been discussed widely (see, e.g., Ross 1974; Crisci and Stuessy 1980; de Jong 1980; Stevens

1980; Watrous and Wheeler 1981; Wiley 1981, 1986a; Farris 1982; Patterson 1982; Donoghue and Cantino 1984; de Queiroz 1985). We will present the case developed by Maddison, Donoghue, and Maddison (1984) for groups in which sister-group relations have already been determined. Before proceeding with this discussion, however, we need to add some more terms to our phylogenetic vocabulary.

The ingroup is the group on which we are working. For purposes of character polarization arguments, the ancestor of the ingroup is depicted as an **ingroup node** (fig. 2.19a; note: the ingroup node = the ingroup, or ancestral, internode). *We are basically on a quest to determine what the character looked like in this ancestor because it represents the plesiomorphic (or ancestral) condition of the character for the ingroup.* The **outgroup node** is the node immediately below the ingroup node (fig. 2.19a). Characters are placed where taxa are usually labeled. Letters are used purely as a heuristic device to avoid connotations of "primitive" and "advanced." The ingroup is indicated by a polytomy since we presume that the relationships among these taxa are

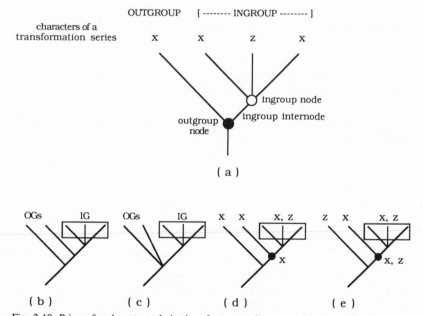

Fig. 2.19. Primer for character polarization. *Lowercase letters* = characters. Members of the ingroup (*IG*) are enclosed in a box. (*a*) Examples of some general terms used in character polarization. (*b*) Outgroup (*OG*) relationships are known. (*c*) Outgroup relationships are unknown. (*d*) State of the character at the outgroup node is known. (*e*) State of the character at the outgroup node is unknown.

Table 2.3 Matrix for characters found within the ingroup (Annidae), its sister group (the monophyletic group P + Q + R), and other outgroups.

Character TS	M	N	O	P	Q	R	Annidae
				Taxon			
1	b	a	a	b	b	a	a, b
2	b	b	a	b	b	a	a, b
3	a	b	b	b	b	a	a, b
4	a	a, b	a	b	b	a	a, b

Note: In this matrix the taxa are in columns and characters are in rows.

unknown. Note that this method applies only when the study group has both characters (otherwise there would be no need to polarize the character states). The relationships among outgroups can be either **resolved** (fig. 2.19b) or **unresolved** (fig. 2.19c). A decision regarding the character found at the outgroup node may be either **decisive,** which provides us with a best estimate of the condition found in the ancestor of our ingroup (fig. 2.19d: x is plesiomorphic and z is apomorphic in the ingroup), or **equivocal,** which does not allow us to postulate the direction of the character change (fig. 2.19e: we are not sure whether x or z is plesiomorphic).

Maddison, Donoghue, and Maddison (1984) began the quest for character polarities by attempting to determine the character state at the outgroup node. They reasoned that because this precedes the ingroup node, it will give us information about the character state in the common ancestor of the ingroup. Simple parsimony arguments are used in conjunction with an optimization routine developed by Maddison et al. (modified from earlier routines of Farris and Fitch). There are two cases: the relationships of the outgroups are known relative to the ingroup, and the relationships of the outgroups are either unknown or only partly resolved. The first case is the simplest, so we will begin by examining character evolution within the hypothetical taxon Annidae, its sister group P + Q + R, and other outgroups M, N, and O.

Character polarity in the group Annidae

1. Draw the phylogenetic tree of the ingroup and outgroups. We cannot reconstruct the tree on the basis of the characters in table 2.3, because these characters apply to the resolution of relationships within the ingroup, not to the relationships between the ingroup and the taxa in the outgroup. Presumably, then, we have access to a phylogeny for these taxa before beginning our investigation (fig. 2.20).

2. Inscribe each of the branches with its corresponding character from the first transformation series (i.e., the first row in the data matrix: fig. 2.21),

and indicate the six nodes. The lowest node on the tree is the **root node,** while the node connecting the Annidae to its sister group (PQR) is the **outgroup node.**

3. Label the two nodes, other than the root node, that are farthest from the outgroup node in the following manner: (1) Label the node "a" if the two closest nodes or branches are either both a or a and a, b. Notice that "closest nodes" refers to the nodes that are adjacent to the node in question, while "closest branch" is defined as any adjacent branch at the equivalent level or lower on the tree than the node in question. (2) Label the node "b" if the two closest nodes or branches are either both b or b and a, b. (3) If the closest branches/nodes have different labels (one a and the cther b), label the node "a, b." Note that the lowest node, termed the **root node,** is not labeled. In

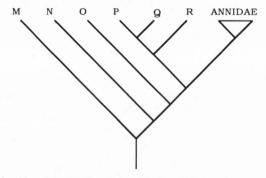

Fig. 2.20. Relationships of the Annidae clade to its closest relatives.

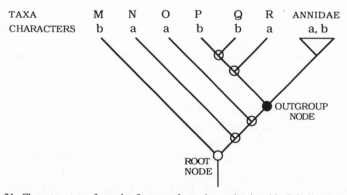

Fig. 2.21. Character states from the first transformation series in table 2.3, listed under the appropriate taxon. In order to determine the state of the character (either a or b) at the outgroup node, we must work towards it based on the information available at other, more-accessible nodes (depicted on this tree as *open circles*).

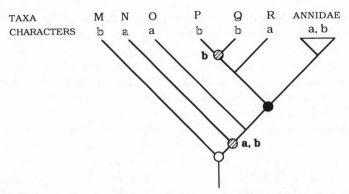

Fig. 2.22. First and second polarity decisions for character transformation series 1. *Outlined letters* = closest branches and nodes; *bold letters* = polarity decisions. The top node is labeled "b" because its closest branches are both b; the bottom node is labeled "a, b" because one of its closest branches is a and the other is b.

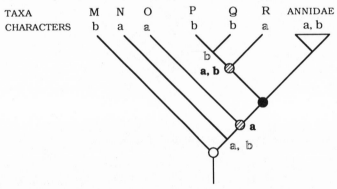

Fig. 2.23. Third and fourth polarity decisions for character transformation series 1. *Outlined letters* = closest branches and nodes; *bold letters* = polarity decisions. The top node is labeled "a, b" because its closest branch is a and its closest node is *b;* the bottom node is labeled "a" because its closest branch is a and its closest node is a, b.

order for us to label the root, we would need another outgroup. Since we are primarily interested in relationships within the ingroup, we will forget about this node. So, label the furthermost two nodes according to this procedure. The node connecting P + Q is given the notation b, because both P and Q have the character b. The node connecting N with O + P + Q + R + Annidae is given the notation a, b because M has character b and N has character a (fig. 2.22).

4. Continue working towards the outgroup node in the same manner (fig. 2.23).

5. The analysis is over when we reach a decision concerning the outgroup node. In this example, the assignment is a decisive **a** (fig. 2.24).

Repeating this procedure for character transformation series 2 of the matrix eventually produces an equivocal decision at the outgroup node (fig. 2.25).

One last thing. *Each of these decisions is made one transformation series at a time, and thus the polarization of every character occurs independently of every other character.* This does not mean that equivocal decisions based on single characters examined in vacuo will remain equivocal at the end of the analysis. The final disposition of character states is subject to an overall parsimony analysis which combines all of the information gleaned from each

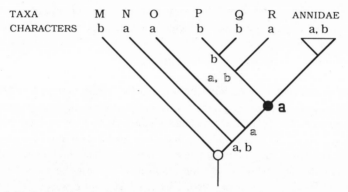

Fig. 2.24. Assignment of polarity to the outgroup node for character transformation series 1. *Outlined letters* = closest branches and nodes; *bold letter* = polarity decision. The outgroup node is labeled "a" because one of its closest nodes is a and the other is a, b.

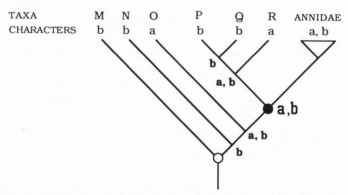

Fig. 2.25. Assignment of polarity to the outgroup node for character transformation series 2. The outgroup node is labeled "a, b" because both of its closest nodes are a, b.

of the separate character analyses according to the grouping and inclusion/exclusion rules.

Maddison, Donoghue, and Maddison (1984) discussed situations in which the relationships between the outgroups are either not resolved or only partly resolved. Since this watershed article is readily available, we will only mention two important observations: (1) Whatever the resolution of the outgroup relationships, the sister group is always dominant in its influence on the polarity decision. If the sister group is decisive for a particular state (i.e., a), no topology of outgroups further down the tree can result in a decisive b. (2) If you are faced with no sister group, but only an unresolved polytomy below the group you are working on, the frequency of a particular character among the outgroups in the polytomy has no effect on the decision at the outgroup node. For example, you could have ten possible, unresolved sister groups with character a and one with character b and the decision would still be equivocal at the outgroup node. Thus, *"common" does not equate with "plesiomorphic" in the phylogenetic system.*

Hennig argumentation was the original phylogenetic algorithm, and its application is still widespread. Working through the argumentation process for each character in a simple phylogenetic reconstruction is not overly difficult. But, even with relatively few taxa and characters, it if often laborious to reconstruct all of the character trees and the logically incompatible alternatives. We have good news and bad news about this situation. The good news is that computer algorithms have been designed to perform this laborious task (Kluge and Farris 1969; Farris 1970; Farris, Kluge, and Eckardt 1970). The bad news is that the ease with which computers perform this task may seduce us into forgetting that the results coming out are only as good as the data going in (Presch 1989). Increasing the amount of homoplasy in your data set increases the number of equally parsimonious pathways open to explain the evolutionary origins of these characters (i.e., increases the number of equally parsimonious trees). By contrast, increasing the number of homologous characters in an analysis decreases the number of equally parsimonious trees, because homologous characters correspond to a single phylogenetic tree. If the number of homologies has been greater than the number of covarying homoplasies in evolution, then the most biologically robust method of dealing with multiple equally parsimonious trees is to collect more data.

Character Coding for Building Trees

A **character code** is a numerical or alphabetical symbol that represents a particular character. We have already encountered codes in the preceding section. Using these characters and their codes has taught you something about the basics of tree reconstruction using classical Hennig argumentation and some of the approaches to determining the polarity of characters through outgroup character argumentation. In this section you will be introduced to some of the different kinds of derived (apomorphic) characters encountered in phylogenetic research and some of the problems associated with assigning codes to these characters. *The ultimate goal here is to formulate rules for coding characters so they can be used to reconstruct phylogenetic relationships among taxa.*

Multistate Transformation Series

All of the derived characters we have considered thus far are qualitative characters and are parts of binary transformation series. A **binary transformation series** consists of a plesiomorphy and its single derived homologue. Binary transformation series present no problem in coding. You simply code each character according to the information available from outgroup argumentation and produce a matrix full of zeroes and ones. Complications can arise if you encounter **polymorphic taxa,** taxa that have both the plesiomorphic and the apomorphic characters; however, this problem is critical only when both characters are found in a single species. Considerable controversy surrounds the coding of such characters, especially when biochemical traits are used. There are two ways of handling this: code the taxon as displaying only the apomorphy and discount the plesiomorphy, because the plesiomorphic trait arose in an ancestor of the taxon (**qualitative coding**), or code according to the frequency of each character. There is considerable controversy about which of these two approaches is appropriate for phylogenetic analysis. (For excellent recent discussions of this problem, see Buth 1984, Swofford and Berlocher 1987, and Murphy 1988.)

Investigators working on a large group, or even a small group that has undergone considerable evolution, may discover that there are several different homologous characters in one transformation series. For example, if you were working on the phylogenetic relationships of fossil and recent horses, the transformation series for the number of toes on the hind foot would contain a goodly number of characters: four toes, three toes, and one toe in the ingroup, and five toes in the outgroup. Such a large transformation series, encompassing a plesiomorphic character and two or more apomorphic characters, is termed a **multistate transformation series.**

Multistate transformation series can be grouped according to the amount of information available concerning the pathway and direction of evolutionary change. As discussed earlier, this produces four possible types of multistate transformation series (see fig. 2.6): ordered, unpolarized; unordered, unpolarized; ordered, polarized; and unordered, polarized. Binary characters are much simpler because they are all automatically ordered (but not neces-

(**X**, A)   (B, C, D)   (E, F)   (G, H)

Fig. 2.26. A simple linear character tree for four characters. Taxa displaying each character are listed in parentheses below the character. X = outgroup; A–H = ingroup taxa.

polarized). A **linear transformation series,** consisting of characters related to one another in a straight-line fashion, is the simplest multistate transformation series (fig. 2.26). The relationships of these characters can be termed a **character tree,** and in this particular case you can see that there are no branches on the tree. Notice that this character tree is both ordered and polarized. At the moment we are not concerned with how to reconstruct the character tree, but rather with how to code the information presented in this tree for use in reconstructing a phylogenetic tree (we will return to the question of just how these transformation series are derived at the end of this section). *It is important to understand that a character "phylogeny" is not the same as a phylogeny of taxa. A character tree only contains information about the relationships among characters. It is not a hypothesis of the underlying phylogenetic relationships of the taxa.* This is a critical, and often misunderstood, distinction.

A simple linear transformation series

The data matrix for the linear transformation series (fig. 2.26) can be coded in one of two basic ways. First, we can simply code the characters in a linear fashion, assigning a value to each character based on its position in the sequence. For example, we have chosen to code rounded square as zero, square as one, black-and-white square as two, and black square as three. Each value is placed in the data matrix in a single column (table 2.4). Every apomorphy will contribute to the length of our forthcoming phylogenetic tree in an additive fashion. We use the term "additive" because each instance of evolutionary modification requires one step along the tree and counting all of the steps in a straight line shows exactly how much the transformation series added to the overall tree length. Indeed, such transformation series are often termed additive multistate characters.

We could also use **additive binary coding,** a method that breaks the character down into a number of binary subcharacters that are each represented by their own column of information. For example, we can consider both black-and-white square and black square as subsets of white square, since each is ultimately derived from white square. The first additive binary column in the data matrix reflects this fact, coding rounded square as the plesiomor-

Table 2.4   Data matrix for the four characters and nine taxa in the linear transformation series in figure 2.26.

Taxon	Linear Coding	Additive Binary Coding		
X (outgroup)	0	0	0	0
A	0	0	0	0
B	1	1	0	0
C	1	1	0	0
D	1	1	0	0
E	2	1	1	0
F	2	1	1	0
G	3	1	1	1
H	3	1	1	1

phic character (0) and white square plus all of its descendants as apomorphic (1). Now, black square is a subset of black-and-white square. The second additive binary column contains the codes for this next level of comparison: both rounded square and white square are plesiomorphic relative to black-and-white square, so they receive a zero, while the black-and-white square and its descendant black square receive a one. Finally, on the last level, black square is apomorphic relative to all the other characters in this linear series, so it is coded one in the third column and everything else receives a zero. Three columns now represent the transformation series. You can double-check your binary coding by adding all the ones together in each row and placing them in a single column. You should find that you have replicated the original linear coding column. Either method of coding produces exactly the same character sequence and thus contributes the same information to a phylogenetic reconstruction (fig. 2.27).

If you choose additive binary coding, it is important to remember that you are dealing with one homologous character series, and not three independent characters. If you count each state as an independent character, you may artificially inflate the consistency index and overestimate the degree of support for your tree. So bear this in mind if you are using a computer program to analyze your data, because computers cannot differentiate between a single multistate transformation series and independent binary transformation series for each of the derived states in the multistate transformation series.

A branching transformation series

A **branching transformation series** (nonadditive or complex transformation series) contains characters that are related to each other in a branching, rather than a straight-line, fashion (fig. 2.28). Notice that this character tree is both ordered and polarized. Since the characters are not related in a linear

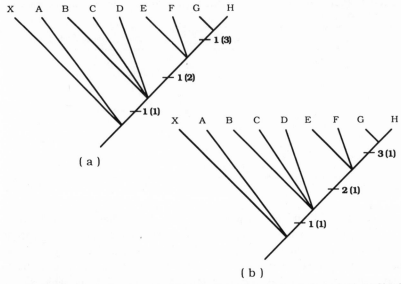

Fig. 2.27. Phylogenetic tree constructed from the linear character tree in figure 2.26. (*a*) Simple linear coding breaks the character into four parts, the plesiomorphic state (0), which need not be represented on the tree, and three apomorphic states (1, 2, and 3; one character, four states). (*b*) Additive binary coding breaks the linear series into three independent characters, then groups according to the apomorphic states (1) of those characters (three characters, two states each). Both methods of coding preserve the phylogenetic information in the character intact and thus will produce the same phylogenetic tree.

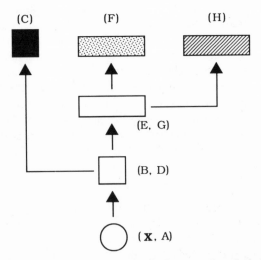

Fig. 2.28. A complex branching character tree for six characters. Taxa displaying each character are listed in parentheses.

fashion, simple additive coding will result in errors in translating the transformation series into a phylogenetic tree.

Labeling the characters of a branching transformation series in a simple linear fashion will obviously result in some misinformation, so how can we display these complex relationships? There are two basic methods: additive binary coding and mixed coding. Since we are already proficient at binary coding, let's turn to this method first.

We arrived at the additive binary codings in the data matrix (table 2.5) by working through the following steps. First, code the first additive binary column: square, which is ancestral to all other characters in the character tree, is apomorphic relative to circle; therefore, square and all of its descendants are coded as apomorphic (1), while circle is coded as plesiomorphic (0). The relationship between square and circle is a simple linear one. Moving to the next comparison, the relationship between square and its most immediate modification brings us to a branch point on the tree, because both black square and white rectangle are directly derived from square. We deal with branch points by examining the derived states one at a time. Second, code the second additive binary column: black square is derived directly from square; therefore square is coded as plesiomorphic (0) and black square as apomorphic (1). Notice that *only the taxa displaying either the apomorphic character or any of its modifications are assigned a code of one; all other taxa receive a zero*. In this case, black square has no modifications and is only present in taxon C (it is an autapomorphy). Third, code the third additive binary column: rectangle is derived directly from square; therefore square is coded as zero and rectangle as one. Again, the taxa bearing either the apomorphic condition or any of its modifications are assigned a code of one, so taxa E and G (apomorphic state rectangle) and F and H (bearing the descendants of rectangle) receive a one; all other taxa receive a zero. Fourth, code the fourth additive binary column: speckled rectangle is directly derived from rectangle and is an autapomorphy for taxon F; therefore F is assigned a one

Table 2.5 Data matrix for the six characters and nine taxa in the branching transformation series in figure 2.28.

Taxon	Additive Binary Coding					Mixed Coding		
X	0	0	0	0	0	0	0	0
A	0	0	0	0	0	0	0	0
B	1	0	0	0	0	1	0	0
C	1	1	0	0	0	1	1	0
D	1	0	0	0	0	1	0	0
E	1	0	1	0	0	2	0	0
F	1	0	1	1	0	3	0	0
G	1	0	1	0	0	2	0	0
H	1	0	1	0	1	2	0	1

for this character and everyone else gets a zero. Finally, code the fifth additive column: striped rectangle is directly derived from rectangle and is an autapomorphy for taxon H; therefore, only H receives a one for this character.

Mixed coding, also called **nonredundant linear coding,** is a hybrid between additive binary coding and linear coding. By convention, the longest straight-line branch of the character tree is coded in a linear fashion, then branches off this linear tree are coded in an additive binary fashion. Depending on the asymmetry of the character tree, this strategy can substantially reduce the number of character columns. Returning to the character tree, the longest sequence of character modifications is circle, square, rectangle, speckled rectangle (note: we could also have chosen the sequence circle, square, rectangle, striped rectangle; however, since nodes can be freely rotated on a tree, the choice between the "sister characters" speckled rectangle or striped rectangle is completely arbitrary and does not change the outcome of the analysis). We place the codes for this section of the tree (0, 1, 2, 3) in the first mixed coding column of the data matrix (table 2.5). Branches from this sequence, the autapomorphies black square and striped rectangle, are then each assigned a code of one in a separate column as described in the preceding section.

Polarization Arguments and Multistate Transformation Series

As promised above, we will now return to the methods involved in determining the evolutionary sequence of multistate characters. Sometimes the ordering of these transformation series can be determined using biological data, such as information about developmental sequences (Nelson 1978; Nelson and Platnick 1981; Patterson 1982; Voorzanger and van der Steen 1982; Brooks and Wiley 1985; de Queiroz 1985; Kluge 1985) or directions of biochemical pathways (Seaman and Funk 1983). More often, though, these data are not available or are not reliable.

What can phylogenetic systematic analysis tell us in the absence of this information? For example, suppose we wish to examine a group of mammals that includes species that do and do not have forelimbs modified into wings. Furthermore, extensive research has revealed that there are two types of wings (let us say short and long, just for this example). Using all other mammals as outgroups reveals that "no wings" came first (is plesiomorphic), but from that point, the evolutionary pathway of wings could have been (1) short to long, (2) long to short, or (3) short and long arising independently from the "no wings" condition. For these species, simple outgroup comparison does not resolve the polarity of the apomorphic states because they do not occur among members of the original outgroups. In order to resolve the transformation series for apomorphic states restricted to the ingroup, it is necessary to find outgroups that possess at least one of these states. In such cases, phylogeneticists determine character polarities by using a method that preserves the logic of outgroup comparisons. This method, developed by Watrous and Wheeler (1981), is called **functional outgroup analysis.**

Let us begin with an ingroup (taxa A-E), a set of outgroups, and three characters (table 2.6). One binary character supports the monophyly of the ingroup. Two characters, one binary and the other multistate, help us resolve relationships *within* the ingroup.

Table 2.6   Distribution of states for one multistate and two binary characters among five members of an ingroup (A–E) and a set of outgroups.

Taxon	Characters		
	Binary		Multistate
Outgroups	0	0	x
A	1	0	x
B	1	0	y
C	1	+	y
D	1	+	z
E	1	+	z

Note: The first binary character supports the monophyly of the ingroup.

1. Use the second binary character to produce a partial phylogenetic tree for the ingroup. Figure 2.29 depicts the distribution of states for this character among the outgroup and ingroup taxa.

2. State 0 is clearly supported as the plesiomorphic condition by outgroup comparison. Hence, taxa C + D + E are united as a group within the ingroup by the shared possession of the derived (synapomorphic) state + (fig. 2.30). Because C + D + E is a putative clade, we can consider it to be a "functional ingroup." A and B then form the "functional outgroups," because their logical relationship to C + D + E is the same as the relationship of the outgroups to A + B + C + D + E.

3. Use the original outgroups and the functional outgroup to polarize the multistate transformation series. First, place the corresponding character states x, y, and z at the tips of the branches for the outgroup and ingroup taxa (fig. 2.30). Polarization via the "outgroup" taxa indicates that character state x is plesiomorphic to either y or z. Now, examine the functional outgroups. One of them (A) exhibits state x, while the other (B) exhibits state y. One of the functional ingroup members (C) exhibits state y, while the other two (D and E) exhibit state z. Because state y occurs in **both** the functional ingroup and the functional outgroup, we conclude that state y is plesiomorphic to state z. The original outgroups tell us that x arose first, while the functional outgroup tells us that y arose next. Consequently, state z arose last, and the

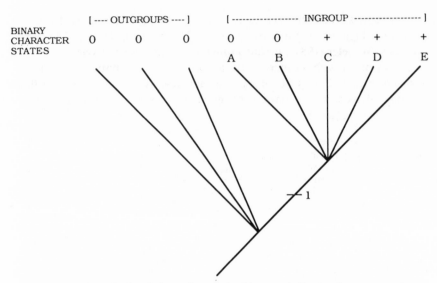

Fig. 2.29. Starting point for phylogenetic analysis of ingroup A–E using three outgroup taxa. Synapomorphy 1 supports the monophyly of the ingroup. Distribution of states (0 and +) for the binary character are indicated at the tips of each branch.

transformation series for this multistate character is the linear sequence x to y to z.

4. Combine the information from the binary and multistate characters. Outgroup comparisons for the binary character support the interpretation that state 0 is plesiomorphic and state + is apomorphic, linking taxa C + D + E as a clade. Functional outgroup analysis suggests that state x is plesiomorphic for the ingroup, that state y is apomorphic to x, linking B + C + D + E as a clade, and that state z is apomorphic to y, linking D + E as a clade. Invoking the inclusion/exclusion rule results in a fully resolved phylogenetic tree (fig. 2.31).

To return to our example of winged mammals, we use some characters to tell us that one group of mammals (ingroup) is distinct from other mammals (outgroup). We then use other characters to give us enough divisions among the winged members of this ingroup (functional in- and outgroups) to establish the sequence in which short and long wings evolved.

The use of functional outgroup comparisons (the logical justification of which is called "reciprocal illumination": Hennig 1966; Wiley 1981) strikes many people as being circular. Although the distinction is a fine one, we believe that the curse of circulatory is avoided because *the groupings within the original ingroup that allow us to determine polarities for the multistate character are determined a priori by reference to other traits.* Of course, if we do not have very many other characters, the robustness of our assessment of multistate polarities is not very great. Consequently, many phylogeneticists worry about using multistate characters at all. In a complementary manner, other phylogeneticists worry about problems associated with treating states of an evolutionary sequence as independent characters. For example, if we treated x, y, and z as three binary characters, we would be able to use the original outgroups to determine that "x present," "y absent," and "z absent" were plesiomorphic conditions. The states "y present" and "z present" would be interpreted as synapomorphies of B + C and of D + E, respectively. Considering "z present" as a synapomorphy of D + E would not conflict with other data (i.e., the second binary character), but considering "y present" to be a synapomorphy of B + C would conflict. In this case, breaking x, y, and z into independent characters would result in unnecessary postulates of homoplasy (violating Hennig's auxiliary principle). This is the reason phylogeneticists study each character carefully, argue transformation series as independently as possible, often considering both binary and multistate options, and constantly seek additional data in an effort to provide the best possible hypothesis at any given time.

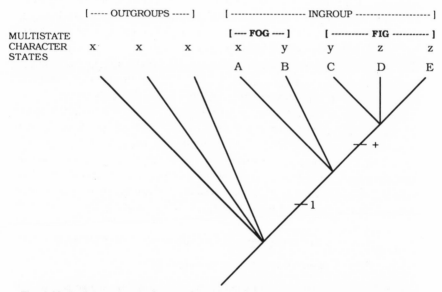

Fig. 2.30. Partial resolution of phylogenetic relationships among members of ingroup A–E, based on distribution of states for one binary character. Outgroup comparisons support the interpretation that state + is apomorphic, linking taxa C + D + E into a clade within the ingroup. This clade now serves as a functional ingroup (*FIG*) within the original ingroup, with the other members of the original ingroup (A and B) serving as functional outgroups (*FOG*). The distribution of states for a multistate character (x, y, and z) among members of the ingroup and outgroup is indicated above the taxa.

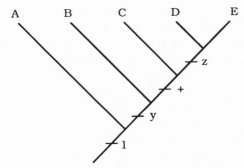

Fig. 2.31. Phylogenetic tree for taxa A–E, based on outgroup comparisons for two binary characters, and outgroup and functional outgroup comparisons for one multistate character.

Summary

Operationally, phylogenetic systematics can be summarized in four steps. First, use nonphylogenetic criteria to assess similarities in traits among species and assume homology whenever possible. Second, use outgroup comparisons to distinguish shared general, shared special, and unique homologies. Third, group according to shared special homologies, mapping all traits onto the resultant tree to show the distribution of the total data base. Fourth, interpret inconsistent traits, post hoc, as instances of homoplasy (parallel and convergent evolution); that is, as cases in which the initial nonphylogenetic homology criteria were misleading. The result of this procedure is a hypothesis of the phylogenetic relationships within the study group.

The number of branching diagrams appearing in the evolutionary biology literature is growing. Many of these are called "cladograms"; however, not all of these diagrams are constructed using phylogenetic systematic methods. The information presented in this chapter will help you critically evaluate these diagrams, for *only trees produced in accordance with phylogenetic systematic principles provide the robust estimates of genealogical relationships that are the necessary precursors for historical ecological studies.*

Answers to Some Common Questions and Misconceptions

What happens if you change the outgroup?

The outgroup's function is to identify the plesiomorphic character state for characters in the ingroup. So changing the outgroup could potentially cause one of the following things to happen to your tree.

1. There might be no change. This occurs when all the outgroups share with the ingroup the same plesiomorphic state for a given character.

2. There might be a change in the polarity of some transformation series.

This would happen if the outgroup and the ingroup share *numerous* homoplasious, apomorphic characters. In this case you look for a consensus state among outgroups and hope that you have enough other characters to provide a tree that highlights the homoplasy. If you do not, you will get multiple trees, and this, in turn, indicates that you need to collect more data.

3. There might be a loss of resolution. If the new outgroup is distantly related to the ingroup, the groups might not share enough similarities for comparison, and this will lead to an increase in the number of unresolved polytomies in the tree. This produces an incomplete, rather than an "incorrect," estimate of phylogeny.

That changing outgroups will change the phylogenetic tree is an important and common misconception about phylogenetic systematics. Figure 2.32 shows a series of trees based on twenty-seven behavioral characters for the gasterosteid fishes (sticklebacks). Characters were polarized in the following manner: first, using the sister group of the Gasterosteidae, the Aulorhynchidae, or tubesnouts (fig. 2.32a); second, using a more distantly related species, the pipefish. *Sygnathus typhle* (fig. 2.32b); and third, using a very distantly related species, the salmonid *Onchorhynchus nerka* (fig. 2.32c). Interestingly, although all outgroups polarized some characters differently, the three tree topologies are consistent with one another; in fact, the trees based on the tubesnouts and the salmonid are identical. Although the phylogenetic relationships are retained, the number of informative characters decreases with use of the other outgroups. Overall, then, the distinction is not one of "correct" versus "incorrect" but rather one of the relative degrees of information available from the particular outgroup chosen. The more robust a tree proves to be in response to changing the outgroup, the greater confidence is instilled in it.

By now you are probably asking, "But even if you don't change the tree topology, haven't you changed some of the character transformation hypotheses by using different outgroups?" This is a critical question because, as the distribution of traits in figure 2.32 demonstrates, changing outgroups can change the hypothesized transformation series for some characters, and this, in turn, will affect our evolutionary hypotheses for those characters. Because of this, it is important to follow one cardinal rule: *never use the characters that are part of the evolutionary hypothesis under investigation to build your phylogenetic tree.* Rather, these characters should be mapped onto an existing tree. We will discuss methods for mapping characters (optimization) in chapter 5.

Is outgroup comparison an exercise in circular reasoning?

No. You do not have to have the sister group to polarize characters, so you do not have to have a phylogenetic scheme a priori in order to determine the

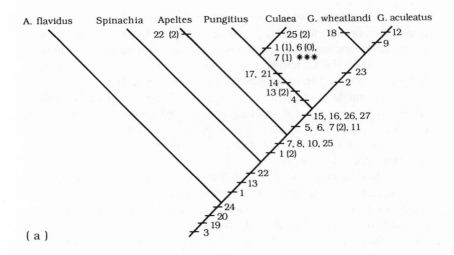

(a)

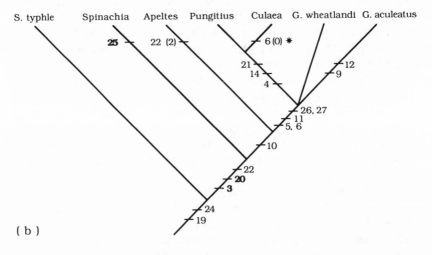

(b)

outgroups, and the polarizations for each character are argued independently based upon comparisons with the outgroups.

> *Does the method assume or rely on any particular evolutionary mechanism?*

No. It assumes only that evolution has occurred. In fact, the utility of phylogenetic systematics in evolutionary biology stems from its independence of any particular evolutionary mechanism.

Fig 2.32 *(continued)*

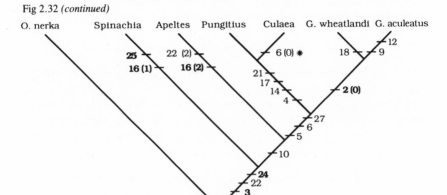

(c)

Fig. 2.32. What happens if you use a different outgroup? (*a*) Polarizations using the sister group, the tubesnout *Aulorhynchus flavidus*. (*b*) Polarizations using a member from the next most closely related order, the pipefish *Sygnathus typhle*. Characters in *bold type* are polarized differently using this outgroup. Characters 1, 2, 7, 8, 13, 15, 16, 17, 18, and 23 cannot be polarized using this outgroup; therefore, they are eliminated from the analysis. Note that although there is a loss of resolution in the polytomy among *Pungitius/Culaea* + *Gasterosteus wheatlandi* + *G. aculeatus*, the overall topology of this tree is compatible with the first tree. (*c*) Polarizations using a distantly related teleost, the salmonid *Onchorhynchus nerka*. Characters 1, 7, 8, 11, 13, 15, 23, and 26 cannot be polarized. This tree is identical to the first tree. * = homoplasious characters; *numbers in parentheses* = the state of a multistate character.

Isn't parsimony an assumption of evolutionary mechanism?

Parsimony is a scientific principle used by scientists to make decisions about ambiguous data. Basically this principle can be stated in the following manner: when there are conflicting hypotheses for a given data set, accept the hypothesis that is support by the greatest amount of data. The use of parsimony in phylogenetic systematics is no different from its use in any other branch of biology or any other science, does not invoke any particular evolutionary mechanism, and does not mean that systematists believe that evolution is parsimonious or that "parsimony" equals "truth." Invocation of this principle simply gives us a starting point for comparative studies. From there, any author who prefers a less parsimonious tree must justify this choice by providing corroborating *biological* evidence.

Are phylogenetic trees tests of evolutionary mechanisms?

Phylogenetic trees are *not* tests of evolutionary mechanisms, they are descriptions about patterns in nature. From these patterns we can obtain critical evidence about some evolutionary principles, which, in turn, may help us to design an experiment to test the existence of a hypothesized mechanism.

When do you have enough characters?

The reconstruction of phylogeny is an open-ended process, so in principle you never have enough characters. In practice, you stop when you stop getting different answers or different resolutions when you add data. Even then it is possible for someone else to come along and modify what you have done.

Is there any inherently preferred data type?

No. It is expected that any data, analyzed phylogenetically, will give the same or highly similar answers because organisms are the result of a dynamic interaction among genetic, developmental, physiological, morphological, and behavioral systems through time. There is no a priori reason to believe that one system contains more information about evolutionary relationships than any other.

If you rotate the branches about a node on a phylogenetic tree, do you change the implied phylogenetic relationships?

No.

What's wrong with equating a taxonomic classification with phylogeny?

It is inappropriate to use a taxonomic classification as a phylogeny because many classifications portray paraphyletic (or polyphyletic) taxa as monophyletic groups. Evolutionary explanations based on such classifications will overestimate the importance of adaptive plasticity, because diagnoses for paraphyletic groups list synapomorphic traits more than once. This gives the impression that these traits are actually examples of parallel or convergent

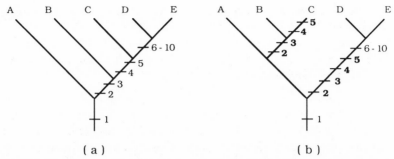

Fig. 2.33. Why is it incorrect to equate "classification" (or taxonomy) with "phylogeny"? (a) Phylogenetic tree reconstructed for species A–E based upon phylogenetic systematic methods. (b) Tree reconstructed from a taxonomic classification scheme that includes the paraphyletic group A + B + C. This forces us to postulate that characters 2, 3, 4, and 5 evolved twice.

evolution, and such homoplasy, in turn, is often considered strong evidence of adaptive evolution.

Consider the following example. Figure 2.33a depicts the phylogenetic tree for a group of hypothetical taxa. Although the presence of five synapomorphic traits (characters 6–10) distinguishes species D and E from species A, B, and C, these taxa are all members of a monophyletic group characterized by the presence of traits 1–5. Now suppose we have a classification scheme that places species D and E in one taxon because they are so distinct, and places species A, B, and C in another taxon. Reconstruction of phylogenetic relationships based on this classification will produce the tree in figure 2.33b. This arrangement forces us to postulate that characters 2–5 evolved twice, overestimating the amount of adaptive evolution. Since most commonly accepted classifications include paraphyletic groups, they cannot serve as independent templates for estimating the origins, elaborations, and associations of ecological characters through evolutionary time. Unfortunately, given the current dearth of available phylogenetic trees, many researchers have been forced to utilize classifications in their preliminary analyses of behavioral/ecological evolution.

Cladograms only represent branching points in evolution and cannot represent the relative degree of evolutionary divergence among lineages, can they?

If we have a large sample of characters, it is possible that the relative degree of evolutionary divergence among lineages can be estimated. This could be obtained by comparing the relative numbers of apomorphic transformations between sister groups on the phylogenetic tree. What we lack at the present are protocols for letting us know when we have a proper sample of characters.

> *Why aren't there primitive and advanced species, just like there are primitive and advanced characters?*

Because not all characters evolve at the same rate and to the same degree in different lineages. As a consequence, all species are mosaics of plesiomorphic and apomorphic traits, and it is inappropriate to speak of plesiomorphic and apomorphic species. One can speak of sister-group relationships or of relative position in a phylogenetic tree.

> *Can't you just manipulate the data to get the answer you "want"?*

Of course, but this problem is one of scientific ethics and not unique to either phylogenetic systematics in particular, or biology in general. In fact, phylogenetics may be slightly more open to scrutiny because you have to report both the character descriptions and codings in each publication; therefore, the original data are more accessible to reanalysis than, say, a table of p values.

PART TWO

Phylogeny and the Evolution
of Diversity

3 Preamble to Speciation and Adaptation

"It is interesting to contemplate a tangled bank, clothed with many plants of many kinds, with birds singing on the bushes, with various insects flitting about, and with worms crawling through the damp earth, and to reflect that these elaborately constructed forms, so different from each other, and dependent upon each other in so complex a manner, have all been produced by laws acting around us" (Darwin 1872:403). This sentence, with its evocation of a diverse, intertwined web of life, can never fail to rouse even the most reductionist of biologists.

The "question of diversity" has been approached on two levels by evolutionary ecologists: description and explanation. Descriptions of diversity are concerned with the total number of species and the relative abundance of species in a given area. These descriptions, in turn, rest upon our ability to perceive, collect, and classify organisms and to depict the data mathematically in the form of diversity indices. Explanation, or the search for processes underlying observed biological diversity, is in itself a complicated process that has blazed a trail of discovery and controversy through the ecological literature. Evolutionary ecologists originally attempted to incorporate both population-level and community-level factors into their explanations of ecological structure. They examined biotic interactions based on traits that were **fixed within** each species but **variable among** species. This permitted researchers to filter out the confounding influence of intraspecific variability and thus formulate hypotheses concerning the influence of species composition on differences in interspecific interactions. General rules for the production and structuring of diversity were sought by investigating (1) specific components of the system, such as patterns of colonization and extinction of species (MacArthur 1965; MacArthur and Wilson 1967); the age (Wallace 1878), productivity (Connell and Orias 1964; Brown 1973; Connell 1978), structural complexity (Hutchinson 1959), and stability (Slobodkin and Sanders 1969) of the system under investigation; predation (Paine 1966; Parrish and Saila 1970); and interspecific competition (Dobzhansky 1951; Williams 1964); (2) patterns of energy flow through the system (Lindeman 1942; Odum 1969; Brown 1981; Wright 1983; Glazier 1987); (3) organism/environment as an inseparable whole (Patten 1978, 1982); and (4) the impact of "stochastic factors" (Simberloff 1971). All approaches have contributed some pieces to the intricate puzzle of biological diversity (for excellent reviews see Brown 1981; McIntosh 1987).

In the past two decades, however, the amount of confounding complexity has led an increasing number of evolutionary ecologists to abandon the search for multispecies, large-scale regularities in favor of single-species studies. This has produced a research program founded on the assumption that an understanding of the population biology of each species within a biota will lead to an understanding of the evolution of that biota. As population studies gained increasing ascendancy, the research emphasis shifted to traits that are variable *within* species. These studies have contributed a wealth of detailed information concerning microevolutionary oscillations at the demic level. Two things are now required: an equivalent data base for ecological systems at the macroevolutionary level and a way to incorporate information from both levels into a comprehensive theory of evolution. We will thus confine our discussions to characters that are relatively fixed within species. This does not mean that variable intraspecific traits are uninteresting or even without phylogenetic components. On the contrary, in chapter 9 we will discuss the potential for dovetailing historical ecology with an emerging research program that uses a combination of phylogenetic and statistical information to discern phylogenetic constraints on the evolution of variable life-history traits. Overall, it is our contention that contemporaneous ecological complexity is embedded in a well-organized historical matrix and that, as a consequence, much of the confounding data compiled by evolutionary ecologists is due to the inability to distinguish between effects stemming from historical background and those stemming from proximal dynamics.

Historical ecologists examine diversity from an explicitly phylogenetic, or historical, perspective. In chapters 4 and 5 we are going to investigate the influence of two evolutionary processes, **speciation** and **adaptation,** on the diversification of individual clades. We believe that if some aspects of diversity represent persistent ancestral conditions, then the study of diversity is inherently a macroevolutionary research program. The term "macroevolution," like many terms in science, is the multifaceted result of contributions from many different people. Of these facets, Cracraft's (1985b) distinction between the "transformational view" and the "taxic view" of macroevolution (see chapter 1) is especially helpful in understanding this section. The transformational approach emphasizes the origin of key innovations and adaptive radiations, and the evolution of large-scale trends in character modifications. The taxic view, by contrast, is concerned with uncovering patterns in the distribution of numbers of species, then examining the underlying mechanisms controlling rates of speciation and extinction. We believe that the phylogenetic analysis of speciation approaches macroevolution from the taxic perspective, whereas the phylogenetic analysis of adaptation represents the transformational view. Although we are presenting them separately, using somewhat different approaches in each case, we hope to show that these two aspects of macroevolution complement each other well, and that robust macroevolutionary explanations often require information about both speciation and adaptation.

Speciation

In chapter 4 we will address macroevolutionary questions concerning the **numbers of species** in clades, both total and relative. To understand this, we must investigate the ways in which new species come into existence; this, in turn, means that we must first know something about the nature of species (Wiley 1981; Cracraft 1989; Nelson 1989; Templeton 1989).

What Is a Species?

Speciation has always been a central process in evolutionary theory. It follows, then, that if speciation is a "real" process, species must be "real" in some sense relevant to evolution. Nineteenth-century philosophers of science argued that the only "real" entities were those that had immutable spatiotemporal existence. Because of their unchangeable nature, such bits of reality could be grouped into "classes" defined by the fixed properties of their components. Classic examples of such "real" entities, sometimes called "species," are "hydrogen" and "gold." Darwin (1859) threw a monkey wrench into this system by suggesting that organisms could be grouped into biological species, but that these species were not immutable. Although this proposal is not considered controversial today, it took two major philosophical revolutions to forge an understanding of just what a biological species really is. The first step was the emergence of what Mayr (1963, 1988) has called "populational thinking" as opposed to "typological thinking." Proponents of the typological approach treat biological species as classes, groups of organisms sharing unique features that define the species. In other words, just as one atom of gold is interchangeable with any other atom of gold, one tiger is interchangeable with any other tiger. This static concept of species makes it difficult to understand how biological species can evolve. In order for speciation to occur, an ancestral species must be variable, and in order for variation to occur, the species must include organisms that do not conform completely to the "definition" of the species. Each new variant that arises in a species must therefore create its own class, so we must either equate "species" with "individual organism," in which case we equate evolution with development, or we must give up the notion of species as real evolutionary entities, if we adopt this perspective.

Advocates of populational thinking treat species as assemblages of organisms held together by reproductive bonds that are exclusive to them, that can develop like an individual organism (but do not have to die of old age), and that can "reproduce" by something analogous to binary fission. This approach allowed biologists to slip comfortably into a transformational or evolutionary mode, because it viewed species as collections of organisms characterized by both common traits and variable traits. Mayr (1942, 1963) called this the biological species concept. The first major step towards an evolutionary species concept had been taken. The next step awaited resolution of the observation that, although species cohesion is provided by reproductive bonds, reproductive structure in many species lies at the level of local breeding units (demes). This leads us in a completely different direction conceptually, to a worldview in which only demes and populations are real, and species are relegated to the role of artificial constructs. Two problems arise from this perspective: demes and populations might be typological constructs themselves, and, if species are not real, then neither is speciation, and evolution is reduced solely to processes involved in reproductive exchange within individual demes. This purely populational view equates evolution with changes in gene frequencies in populations, and this, in turn, construes evolution as a reversible phenomenon, in contrast to all of our empirical evidence to the contrary. Interestingly, speciation, like evolution, is irreversible. It is a property of collections of demes that is not manifested by the demes themselves because, although changes are always occurring within populations, the coherent structure of the species is not affected unless gene flow is severed between populations (fig. 3.1). Not surprisingly, then, the collection of demes construed as representing a species often exhibits more geographical and ecological coherence than the demes themselves (i.e., demes can disappear and re-form without destroying the species).

For those, like Mayr, who have always felt that species and speciation were important aspects of evolution, the biggest problem has been to determine just how to consider those groups of demes as real without being typological. Michael Ghiselin (1974) provided the solution to the problem by considering species as if they were individual, rather than collective, entities (see also Hull 1976, 1978, 1980; Wiley 1978, 1980a,b; Mishler and Donoghue 1982; Cracraft 1983b; Donoghue 1985; McKitrick and Zink 1988). Biological species are real, but not in the same sense that "hydrogen" is real. A molecule of hydrogen found anywhere, and formed at any time, in the universe would be a member of the class hydrogen. By contrast, an organism that looks like a tiger on this planet would not be part of the same species as an organism that looks like a tiger on another planet unless the two organisms shared a common ancestor. Viewed in this way, we can see that classes are defined by convergences, whereas individuals are defined by homology. So the typological view and the purely populational view of species are nonevolutionary,

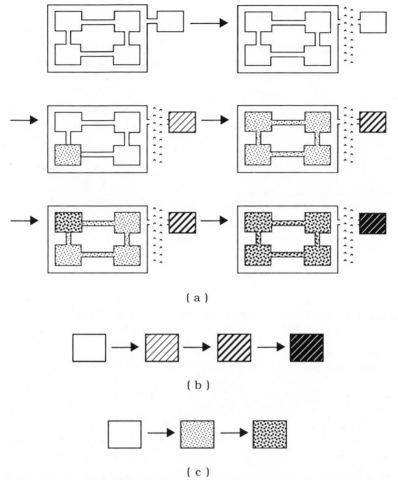

Fig. 3.1. Heuristic diagram depicting the difference between speciation and population differentiation. (*a*) The ancestral species is initially composed of five populations (represented by small boxes) connected by some level of gene flow. One population is isolated (gene flow is severed or severely restricted); the remaining populations continue to experience gene flow. Character changes, depicted by *different patterns in the boxes*, occur in both the ancestral populations and the isolated population. (*b*) Character changes in the isolated population. (*c*) Character changes in the ancestral species' populations. It is impossible to determine which scenario, *b* or *c*, represents the speciation event *without reference to other ancestral populations*. In this situation, the isolated population has established an independent evolutionary trajectory (speciated), while the remaining populations, although they have changed, remain part of the original evolutionary lineage.

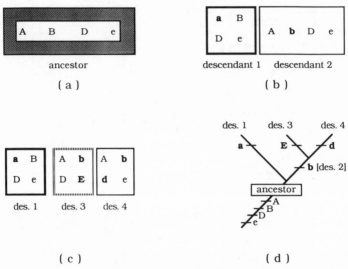

Fig. 3.2. Species are a mosaic of ancestral and derived traits, but only synapomorphies distinguish phylogenetic relationships. *Bold letters* = derived traits. (*a*) Ancestral species. (*b*) Ancestral species divided, descendant species 1 and 2 produced. (*c*) Descendant species 2 divided, species 3 and 4 produced. (*d*) Phylogenetic tree depicting the relationships among the ancestor and all its descendants. Notice that descendants 1 and 2 are sister species (possession of character e); descendants 3 and 4 are sister species (possession of character b). Autapomorphies (a, d, E) distinguish individual species, but not phylogenetic relationships.

because they are based on homoplasy and evolutionary descent involves homology (common ancestry; Wiley 1989).

Under this "species as individuals" view, the most important characteristic of a species is that its members are bound together by unique common ancestry, and not that its members are reproductively isolated from members of other species. The evolution of a single species is analogous to the development of a single organism; just as an organism changes its appearance without losing its identity during development, so a species can change its appearance without losing its identity during evolution (fig. 3.1a and c). The formation of new species is analogous to asexual reproduction, in which new individuals are distinct from the old individual because they form independent evolutionary lineages (fig. 3.1a and b). Over time, distinct historical trajectories emerge from the speciation process, each differing to some degree from its ancestor and closest relatives, but retaining some of its ancestry in the form of synapomorphies. We take advantage of this historical mosaic nature of the attributes of organisms that comprise species when we use synapomorphies to reconstruct phylogenetic trees (fig. 3.2). The adoption of the evolutionary species concept (Wiley 1978, 1980a) appears to have finally exorcised the persistent spector of typology and also freed us from a reductionist view of evolution (Mishler and Donoghue 1982; Cracraft 1983b; Donoghue 1985; McKitrick and Zink 1988). All species that conform to the biological species concept are evolutionary species, but not all evolutionary species need conform to the biological species concept (see also Endler 1989; Templeton 1989).

Understanding what species are conceptually remains an interesting issue in the philosophy of biology (see, e.g., articles in *Biology & Philosophy,* vol. 2, 1987 and for an enlightening counter-point, see Powers 1909). What is important to most biologists, however, is not definition but identification. Wiley (1981) discussed a series of criteria that can be used to make decisions about species membership. Despite these empirical tools, delimitation of species boundaries remains a problem in some groups, and may limit our ability to study speciation processes in those organisms. Nonetheless, students of such groups should take heart. As David Hull (pers. comm.) has pointed out, if all species had sharply distinct boundaries, we would have no reason to suppose that any of them could evolve into other species. Hence, those groups for which it is difficult to delimit boundaries represent some of the strongest evidence we have for the evolutionary potential of biological species.

How Species Are Produced: Uncovering Patterns and Processes of Speciation

Mayr (1963) recognized three general classes of speciation. The first is *reductive speciation,* in which two existing species fuse to form a third. Harlan and DeWet (1963) proposed the term "compilo-species" for cases in which one species absorbs another; however, examples of this phenomenon have not been documented to date. The second is *phyletic speciation,* in which a gradual progression of forms within a single lineage is assigned species status at different points in time. As noted above, we consider each individually evolving lineage to be a single species; therefore, phyletic speciation represents evolutionary change within a single species, usually termed **anagenesis,** rather than a mode of species formation, or **cladogenesis.** The third class, *additive speciation,* is characterized by an increase in the number of species. The majority of speciation models, although based on several different mechanisms, are models of additive speciation. The most important thing to remember about speciation is not that it produces species, but that it produces *sister species,* so you cannot formulate explanations about speciation modes based on analysis of a single species. Rather, you need to examine sister species and clades.

Wiley (1981; see also Felsenstein 1981; Templeton 1982; Wiley and Mayden 1985) suggested that various models of additive speciation could be studied if phylogenetic, biogeographic, and population biological data were available, and if three assumptions could be met. First, character evolution must provide a reliable basis for reconstructing sequences of speciation events; that is, speciation has left a trace of its actions that we can detect. This assumption requires the operation of one of two processes, character evolution tightly coupled with speciation, or character evolution occurring at the same or higher rate than lineage splitting. So long as one of these processes is operating, even if the divergence of particular characters is not driving the divergence of the lineages, there will be a historical trail of character transformation highlighting speciation events. The second assumption is that there have been no extinctions in the clade. If we are to use phylogenetic trees to study particular modes of speciation, we must have confidence that sister species are each other's closest relatives and not, in reality, more distantly related due to the extinction of several unknown intermediate species.

The third assumption postulates that, if speciation has been associated with geographical changes, then we can reconstruct the original background of speciation events because the current distributions of descendant species do not differ dramatically from their original distributions. In other words, we assume that the dispersal of descendant species has not obscured the geographical context of the speciation events. This assumption requires closer inspection because of the confounding ways in which the movements of organisms and populations have been described in evolutionary biology. One of the most contentious words in biogeography and speciation has been *dispersal*. To some, the word refers to localized movements of organisms during short periods of time and is associated with concepts such as home range. To others, dispersal refers to the expansion and contraction of populations and species over longer periods of time. This form of dispersal is a necessary part of one class of speciation model (what we will call "allopatric speciation mode I" in chapter 4), because range expansion results in the widespread distribution of a species through space. This, in turn, increases the probability that geological changes will isolate large enough populations to form descendant species. A third meaning of the term dispersal is associated with another class of speciation models (what we will call "allopatric speciation mode II" in chapter 4), in which speciation is initiated by the dispersal of organisms into new areas. Thus, in some cases dispersal is not associated with speciation, in other cases it establishes the conditions under which speciation can be initiated, and in still other cases it is responsible for initiating speciation.

This highlights an important and often misconstrued aspect of the relationship between speciation and dispersal. The third assumption does not state that dispersal is unimportant, only that postspeciation dispersal does not overwhelm speciation patterns. Like many biological assumptions, this is a necessary starting point because without it we have no a priori justification for attempting to reconstruct speciation patterns, and thus no hope of studying the process. Unlike many biological assumptions, data are available to examine the validity of this supposition. For example, Lynch (1989) examined the distributions of sixty-six pairs of sister species and concluded that since "only 3–5 cases out of 66 possibilities reflect appreciable dispersal, then significant dispersal should not be envisioned as an important hypothesis [*in studies of speciation*]."

Speciation processes can be characterized in a number of ways. At the coarsest level there are two categories of models, those involving the physical disruption of gene flow by geographical isolation (allopatric modes), and those which do not require isolation for speciation to occur (nonallopatric modes). The allopatric category can be further subdivided depending on whether disruption of gene flow is accomplished through geological alteration (passive allopatric, or vicariant, speciation) or through movements of members of the ancestral species that eventually result in their geographical isolation (active allopatric speciation). In a different, but complementary, vein adaptive changes within populations play different roles in each of these three general classes of speciation processes. Adaptive changes are not required to initiate passive allopatric speciation, although they may accompany such speciation events; conversely, adaptive changes are often postulated to accompany active allopatric speciation and are a necessary component in initiating nonallopatric speciation.

In chapter 4 we will use "passive allopatric speciation" as the null hypothesis because, being independent of any particular underlying biological properties, it is a mode of speciation that could occur in any group of organisms. All that is required is for an ancestral species to "get separate and get different." Since most models were developed to explain the breakdown of a single ancestral species into descendants, an entire clade is not necessarily expected to be the product of a single speciation process, unless it is the "null" mode. From a conceptual standpoint then, uncovering incidents of active allopatric and nonallopatric speciation is just cause for celebration, because these modes represent departures from the historical background of vicariance and give us insights into the possible roles of a variety of environmental processes in speciation. As a consequence, historical ecological researchers must delve into the minutest details of each putative case of speciation within a clade in the quest to delineate and understand the patterns and processes composing the important evolutionary force of "speciation."

The Frequency of Different Speciation Modes

Theoretical studies of speciation have produced a plethora of models that are variations on the three themes of passive allopatric, active allopatric, and nonallopatric speciation. Now that theoretical biologists have delineated these models by mathematical and deductive reasoning, the next question is, How often does each of these speciation modes actually occur in nature? Unfortunately, investigating this question is hampered by the paucity of explicit species-level phylogenetic hypotheses. There has been only a single such study undertaken to date, but, as we will discuss in chapter 4, the results of that study are intriguing. In brief, Lynch (1989) discovered that vicariance (passive allopatric speciation), the most plausible speciation mode on theoretical grounds, also seems to have been the most prevalent on empirical grounds. Vicariant speciation requires only the physical disruption of gene flow; it does not require that speciation be initiated by adaptive processes (although it may be accompanied by such responses). Speciation modes, such as parapatric and sympatric speciation, that require adaptive changes to initiate and/or complete the process are relatively unlikely on theoretical grounds, and very few putative examples of these modes have been documented. This implies that speciation and adaptation need not always be tightly coupled evolutionarily. If that is true, the relationship between speciation, adaptation, and diversity may not be as straightforward as previously thought.

Macroevolutionary Trends in Diversity: Species Numbers

Other questions concerning numbers of species in evolutionary ecology have been formulated within two contexts. The first of these is comparison of species numbers in different environments; for example, Why are there more species in the tropics than in the polar regions? The second involves comparison of species numbers between different groups of organisms; for example, Why are there more species of insects than species of birds? The first approach has generally sought nonhistorical explanations, so there are, at present, no empirical studies available for discussion in a book dedicated to "historical ecology," although we will discuss a phylogenetic perspective on the question in chapter 4. We hope researchers will eventually become interested in examining the phylogenetic component of species distribution in different environments (see also Ricklefs 1989). The second question has traditionally incorporated some minimal concept of "taxonomic relatedness" into its explanations. The goal of research comparing species numbers between groups has been to determine if there are clades of "unusually high" or "unusually low" species number, and to attempt to explain those unusual groups. Although there have been numerous discussions of this problem, the lack of rigorous, objective criteria has proved to be a major impediment to its resolution. For example, how much asymmetry in species numbers can be considered unusually high or unusually low? Or, since the "number" of species within a given clade is strongly dependent upon the window of time through which one is viewing, on what temporal scale should an investigation be conducted? Answering this question may require more information than is readily available for most clades; therefore, there has been a tendency to focus only upon the diversity of extant species. Nonetheless, there is a perception that some groups are "more species-rich" than other groups that are somehow "equivalent," and that there is an evolutionary explanation for this difference. It is tempting to postulate that highly species-rich groups are, or have been, "better" or "more successful" than average, whereas groups of low species number are, or have been, somehow "less successful" than average.

Mayden (1986) took the first steps towards establishing a more rigorous foundation for assessing differences in species number between or among groups. He suggested that two standards were required. First, each group under examination must be monophyletic, and second, the groups being compared must be of equal antiquity. Because phylogenetic systematics stresses the recognition of monophyletic groups, an appropriate starting point for investigations into this component of diversity should be a phylogenetic analysis of the study organisms. A number of methods have been proposed for documenting the relative ages of clades. Stratigraphic and biogeographic analyses attempt to use environmental parameters as independent indicators of age. The use of "genetic distances" attempts to use a hypothetical "internal clock" that is universally informative about evolutionary rates. Phylogenetic systematics uses yet another criterion, sister-group relationships, because, by definition, each of two sister groups is the same age. If Mayden's conditions are met, groups of equal age may differ in species number either because of unequal rates of speciation or because of unequal rates of extinction. Assessing this component of diversity requires comparative analysis of speciation patterns and mechanisms.

Adaptation

We will address macroevolutionary questions concerning the degree of functional (ecological and behavioral) differentiation within and among groups of species in chapter 5. Since the beginning of modern evolutionary theory, this aspect of diversity, traditionally examined within a nonhistorical framework, has been a central focus of adaptation theory. We hope to show that there is generally a coherent phylogenetic sequence by which complex behaviors and ecologies are "assembled." Rarely do these character complexes arise de novo in the "lifetime" of any single species. We will also show, as a correlate, that current data suggest that ecological and behavioral diversification within groups is generally more conservative than morphological diversification and speciation. As a consequence, we will begin to view adaptation as a conservative, or cohesive, influence on evolution, complementing rather than causing diversification.

Contemporary adaptive explanations refer to an individuals' response to some problem set by nature. Adaptation is thus related to notions of a functional "fit" between an organism and its environment, implying that adaptive characters can be identified by seeking correlations between particular traits, or combination of traits, and the relevant environmental variable (for an excellent discussion of "adaptation," see Dunbar 1982). Traits that are "the same" in two species are postulated to have arisen as a common adaptive response to similar selection pressures. This leads to the assumption that, if the two species do not live in similar environments today, then they must have lived in similar environments in the past. Traits that differ between two species are postulated to have arisen as differential adaptive responses to different selection pressures. This leads to another assumption, namely that, if the species live in similar environments today, then either they lived in different environments in the past, or the current environments have not been partitioned on a fine enough scale to determine that they really are different in some evolutionarily significant way.

One way to characterize this traditional point of view is to contrast the traits ("similar" or "not similar") with the species possessing them ("related" or "not related": table 3.1). The classical explanation for unrelated taxa bearing the same trait is **convergent adaptation** (II), whereas the explanation for related taxa displaying different traits is **divergent adaptation** (III). This leaves two possibilities unexplored. In the case of unrelated taxa bearing different traits (IV) there is nothing of interest to study because there is no reason to believe the taxa have anything in common evolutionarily, either ecologically or genealogically. In the case of related taxa displaying the same trait (I), we are dealing with a phylogenetic scale that is uninformative to traditional evolutionary ecology, that is, there is nothing to study because the trait has not changed evolutionarily. Stabilizing selection is often invoked to explain the absence of change in this sort of situation, based on the assumption that the two related species occur in similar environments (or the environments have changed so recently that there has not yet been "time" for an adaptive response). Note that, in this approach to explanations, there is no

Table 3.1 Adaptation scenarios.

Traits	Taxa	
	Related	Not related
Similar	not interesting I	convergent adaptation II
Not similar	divergent adaptation III	not relevant IV

Note: Only type II and type III patterns are considered to be of interest to students of adaptation (Tinbergen 1964).

direct comparison between the species and its environment; the functional fit is assumed, the adaptation is assumed to be the explanation for the fit.

Historical ecology allows us to expand our evolutionary perspective to encompass macroevolutionary patterns (patterns that appear on protracted temporal scales) and thus dispense with the assumptions of adaptation theory listed above, without losing the robustness of adaptive explanations. When the four classes of patterns depicted in table 3.1 are examined within two environmental contexts, eight possible outcomes are produced from the interactions between phylogenetic and environmental information (table 3.2).

Table 3.2 Expanded adaptation scenarios based upon the interaction between phylogenetic and environmental information.

Traits	Environments	
	Similar	Not similar
Similar (homologous)	phylogenetic constraints (stabilizing selection?) I	phylogenetic constraints II
Similar (homoplasious)	convergent adaptation III	convergence not due to environment IV
Not similar (homologous transformation series)	divergence not due to environment V	divergent adaptation VI
Not similar (nonhomologous)	not relevant VII	not relevant VIII

Under this view, there are two manifestations of phylogenetic constraints. Patterns of type I may provide weak evidence of adaptation if a researcher wishes to invoke stabilizing selection to explain the absence of evolutionary change in a trait displayed by related species inhabiting similar environments (x_o in fig. 3.3a). This is a weak test of adaptation because there is no macroevolutionary evidence of a functional change involved with an environmental change. We need to increase the level of the phylogenetic analysis until the origin of x_o is uncovered. If there is a macroevolutionary association between a shift in the environment and a change in x_o, then we have strong support for the adaptational hypothesis for the trait. The type II pattern does not corroborate hypotheses of adaptive evolution because x_o does not change even though speciation has been accompanied by movement into new environ-

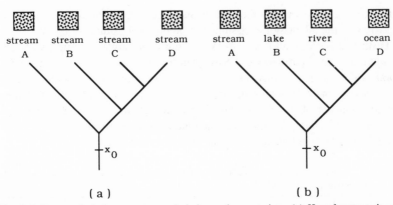

Fig. 3.3. Patterns depicting two types of phylogenetic constraints. (*a*) Homologous trait and similar environments. Postulate that stabilizing selection reinforces the maintenance of trait x_0 in all the members of the monophyletic group A + B + C + D. (*b*) Homologous trait and different environments: *environment changes, trait remains the same*. This does not corroborate a hypothesis of a "functional fit" between trait x_0 and the environment.

ments (fig. 3.3b). Of the eight patterns depicted in table 3.2, this is the strongest "falsifier" of adaptation scenarios. Types I and II are the "absence of evolutionary change" patterns; therefore they will serve as "null hypotheses" in the discussion of adaptational studies presented in chapter 5.

Patterns of type III provide strong support for convergent adaptation. This can be viewed as a form of environmentally driven change (fig. 3.4a). Pattern IV, on the other hand, postulates the appearance of convergent traits in different environments and thus does not support a hypothesis of convergent adaptation (fig. 3.4b). We suspect that this pattern usually represents instances of developmentally driven (or developmentally allowed) evolutionary change. Similar homoplasious characters keep arising regardless of the environment, perhaps due to the presence of a developmental program that is manifested in only a limited number of ways and can be "pushed" down the different pathways quite easily. For example, viviparity is a character that shows up repeatedly within several frog families whose members live in different habitats (Duellman 1985; see also chapter 5).

Patterns of type VI, providing support for divergent adaptation, can also be viewed as a form of environmentally driven change (fig. 3.5a). Pattern V, on the other hand, postulates the appearance of divergent homologous traits in similar environments and thus does not support a hypothesis of divergent adaptation (fig. 3.5b). This scenario depicts the classical case of passive allopatric speciation; that is, random developmental changes which are "allowed" by the developmental program, arise and are then incorporated into the isolated population following cessation of gene flow. Like pattern type IV, then, this incorporates a developmentally driven component into explanations of evolutionary change within groups of organisms.

There are two manifestations of evolutionary changes that are not related to each other either phylogenetically or ecologically. These situations are not pertinent to studies of adaptive change (VII and VIII). As with all complex systems, explanations about adaptational changes within a given clade often contain combinations of the above eight patterns. For example, the traditional story about the adaptive radiation of Galapagos finches invokes among-islands convergent adaptation (III) and within-island divergence in similar environments (IV: Coddington 1988).

We will examine six aspects of the historical ecological approach to adaptation theory within this reference framework in chapter 5: formulating hypotheses of adaptation within a phylogenetic framework, examining the temporal sequence of adaptive changes, examining the evolutionary association of traits ("coadapted traits"), examples of convergent adaptation, examples of divergent adaptation, and examining phylogenetic constraints on ecological diversification of characters and species (adaptive radiations). Reference

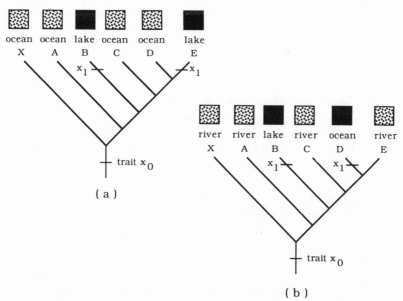

Fig. 3.4. Patterns depicting two types of convergent evolution. (a) Homoplasious traits and similar environments. Postulate that selection drives the modification of trait x_0 to x_1 in the similar lake environment. (b) Homoplasious traits and different environments: *both environment and trait change, but there is no evolutionary association between the changes.* This does not corroborate a hypothesis of a convergent "functional fit" between trait x_1 and the environment.

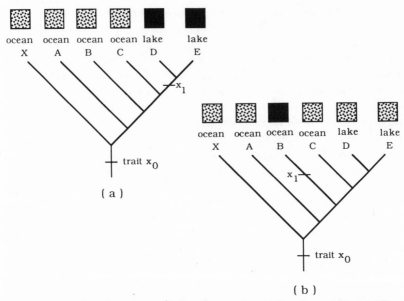

Fig. 3.5. Patterns depicting two types of divergent evolution. (*a*) Homologous traits and different environments. Postulate that selection drives the modification of trait x_0 to x_1 in association with the shift from an ocean to a lake environment. (*b*) Homologous traits and similar environments: *environment stays the same, trait changes*. This does not corroborate a hypothesis of a divergent "functional fit" between trait x_1 and the environment.

to a phylogenetic context will allow us to ask questions relevant to studies of adaptation for single traits, such as: Is the trait unique to one species? If the trait is unique to one species, is it an ancestral remnant, or is it recently evolved? If the trait is found in more than one species, how many times has the trait evolved? If we are concerned with possible evolutionary correlations among two or more traits, we may use phylogenetic analysis to identify cases in which the traits do not co-occur and did not co-originate in any members of the study group, other cases in which the traits co-occur but did not co-originate in some members of the study group, and yet other cases in which the traits both co-occur and co-originated in members of the study group. *One of the most important contributions of phylogenetic analysis to studies of adaptation is that it allows us to unveil traits that are persistent ancestral characters, and not adaptations in the species at hand.* Researchers interested in the dynamics of evolutionary change can thus avoid studying adaptation in species that have simply inherited the trait(s) and focus their attention on associations between character modifications and the original environmental context.

Summary

When biologists explain the evolution of taxa ("taxic evolution"), they refer to speciation. When they explain the evolution of traits ("transformational evolution"), they refer to adaptation. Hence, speciation and adaptation are cornerstones of evolutionary theory. And yet, the relationship between speciation and adaptation in evolutionary theory has always been ambiguous. Darwin (1872) considered the strongest evidence in favor of evolution, as a genealogical process, to be the existence of two related species displaying the same ecological and behavioral traits in different environments. By contrast, he also felt that adaptation to different environments would produce new species, emphasizing the ecological aspects of evolution. We still have that ambiguity between speciation and adaptation in evolutionary theory today. Some modes of speciation require adaptive changes, others do not. Some adaptive changes lead to speciation, others are not accompanied by speciation. In either case, the situation is clouded when the origins of characters and species are confused with the persistence of characters and species in evolutionary explanations. In the following two chapters we will begin to discuss how phylogenetic systematics, employed within a historical ecological framework, offers one pathway out of this particular confusion into a realm of more robust, unified evolutionary theory.

4 Speciation

Historical ecologists have a perspective that speciation does not just produce species, it produces **sister species.** Since this irreversible production of groups that are each other's closest relatives introduces a historical component into the process, speciation cannot be studied without first determining the sister-group relationships within the system of interest. Assuming that two species are or are not "closely related," and basing hypotheses of the speciation model involved in their production on this assumption, will, in most cases, ultimately lead to confusing and contradictory results.

Mayr (1963) recognized three general classes of speciation. In *reductive speciation* two existing species fuse to form a third. In *phyletic speciation* a gradual progression of forms through a single lineage (anagenesis) is assigned species status at different points in time. Although the endpoints of such a continuum may be recognizably "different," separation of the intermediate forms into distinct groups is an inherently arbitrary exercise (Hennig 1966; Wiley 1981). Additionally, since we consider each individually evolving lineage to be a single species, "phyletic speciation" can only represent intraspecific evolutionary change, that is, change preceding or following, but not correlated with, speciation. *Additive speciation* involves lineage splitting (cladogenesis) and reticulate evolution. The majority of examples of speciation represent cases of additive speciation. No single mechanism is responsible for the initiation of additive speciation (Wiley 1981). Several mechanisms have been proposed, and we will consider each as a distinct "model" of speciation. Our discussion will rely heavily on the methodological framework incorporating phylogenetic patterns with biogeographic and population information provided by E. O. Wiley (see especially Wiley 1981; Wiley and Mayden 1985). Wiley's evolutionary detective work, in turn, has been based on the pioneering studies of Mayr (e.g., 1954, 1963), Bush (1975a,b), Endler (1977), White (1978), Wright (1978b), Lande (1980a, 1981), Templeton (1980, 1981, 1982), and Felsenstein (1981).

Assumptions of a Speciation Study

Wiley (1981; see also Wiley and Mayden 1985) suggested that the various models of additive speciation could be studied by establishing phy-

logenetic, biogeographic, and population biological predictions corresponding to each model. In order to begin such extensive studies, three assumptions
concerning the nature of the data must be satisfied. First, character evolution
must provide a reliable basis for reconstructing sequences of speciation
events, that is, speciation has left a trace of its actions that we can detect.
This assumption requires that one of two processes is occurring: either character evolution is tightly coupled with speciation, or character evolution occurs at the same or higher rate than lineage splitting. Thus, even if the divergence of particular characters is not driving the divergence of the lineages,
there will be historical trail of character anagenesis highlighting speciation
events. Although the second condition represents the traditional perspective
of evolutionary biologists, the recent advent of punctuated equilibrium models (Eldredge and Gould 1972) has strengthened the proposition of a causal
relationship between character modification and speciation. The first assumption is violated if gene flow is halted permanently between populations at a
faster rate than character change is occurring. If this happens, the traits present in each species will represent a combination of (1) characters that existed
prior to the isolation of the populations, providing information about common ancestry (symplesiomorphies), and (2) evolutionary modifications that
occurred subsequent to the population's isolation, providing information
about the unique status of the population (autapomorphies). Since derived
traits are not shared between populations under these circumstances, sequences of speciation events will be difficult or impossible to determine (we
will discuss this more in a later section) (fig. 4.1).

The second assumption is that there have been no extinctions in the clade.
If we are to use phylogenetic trees to study particular modes of speciation,
we must have confidence that sister species are each other's closest relatives
and not, in reality, more distantly related due to the extinction of several
unknown intermediate species. Consider the following hypothetical example.
Two groups of fish, demonstrated to be sister species on the basis of a phylogenetic analysis, are located on either side of a mountain range (fig. 4.2a).
Based on these observations, we might hypothesize that the disjunct distribution was caused when the upheaval of the mountains separated the ancestral
species into two populations, which subsequently diverged in isolation (fig.
4.2b). Unfortunately for our theory, a group of enthusiastic paleontologists
discover an abundance of fossil evidence suggesting that at least two other
species fall between the extant representatives (fig. 4.2c). Hence, the current
disjunction of fishes B and C was probably derived through a series of speciation **and extinction** events, only one of which need have been associated
with the tectonic activity (fig. 4.2d).

The third assumption postulates that the influence of geographical separation during the evolutionary divergence of a clade has not been obscured by
rampant dispersal of the descendant species (fig. 4.3). Pairs of sister species

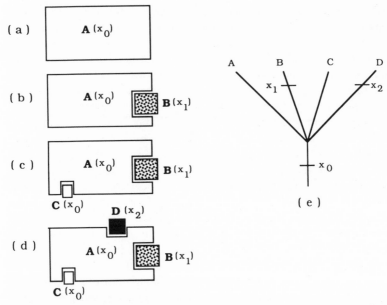

Fig. 4.1. Problems arising when gene flow between populations is severed faster than character change occurs. (*a*) Ancestral species A bearing character state x_0. (*b*) Speciation producing B accompanied by the divergence of x_0 to x_1. (*c*) Speciation of C, no character change. (*d*) Speciation of D accompanied by the divergence of x_0 to x_2. (*e*) Phylogenetic tree reconstructed by changes in character x. Note that there is no way to differentiate between A and C, and that it is impossible to reconstruct the temporal sequence of speciation because the derived characters x_1 and x_2 are autapomorphies.

or clades that show such dissemination may be identified by large-scale sympatry; however, uncovering such sympatry creates a problem because it is difficult to determine whether the current distribution pattern existed during the speciation of the group, or whether it represents widespread dispersal following speciation in isolation.

It is probably true that many groups will not satisfy all the assumptions; however, until a larger data base is established, it is impossible to determine whether these nonconformists need be accorded the status of an overwhelming majority or a confounding minority. We are confident that numerous clades will emerge in which phylogenetic patterns and distribution patterns are congruent with predictions from particular speciation models.

Phylogenetic Patterns of Speciation

Allopatric Speciation

"Allopatric" speciation is a generic term for models that invoke the complete geographical separation of two or more populations of an ancestral spe-

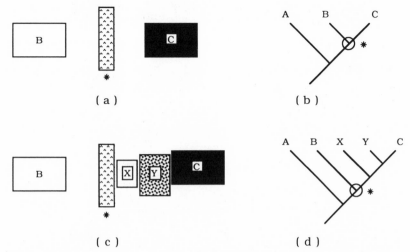

Fig. 4.2. Problems arising from extinctions. (a) Two fishes, B and C, are located on either side of a mountain range. (b) Phylogenetic tree for the genus containing species B and C. *Open circle* = the speciation event; * = the upheaval of the mountains. (c) Fossil evidence of extinct species X and Y on the same side of the mountains as extant species C. (d) New phylogenetic tree incorporating fossils. According to this new hypothesis, the mountains may have played a role in the production of species B and the ancestor of the X + Y + C clade.

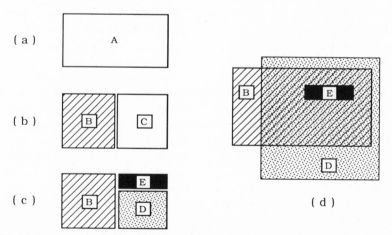

Fig. 4.3. Problems arising from widespread dispersal of descendant species. (a) Ancestral species A. (b) Geographic separation of A produces descendant species B and C. (c) Geographic separation of C produces descendant species D and E. (d) Rampant dispersal of descendants B and D produces a current distribution pattern of widespread sympatry and obscures the original pattern of geographic disjunction.

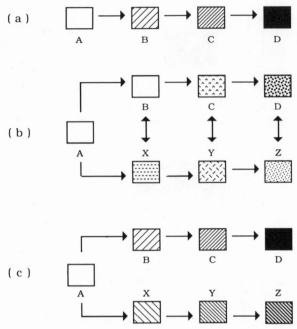

Fig. 4.4. Three types of population structure. (*a*) A single deme evolving through time. (*b*) Two demes, linked by gene flow, evolving through time. (*c*) Two demes with no gene flow, evolving through time.

cies to initiate speciation. Distinguishing the allopatric models from one another requires that we answer three questions: (a) Were the disjunct populations created by the actions of geological processes (*passive role for the ancestor*) or by the dispersal of some ancestral individuals over preexisting barriers (*active role for the ancestor*)? (2) Was gene flow among ancestral populations *present* (fig. 4.4a and b) or *absent/rare* (fig. 4.4c) prior to the isolating event? This is an important question because the rate of speciation will be affected by the interaction between local differentiation in response to selection, which tends to promote speciation (e.g., Fisher 1930; Wright 1931, 1940, 1978a; Haldane 1932; Lande 1980a, 1981; Templeton 1980, 1981, 1982; Coyne and Kreitman 1986), and the *cohesive* forces of persistent ancestral traits and gene flow among demes, which tend to inhibit speciation (Wiley 1981; Wiley and Brooks 1982; Brooks and Wiley 1986, 1988; Templeton 1989). (3) Was the ancestral population equally subdivided, or was only a very small part of the ancestral range "budded off" from the rest of the species range (or was the division somewhere between these two extremes)?

Bearing these questions in mind, let us examine the phylogenetic patterns predicted by the three allopatric models (as summarized by Wiley 1981; Wiley and Mayden 1985; Funk and Brooks 1990).

Allopatric speciation mode I

Usually called **vicariance,** or geographic speciation, allopatric speciation mode I combines gene flow among populations prior to separation with a passive role for range changes in the ancestral species. It occurs when an ancestral species is geographically separated into two or more relatively large and isolated populations, with subsequent lineage divergence by the isolated descendant populations (fig. 4.5). The speciation rate will depend on the degree of variation in the ancestral species prior to isolation and the rate of origin of evolutionary novelties in the subdivided populations. Three predictions from this model are of interest to students of speciation: (1) The phylogenetic tree for the group will be predominantly dichotomous because the fragmentation of the ancestral species and concomitant interruption of gene

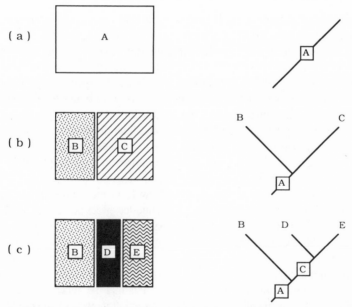

Fig. 4.5. Allopatric speciation mode I. (*a*) Species A extends throughout a geographical area. (*b*) The species is divided by the appearance of some geographical barrier preventing gene flow; the two populations continue to evolve independently of one another, producing new species B and C. (*c*) Species C undergoes another geographical upheaval, gene flow is eliminated, and changes continue in isolation, leading to the eventual production of new species D and E. The outcome of this division of space through time is the production of three extant species (B, D, and E) and the extinction through total speciation of two ancestors (A and C).

flow among the isolated populations makes is unlikely that either of the descendant species will be identical to the ancestor or to each other. In this case, the ancestor experiences "extinction through total speciation." (2) The points of geographical disjunction between sister species will correspond to the historical boundaries established by the geological changes. Based on this, the ancestral range may be estimated by combining the distributions of the descendant species, assuming no substantial range expansion or contraction following speciation (fig. 4.3). (3) A multitude of ancestral species, fragmented in the same way by the same geological event, could all theoretically speciate subsequent to the event, because the mechanism initiating speciation is independent of any particular biological system. Hence, we would expect to find the same biogeographical distribution pattern shared by a number of different clades. The research program called "vicariance biogeography" relies on this mode of allopatric speciation to detect episodes of parallel biological and geological evolution (see chapter 7).

Allopatric speciation mode II

Allopatric speciation mode II, more commonly known as **peripheral isolates allopatric speciation** or **peripatric speciation,** postulates that a new species arises from a small, isolated population usually, but not always, on the periphery of the larger central ancestral population. Gene flow between the peripheral and central populations contributes to species cohesion, because it is initially sufficient to keep novel traits from being fixed; however, it is not strong enough to prevent the establishment of novel phenotypes in the peripheral population. The foundation and final disjunction of the peripheral population that initiates speciation may involve either a passive or active role for the ancestor. When the new species arises in a geographical locality not previously occupied by the ancestor, an active role in invoked. Once geographic separation is complete, gene flow from the central population is stopped. This could happen rapidly, as in founder-effect phenomena (Carson 1975, 1982; Templeton 1980; Lande 1981; Carson and Templeton 1984; Goodnight 1987; Charlesworth and Rouhani 1988; Barton 1989), or it could be a relatively gradual process, such as a gradual environmental change in the peripheral area (Mayr 1954, 1963, 1982; Patton and Smith 1989). Unlike allopatric speciation mode I, this model predicts that the ancestor will persist after the speciation event, because its fragmentation was so asymmetrical. When we say that the ancestor "survives," we mean that it does not exhibit any evolutionary change correlated with the speciation event (i.e., the peripheral descendant shows all the divergence). Because the peripheral population is small and thus released somewhat from the homeostatic constraints of large-scale gene flow, the peripheral descendant will exhibit more autapomorphic traits than the central population, even if that central population experiences its own anagenetic events or eventually speciates again. Hennig (1966) termed the general observation that one of two descendants tended to be more divergent from the ancestor than the other his "deviation rule."

There are three different pathways to this speciation mode, so there are several predicted phylogenetic and biogeographic patterns. In "classic" peripheral isolation (Wiley and Mayden 1985), there are both ecological and phylogenetic components to speciation, and the ancestor is accorded an active role. The assumption is that populations in peripheral habitats are initially free to diverge evolutionarily from their ancestor because of reduced gene flow (Mayr 1963; Hennig 1966; Brundin 1966), and that this divergence, in turn, may be reinforced by local adaptive responses to the new habitat (Hennig 1966) and/or genetic drift and founder effects. Proponents of this speciation mode often assume that peripheral habitats are necessarily marginal in composition. Lynch (1989) pointed out the dangers in equating "different" with "marginal." This is an important distinction because it allows the potential for evolutionary change to occur in response to the presence of a new selective regime, without invoking the assumption that the habitats occupied by the central populations are somehow "better" than the habitats occupied by their peripheral counterparts. If peripheral isolation is due to random settlements of ancestral individuals around the margins of the species' range, we expect to find phylogenetic trees comprising polytomies in which the number of terminal taxa equals the number of peripheral descendants "budded off" from the ancestor, plus the ancestor itself (fig. 4.6). Biogeographically, we would expect to find similar patterns of distribution only among clades with similar ecological requirements.

If isolation is due to ancestral individuals dispersing into a new habitat, experiencing peripheral isolation, and speciating, followed by movement of some members of the new species into another peripheral area, repeating the process, we would expect to find dichotomous phylogenetic patterns reflecting the alternating episodes of dispersal and isolation (Hennig's "progression rule"; fig. 4.7). An excellent example of this type of pattern might be found among organisms that have speciated repeatedly during progressive colonization of island archipelagos. Unlike allopatric speciation mode I, which predicts that all clades segregated by a vicariance event could theoretically show the same biogeographical patterns, similar patterns are expected here only for clades demonstrating analogous ecological requirements and dispersal abilities.

The final form of peripheral isolates allopatric speciation, "vicariant" peripheral isolation (Wiley and Mayden 1985), invokes a passive role for the ancestral species. Specifically, when large chunks of ancestral populations are subdivided geographically, we speak of vicariant speciation (allopatric spe-

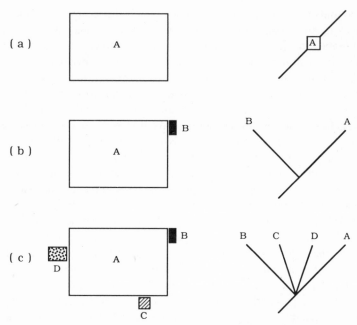

Fig. 4.6. Allopatric speciation mode II, peripheral isolates allopatric speciation via random dispersal. (*a*) Species A extends throughout a geographical area. (*b*) Some individuals disperse into a new area; gene flow is severed, producing species B. (*c*) Random dispersal and severing of gene flow results in the budding off of descendant species D and C. The traits present in each species represent a combination of characters that existed prior to the isolation of the population (symplesiomorphies) and evolutionary modifications that occurred subsequent to the isolation (autapomorphies). Since derived traits are not shared between populations under these circumstances, sequences of speciation events will be impossible to determine, and the resultant phylogenetic pattern will be a polytomy.

ciation mode I), and when small chunks are subdivided, we speak of this form of peripheral isolates speciation (fig. 4.8). Lynch (1989) has termed this "microvicariance."

Since the mechanism initiating speciation (vicariance) is independent of any particular biological system, there is a phylogenetic component to speciation (the relationship between ancestor and descendant) but not necessarily an ecological one. If more than one small group is isolated by the vicariance events, we would expect phylogenetic patterns showing polytomies, and would expect to find similar biogeographical distribution patterns only for other species that were fragmented by the same geographic events.

The peripheral isolates model is the traditional favorite of evolutionary biologists (Mayr 1963) because it combines geographically restricted gene flow with the exposure of small populations to new selection pressures from the peripheral habitats, which, in turn, is postulated to reinforce evolutionary divergence. However, Barton and Charlesworth (1984; see also Barton 1989) have recently questioned the likelihood of this mode of speciation. Biologists have been particularly interested in the possibility that an ancestor occupying a large central range can remain relatively unchanged for a long period of time, become extinct, and, in a relatively short time, have its range reclaimed by descendant species that have been evolving in peripheral areas. If this happens, the fossil record will show the long-term persistence of a "static" ancestral species, followed relatively quickly by its extinction and replace-

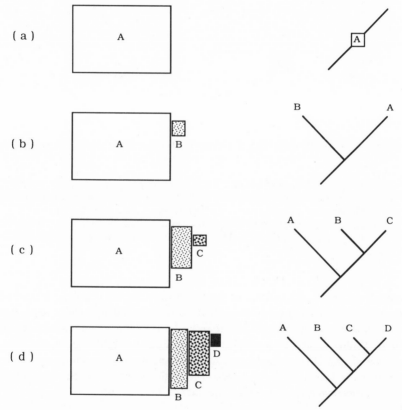

Fig. 4.7. Allopatric speciation mode II, peripheral isolates allopatric speciation via sequential dispersal. (a) Species A extends throughout a geographical area. (b) Some individuals disperse into a new area; gene flow is severed, producing species B. (c) Individuals from species B disperse into a new area; gene flow is severed, producing species C. (d) Individuals from species C disperse into a new area; gene flow is severed, producing species D. In this case, sister species can be identified by the presence of shared derived traits; therefore the resultant phylogenetic pattern will be dichotomous.

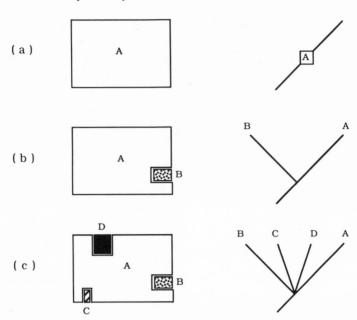

Fig. 4.8. Allopatric speciation mode II, peripheral isolates allopatric speciation via microvicariance. (*a*) Species A extends throughout a geographical area. (*b*) Gene flow is severed in a small area, producing species B. (*c*) Severing of gene flow by a series of microvicariance events results in the budding off of descendant species D and C. The traits present in each species represent a combination of characters that existed prior to the isolation of the population (symplesiomorphies) and evolutionary modifications that occurred subsequent to the isolation (autapomorphies). Since derived traits are not shared between populations under these circumstances, sequences of speciation events will be impossible to determine, and the resultant phylogenetic pattern will be a polytomy.

ment by a descendant bearing novel traits (Mayr 1954). This combination of peripheral isolates allopatric speciation, extinction of the ancestor, and range expansion of the descendant has come to be known as "punctuated equilibrium" (Eldredge and Gould 1972; see also Futuyma 1986) because it looks like rapid speciation pulses are occurring sporadically throughout evolutionary time. In fact, it is the disappearance of the ancestor, not the appearance of the descendant, that is sudden and unexpected.

Allopatric speciation mode III

Some species exist as several disjunct populations without appreciable gene flow among them. Under these circumstances, species cohesion is provided only by the constraining influences of developmental homeostasis (Eldredge and Gould 1972; Gould and Eldredge 1977), as evidenced by the persistence of ancestral traits in the populations. Depending upon the strength of these constraints, the species may remain in a state of relative evolutionary stasis for long periods of time. Speciation occurs whenever a deme becomes fixed for a novel phenotype; therefore, phylogenetic trees will be composed of one branch for each ancestral deme that speciates, plus a branch for the collection of unmodified ancestral demes (fig. 4.9).

Stasipatric speciation (White 1978; see also Key 1968; Patton and Sherwood 1983; Thompson and Sites 1986) is a special case of allopatric speciation mode III that invokes a specific genetic mechanism to explain population differentiation. More particularly, population differentiation is hypothesized to occur via chromosomal mutations, coupled with genetic drift or meiotic drive to fix the mutation in the population. Although Futuyma and

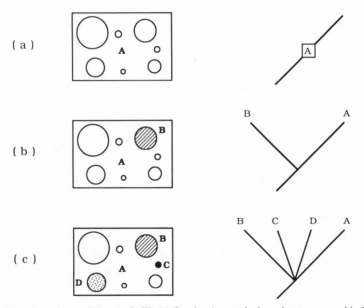

Fig. 4.9. Allopatric speciation mode III. (*a*) Species A extends throughout a geographical area and comprises several populations. Gene flow among the populations is either insignificant or absent. (*b*) Evolutionary change in one of the populations at any position in the species range, due to chance (genetic drift) or selection, produces species B. (*c*) Evolutionary change in two of the populations, due to chance (genetic drift) or selection, produces species D and C. The traits present in each species represent a combination of symplesiomorphies and autapomorphies. Since derived traits are not shared between populations under these circumstances, sequences of speciation events will be impossible to determine, and the resultant phylogenetic pattern will be a polytomy.

Mayer (1980) have questioned the reality of this model, based on population genetical arguments, groups that are suspected of having been produced by stasipatric speciation have not yet been investigated phylogenetically.

Aside from the recognition that the ancestral species is a collection of allopatric demes, there are no biogeographical correlates for this model, because speciation occurs within demes that are already disjunct. If the ancestral species is geographically widespread, the distance between disjunct populations, combined with their phenotypic similarity, is guaranteed to provide an inordinate number of sleepless nights for systematists. For example, despite having no demonstrable qualitative or quantitative phenotypic differences, two species of tropical plants, *Acacia heterophylla* in the Mascarene Islands and *A. koa* in the Hawaiian Islands, have been assigned separate species status on the basis of their vast geographical separation (see Geesink and Kornet 1989 for a discussion of this and other examples). Although some sexually reproducing species may show this type of spatially disjunct population structure, this is primarily a speciation model for asexual organisms. Given the number of asexual species on this planet, it is unfortunate that studies based on allopatric speciation mode III are so scarce.

Parapatric and Alloparapatric Speciation

Parapatric speciation (see Endler 1977; Lande 1982; Barton and Charlesworth 1984) occurs when two populations of an ancestral species differentiate into descendant species despite the maintenance of some gene flow and geographical overlap during the process (fig. 4.10). Stochastic events (e.g., drift) and/or adaptive responses to local selection pressures initiate the differentiation, which is then promoted by low vagility among members of the populations (decreasing gene flow even when sympatric) and/or a decrease in heterozygote/hybrid fitness leading to positive assortative mating.

A related mode, alloparapatric speciation (Mayr 1942; Dobzhansky 1951; Key 1968; Endler 1977), occurs when allopatric populations of an ancestral species begin to differentiate during the period of allopatry, become sympatric

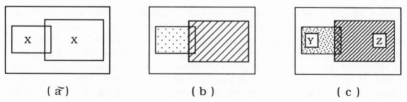

(ã)   (b)   (c)

Fig. 4.10. Parapatric speciation. (*a*) Overlap of two populations of ancestral species X. (*b*) Differentiation of populations begins while they are still in contact. (*c*) Speciation of Y and Z is completed despite maintenance of the contact area.

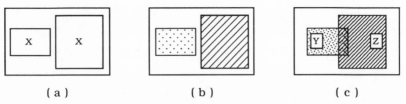

(a)   (b)   (c)

Fig. 4.11. Alloparapatric speciation. (*a*) Two populations of ancestral species X are separated geographically. (*b*) Differentiation of populations begins while they are allopatric. (*c*) Speciation of Y and Z is completed when contact is established between the diverging populations.

over a limited area, and complete their divergence because of interactions between the differentiated populations in the zone of sympatry that reinforce the differentiation (fig. 4.11).

These two speciation modes differ primarily in two ways. First, in parapatric speciation the zone of sympatry between two sister species is a primary zone of contact (i.e., the species have always been in contact at that point), whereas in alloparapatric speciation the zone of sympatry is an area of secondary contact. Second, in the parapatric model, population differentiation begins *in spite of* any interactions between populations, while in the alloparapatric model, differentiation begins in isolation. Once contact is established, both models postulate that speciation is completed *because of* the interactions among the differentiating populations in the areas of overlap. These models are difficult to study because, with the exception of the zone of sympatry, their phylogenetic and biogeographic predictions do not differ from those for allopatric speciation.

Endler (1977) presented a detailed defense of parapatric speciation in his extensive treatise on the microevolutionary aspects of geographical and clinal variation. He suggested that members of (1) the anuran *Rana pipiens* group, (2) the mosquito-fish genus *Gambusia*, (3) the fruit-fly genus *Drosophila*, (4) the plant genus *Gilia*, (5) the frogs *Hyla ewingi* and *H. verreauxi*, and (6) the frogs *Pseudophryne dendyi*, *P. bibroni*, and *P. semimarmorata* might all be examples of parapatric speciation. In the case of *H. ewingi* and *H. verreauxi*, hybrids from the zone of sympatry showed depressed fitness relative to hybrids from parents taken from allopatric portions of the species population. This satisfies one of the conditions of both parapatric speciation models; however, since Endler did not present evidence of the phylogenetic relationships of the groups under investigation, we are not certain that any of the species pairs are sister species. There are also inherent problems in attempting to discern primary from secondary contact zones. For example, Woodruff (1972) suggested that the three species of *Pseudophryne* frogs originated allopatrically and became secondarily parapatric. In this case it is difficult to ascertain whether the parapatric contact had anything to do with the comple-

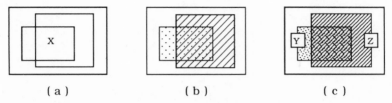

Fig. 4.12. Sympatric speciation. (*a*) Extensive overlap of two populations of ancestral species X. (*b*) Differentiation of populations begins while they are still in contact. (*c*) Speciation of Y and Z is completed despite maintenance of the contact area.

tion of speciation. If it did, this favors an interpretation of alloparapatric, rather than parapatric, speciation.

Sympatric Speciation

Sympatric speciation (Maynard Smith 1966; Dickinson and Antonovics 1973; Felsenstein 1981; Gittenberger 1988) occurs when one or more new species arise without geographical segregation of populations (fig. 4.12). Unlike the allopatric models, which postulate that gene flow between populations is initially severed by factors extrinsic to the biological system, sympatric speciation requires the involvement of biological processes intrinsic to the system, for example, hybridization, ecological partitioning, the evolution of asexual or parthenogenetic populations, or a change in mate recognition. Additionally, differentiation must occur "within the dispersal area of the offspring of a single deme [the cruising range]" (Mayr 1963:257).

Although this was the mode originally preferred by Darwin (1859), support for sympatric speciation wavered when population geneticists demonstrated that the effects of gene flow among populations would tend to swamp out or homogenize any novel traits arising within a population. If gene flow were restricted or interrupted, as in the allopatric or parapatric speciation models, the novel trait would have a better chance of becoming fixed within a deme, and the whole process would operate much more smoothly. The work of the population geneticists was coupled with the earlier recognition that most "related" species (this usually meant members of the same genus) exhibited allopatric distributions (e.g., Mayr 1942; Wallace 1955), and this combination provided a strong foundation for the hypothesis that most speciation was allopatric. However, in recent years there has been a revival of interest in the possibility of sympatric speciation modes, as researchers have intensified investigations of mechanisms of phenotypic plasticity, disruptive selection, and chromosomal divergence (see, e.g., discussion and references in West-Eberhard 1989; papers in Otte and Endler 1989). One of the most eloquent supporters of this and other nonallopatric models has been Guy Bush (e.g.,

1975a,b, 1982; Diehl and Bush 1989), who suggested that it is unnecessary to postulate a link between speciation and adaptation in allopatric speciation models, while in nonallopatric models (Bush 1982),

> speciation is the direct outcome of adaptation, and divergence occurs as a product of selection for habitat preference, competition, and selection to enhance reproductive isolation.

Sympatric speciation requires observations of the sympatric distribution of sister species that differ in some special ecological or genetic characteristics that could, in themselves, produce independent species. Phylogenetic trees reflecting incidents of sympatric speciation may be either dichotomous or polytomous, depending on how many species have been produced sympatrically from the same ancestor, and depending on whether or not the ancestor persists. Biogeographically, this mode requires that sister species be broadly sympatric today and *at the time of speciation*. Observing that the two species are sympatric today is not sufficient evidence of either sympatry in the past or a sister-group relationship.

Sympatric speciation by ecological segregation

The most controversial form of sympatric speciation proposes that evolutionary divergence has been driven solely by ecological segregation, usually studied in terms of host (habitat) switching (see, e.g., Diehl and Bush 1989; Grant and Grant 1989; Tauber and Tauber 1989). This mode of speciation is problematical because it is at once theoretically attractive and perplexingly paradoxical. The attraction lies in the models' invocation of adaptive processes to drive speciation. The paradox is twofold. First, once colonization of a new type of resource (habitat or host) within the ancestral-species range has occurred, the probability that the new resource will exert strong directional selection pressure on the colonizing population should be higher for species displaying pronounced habitat specificity. These species are more tightly coupled to their resource bases and thus should be more sensitive evolutionarily to changes in that component of their environment than their generalist counterparts. However, the likelihood of a habitat change occurring in the first place is decreased for species that respond to only a small number of cues. Therefore, the species least likely to colonize new habitats (specialists) are the ones most likely to speciate as a result of any such switch, while the species most likely to colonize new habitats (generalists) are the least likely to speciate as a result of the interaction.

The second dilemma arises because, when habitat switching means host switching, the switches can occur only while the hosts (and thus, the initial associate population) are *sympatrically distributed*. However, speciation may *speciation*) or after the hosts have been separated (*allopatric speciation*). When sister species are associated with allopatric hosts, it is often impossible to determine whether the actual speciation event producing those species occurred before or after the hosts's isolation. Just as the observation of current sympatric overlap between two "related" species is a weak test of sympatric speciation without the relevant phylogenetic information (i.e., Are the taxa sister species?), knowledge that two sister species show evidence of host switching in their evolution is not sufficient to invoke sympatric speciation in the absence of current sympatric overlap between them. Of course, it is always possible to invoke the effects of "sympatric speciation in the past," followed by vicariance, to explain the observed disjunct distribution, but this is a weak hypothesis at best. Overall then, the strongest evidence for speciation via host switching is provided by situations in which a sympatric overlap of associate sister species can be demonstrated (fig. 4.13).

There are a variety of population genetical models explaining the origins of reproductive isolation under a host-switching regime (see, e.g., Maynard Smith 1966; Dickinson and Antonovics 1973; Caisse and Antonovics 1978; Wood and Guttman 1983; Sturgeon and Mitton 1986; Thompson 1988). Futuyma and Mayer (1980) explored putative examples of sympatric speciation by host switching within this genetical framework and concluded that these examples were not particularly convincing. They focussed their attention on two paradigm cases involving predatory lacewings of the genus *Chrysopa* (see Tauber and Tauber 1977a,b) and true fruit flies of the genus *Rhagoletis* (see also Bush 1966, 1969, 1974, 1975a,b; Berlocher and Bush 1982). In the lacewing example, there are two species, *Chrysopa downesi* and *C. carnea,* the first a specialist feeder on conifers and the second a generalist feeder in meadows. There is no phylogenetic evidence that they are sister species, or biogeographic evidence that they are primarily sympatric. Hence, the traits by which these species can be distinguished and by which they are ecologically segregated today may, or may not, have evolved in a single ancestral species under conditions of sympatry. The fruit-fly example is somewhat more complex because it is purportedly an example of sympatric speciation in progress. The debate centers on the association of the North American *Rhagoletis pomonella* "host race" complex with hawthorn, apple, and cherry hosts. Futuyma and Mayer amalgamated the results from a variety of studies and concluded that there was no evidence that the putative host races were existing in either behaviorally or genetically differentiated forms. This, however, does not detract from the exciting discovery that the researchers have potentially tapped into an actual speciation event rather than, as is generally the case, being restricted to documenting the historical traces of the process. Determining that the "races" are not genetically distinct at this point in time does not rule out a role of sympatric speciation. There is really nothing to do at the moment but sit back and watch the process.

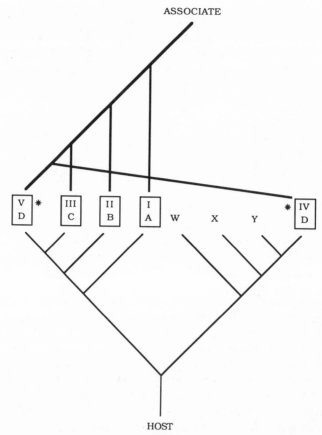

Fig. 4.13. The "best case" scenario for postulating the existence of sympatric speciation via host switching. The *letters* on the host tree refer to the areas where the hosts are found; thus the two different host species marked with *asterisks* inhabit the same area (area D). When the phylogeny for the associates is compared with the phylogeny for the hosts, we find that all the branches match except the relationship between associate IV and its host. Branches that match represent cases of cospeciation between the associate and its host. Branches that are "out of synch" represent cases of host switching (this will be explained in detail in chapter 7). In this case, the switch and subsequent speciation occurred in a host that was sympatric with the historical host group. The observation of sister species in sympatric hosts is strong evidence for sympatric speciation.

Sympatric speciation by hybridization

Regardless of the outcome of the preceding studies of ecological segregation, there are other mechanisms of genetic change that can result in sympatric speciation. One of these, speciation by hybridization (see Harrison and Rand 1989; Hewitt 1989; Wake, Yanev, and Frelow 1989), is a phenomenon of particular importance to diversification among plant groups, certain groups

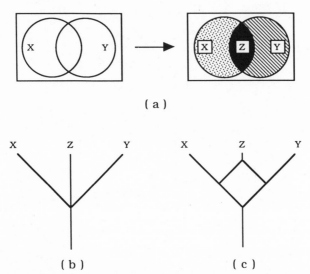

(a)

(b)   (c)

Fig. 4.14. Sympatric speciation by hybridization. (*a*) Two species, X and Y, exist in sympatry in a portion of their range. In this area of sympatry, a third species, Z, arises due to hybridization between X and Y. (*b*) Branching pattern expected if X, Y, and Z are not related by hybridization. (*c*) Reticulate branching pattern that arises when Z is a hybrid of species X and Y.

of freshwater fishes, and lizards (fig. 4.14a; Grant 1981). This mode of speciation produces reticulate (fig. 4.14c) rather than hierarchical (fig. 4.14b) patterns of phylogenetic relationships for a group of species, one of which is a hybrid of two others. (See Funk 1985 and Funk and Brooks 1990 for detailed discussion of phylogenetic protocols for detecting species of hybrid origin and depicting their relationships on a phylogenetic tree.)

One of the most intriguing aspects of speciation by hybridization is that it may lead to three different classes of phenotypic and ecological outcomes. In the first case the hybrids segregate phenotypically and ecologically with one of the parents and are thus subjected to the same selection regime as that parent. Since the parent is already surviving in the environment, it is likely that the hybrids will survive as well, assuming that necessary resources are not limiting. Mixed stands of such "species groups" have been documented. It is also possible that, under conditions of limited resources, interspecific competition will occur, since new genetic information has been added to the system without enlarging the available resource base. In the second case the hybrids display a mixture of parental attributes, some of which may be intermediate in nature. If such hybrids are capable of living in a wider range of habitats than either parent, they should have a good chance of survival because, although subject to a wider range of selection pressures, their flexibility will reduce the likelihood of competition with each parent. If adaptive processes in evolution tend to promote specialization, this is an evolutionary mechanism for producing new generalists. Finally, the hybrids could represent a unique phenotypic and ecological system. In this case survival will be more problematical because the new combination must correspond to one allowed by un- or underexploited resource bases in the local environment. However, such hybrids, if they survive, should face no competition from either parent. Each of these three survival pathways open to species of hybrid origin implies different microevolutionary scenarios. Testing these scenarios requires knowledge of parental identity as well as the degree of phenotypic and ecological similarity between parental and hybrid species.

Sympatric speciation by shifts in sexuality

Another plausible mechanism for sympatric speciation is the evolution of genetic changes resulting in the production of asexual lineages from a sexually reproducing ancestral species. Mechanisms that may be involved in such changes include apomixis, parthenogenesis, and ploidy shifts (Felsenstein 1981; Barrett 1989). Support for this class of explanations requires that we find asexual and sexual sister species occurring sympatrically. Wiley and Mayden (1985) discussed three unisexual fish species that have apparently evolved via a combination of hybridization and subsequent genetic alteration: the gynogenetic species *Poecilia formosa,* found in northern Mexico and southern Texas, is thought to be a result of hybridization between *P. latipinna* and either *P. sphenops* or *P. mexicana; Menidia clarkhubbsi,* a unisexual species, is postulated to have arisen from hybrid-producing interactions between *M. beryllina* and *M. peninsulae;* and finally there are at least five parthenogenetic "forms" of unclear phylogenetic status within the topminnow genus, *Poeciliopsis,* which are potentially the byproducts of hybridization events (see Vrijenhoek 1989). In none of these cases were the parental species each other's closest relatives. This suggests that, while the production of the asexual hybrid species might well have occurred in sympatry, some degree of geographical dispersal was involved in getting the parental species together in the first place.

Sympatric speciation and sexual selection

The observation that some species possess sexually dimorphic traits that appear to decrease the survivability of their bearers was problematical for the theory of natural selection. Darwin sought a way out of this dilemma by reasoning that such extreme characters must confer some sort of advantage to their bearers, which at least balanced, or at best outweighed, their deleterious effects on survival. He looked for this advantage in the second component of natural selection—production of offspring—and proposed from this his theory of sexual selection (Darwin 1872:64).

> This form of selection depends, not on a struggle for existence in relation to other organic beings or to external conditions, but on a struggle between the individuals of one sex, generally the males, for the possession of the other sex.

Fisher (1930, 1958) examined sexual selection in genetic terms and formulated the implications of this selective regime to speciation (Fisher 1930).

> It is, of course, characteristic of unstable states that minimal causes can at such times produce disproportionate effects; in discussing the possibility of the fission of species without geographic isolation, it will therefore be sufficient if we can give a clear idea of the nature of the causes which condition genetic instability.

Sexual selection has been examined from two perspectives, interactions between members of one sex (usually males) to acquire mates (intrasexual selection) and interactions between the sexes to choose mates (intersexual selection). Investigations of the relationship between sexual selection and speciation have generally focussed on the latter form and have followed three pathways. Dobzhansky (1940) emphasized the role of mate discrimination in reinforcing speciation once populations that had diverged in allopatry came back into contact (alloparapatric speciation mode; see reviews in Dobzhansky 1970; Mayr 1970). Muller (1942; see also Paterson 1985) proposed that the appearance of divergent mate-recognition systems could occur and complete the speciation process in allopatry without the need for reinforcement via secondary contact of the diverging populations (allopatric and alloparapatric modes; see discussion in Kaneshiro 1980). Lande (1981, 1982) was the first researcher to develop explicitly genetical models that demonstrated the potentially powerful nature of sexual selection as a mechanism of sympatric speciation (see also West-Eberhard 1983). He concluded (1982),

> Incipient speciation in a population occupying a continuous range is modeled as the joint evolution of geographic variation in female mating preferences and a quantitative secondary sexual character of males. Even in the absence of genetic instability, the evolution of female mating preferences can greatly amplify large-scale geographic variation in male secondary sexual characters and produce widespread sexual isolation with no geographical discontinuity.

The proposal that intersexual selection is a strong driving force in sympatric speciation is a promising new line of research because the theoretical framework is well developed and a plethora of sexual selection studies exist for individual species (see articles and references in Bateson 1983; Thornhill and Alcock 1983; Bradbury and Andersson 1987). What is needed now is an examination of these data within a phylogenetic context. In order to do this, we require three pieces of information: a phylogeny for the group of interest, biogeographical data, and experimental evidence for intersexual selection. Remember, the comparisons of changes in female mating preference and changes in the male character must be made between sympatric sister species.

A Comment on Sympatric Speciation

Although it appears that sympatric speciation often requires unusual genetic and/or ecological circumstances (Futuyma and Mayer 1980), there is evidence that those circumstances occur regularly in some restricted groups of plants and animals. In chapters 7 and 8 we will discuss a number of examples in which host switching is associated with speciation. To some, this might be seen as de facto evidence of sympatric speciation, in which case approximately half of the host-parasite and phytophagous insect-plant associations for which there is phylogenetic information are the result of this speciation mode. However, in the majority of cases we do not know if the host switch had anything to do with the speciation event, because we are not certain that the colonization events represented the invasion of a new resource base or the expansion of an old one. In many cases there is biogeographic evidence suggesting that passive allopatric speciation accounts for the speciation events and that host switching is not coupled with speciation. In any event, we think it is apropos to close this section with the statement by Bush (1982) that

> the future holds many surprises. . . . I suspect that macromutations and rapid nonallopatric mechanisms of speciation will prove to be far more important in many groups of organisms than previously imagined.

Some Sample Studies

Freshwater Fishes of the Mobile basin and the Interior Highlands of southeastern North America

The freshwater stream fishes of southeastern North America represent an excellent model system for studying speciation in a historical ecological framework. Detailed distributional data are available, and explicit phylogenetic hypotheses have been published and are being reinforced or upgraded on a regular basis. Although there is a plethora of examples, we will discuss only a few of these from each region (for a more extensive discussion see Wiley and Mayden 1985 and examples in Mayden 1988, in press; also example in chapter 7).

Darters are small, bottom-dwelling percids. Although distributed in freshwater throughout the Northern Hemisphere, approximately 90% of the species are restricted to locations east of the Rocky Mountains (Moyle and Cech 1982). Killifish are small, brightly colored cyprinodontids that are both geographically and ecologically diverse. They can tolerate a wide variety of habitats, from freshwater streams and desert springs to salt marshes and mangrove swamps. This discussion will focus on speciation patterns within one group of sand darter (*Ammocrypta;* fig. 4.15) and one group of killifish (*Fundulus;* fig. 4.16).

Examination of the disjunct distribution patterns of these fishes reveals that vicariant speciation (allopatric speciation mode I) has been the predominant mode in these groups. Within the sand darters (fig. 4.15), geographic division of ancestor x into two populations produced *Ammocrypta clara* and its sister species, ancestor y, with some apparent extinction of *A. clara,* leaving two disjunct populations of the species today. In addition, geographic division of ancestor y into populations east and west of the Mobile drainage system produced *A. beani* and *A. bifascia.*

The killifish pattern (fig. 4.16) is slightly more complicated. Production of sister species p and q is problematical because, since they no longer exist, we can obtain only an estimate of their distribution by combining the ranges of their descendants. This method assumes that there has not been any widespread extinction or dispersal in the area, assumptions that are tenuous at best for these fishes. The results of such an analysis indicate that the ranges of p and q potentially overlapped near the mouth of the Mobile drainage area. Although fossil evidence of overlap would be illuminating, it would not resolve this quandary because, since p and q are extinct, we can no longer test for the presence or absence of interpopulation interactions in the parapatric area. So, for the present, this speciation event must be tentatively assigned an indeterminate status (parapatric mode?). Division of ancestor p producing *Fundulus blairae* and *F. dispar,* division of ancestor q producing *F. lineolatus* and r, and division of ancestor r into populations east and west of the Mobile drainage system, producing *F. nottii* and *F. escambia,* all represent apparent examples of vicariant speciation (allopatric speciation mode I). Although suggestive of sympatric speciation, the observation that *F. nottii* is located almost entirely within the range of *F. blairae* is not important to a study of speciation modes, because the two fishes are not sister species (see also the slight overlap between *F. escambia* and *F. lineolatus*). It is important to remember that there is a fundamental difference between models of speciation and mechanisms of speciation. Speciation models describe different original conditions, which set the stage for the subsequent divergence of ancestral populations. This divergence, in turn, may be accomplished by a variety of biological processes. For example, it could be envisioned that once ancestor

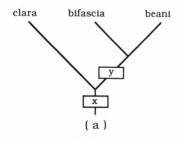

(a)

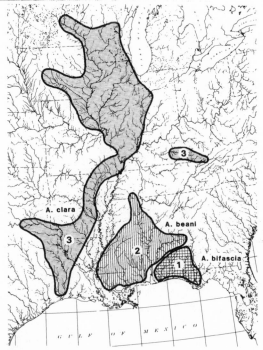

Fig. 4.15. Speciation of sand darters. (a) Phylogenetic tree for sand darters in the *Ammocrypta beani* group. *Names* = species; *letters* = ancestral species. (b) Distribution map for three *Ammocrypta* species. (Modified from Wiley and Mayden 1985.)

r was separated on either side of the Mobile River, interactions between *F. blairae* and the population of ancestor r, isolated on the west side of the basin, were involved in driving the population along the pathway to "specieshood." Nevertheless, these interactions have no effect on the original event that established the potential for the speciation of ancestor r that is, its separation into two disjunct populations.

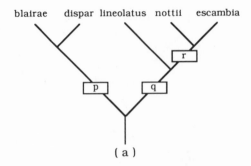

(a)

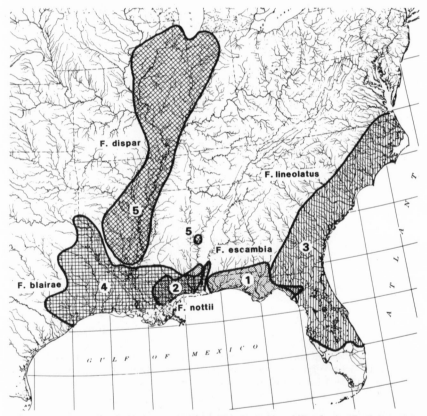

Fig. 4.16. Speciation of topminnows. (*a*) Phylogenetic tree for killifish in the *Fundulus nottii* group. *Names* = species; *letters* = ancestral species. (*b*) Distribution map for five *Fundulus* species. (Modified from Wiley and Mayden 1985.)

Examples of allopatric speciation mode I are widespread among fishes inhabiting the central areas of the Mississippi drainage from the Ozark Plateaus and Ouachita Highlands in the west to the Interior Low Plateau, Ridge and Valley Province, and Blue Ridge Province in the east. Darters are well represented in this area, as are the small, species-rich, silvery cyprinids aptly referred to as shiners. During the breeding season many shiner species don flamboyant red, orange, and/or yellow breeding liveries, and because of this, they have drawn the attention of many researchers. This discussion will be restricted to speciation patterns within two groups of shiners (*Notropis;* figs. 4.17 and 4.18) and one group of darters (*Etheostoma;* fig. 4.19).

Examination of the distribution patterns of these fishes reveals that, like the situation in the Mobile drainage system, vicariant speciation (allopatric speciation mode I) has been the predominant speciation mode in these groups. The simplest pattern occurs in the *Notropis nubilus* clade (fig. 4.17), where two speciation events associated with geographical vicariance have been coupled with the apparent loss of one species' *(N. nubilus)* central populations.

The second *Notropis* example (fig. 4.18) is equally straightforward. The distributions and phylogenetic relationships support the proposal that the four putative speciation events within this clade have been vicariant, via the geographical divisions of (1) ancestor o, possibly through a vicariance event eliminating the central populations, producing sister species p and q; (2) ancestor p producing *N. pilsbryi* and *N. zonatus;* (3) ancestor q producing *N. cerasinus* and ancestor r; and finally (4) ancestor r producing *N. coccogenis* and *N. zonistius.*

The pattern depicted for the darters (fig. 4.19) is slightly more problematical because the current pattern appears to have resulted from a combination of two and possibly three speciation modes. First, there is the possible asymmetrical division of ancestor u, producing *Etheostoma blennius* and ancestor v, plus the asymmetrical division of v, producing *E. sellare* and ancestor w, both examples of allopatric speciation mode II, either peripheral isolates or microvicariance. Second, there are three putative speciation events that indicate vicariant speciation (allopatric speciation mode I): the division of ancestor w, associated with extinction of central populations, producing ancestors x and y; the division of ancestor x, producing *E. tetrazonum* and *E. euzonum;* and the division of ancestor y, producing *E. variatum* and ancestor z. Finally, there is the potential parapatric speciation involving ancestor z, producing *E. osburni* and *E. kanawhae.* The case for parapatric speciation in this instance would be strengthened if interactions between the two species could be documented in the region of overlap.

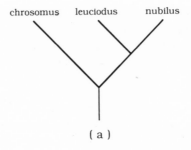

(a)

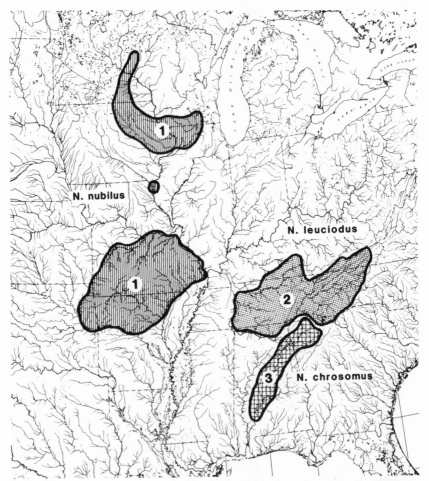

Fig. 4.17. Speciation of shiners. (*a*)Phylogenetic tree for shiners in the *Notropis nubilus* group. *Names* = species. (*b*) Distribution map for three *Notropis* species. (Modified from Wiley and Mayden 1985.)

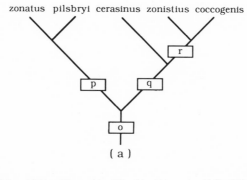

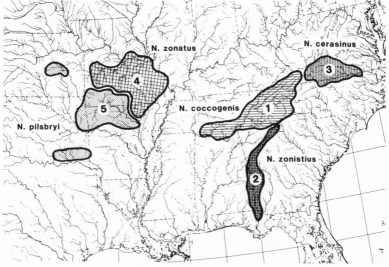

Fig. 4.18. Speciation of more shiners. (*a*) Phylogenetic tree for shiners in the *Notropis zonatus-coccogenis* group. *Names* = species; *letters* = ancestral species. (*b*) Distribution map for five *Notropis* species. (Modified from Wiley and Mayden, 1985.)

South American horned frogs

One of the most diverse and widespread of all frog groups is the family Leptodactylidae. Among South American leptodactylids, the subfamily Ceratophryinae comprises two genera, *Lepidobatrachus*, with three species, and *Ceratophrys*, with six species. *Ceratophrys* species are boldly colored, voracious predators that are well known to aquarists and tropical hobbyists as "horned frogs." They dwell in a variety of different habitats, ranging from neotropical rainforests (*C. aurita* and *C. cornuta*) through grasslands (*C. ornata*) to semixeric (*C. calcarata*) and xeric regions (*C. stolzmanni* and *C. cranwelli*). Lynch (1982) presented a phylogenetic analysis of the six species

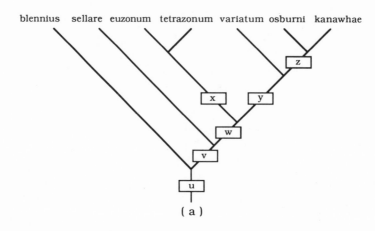

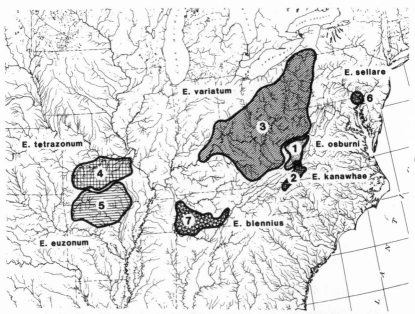

Fig. 4.19. Speciation of darters. (*a*) Phylogenetic tree for darters in the *Etheostoma variatum* group. *Names* = species; *letters* = ancestral species. (*b*) Distribution map for seven *Etheostoma* species. (Modified from Wiley and Mayden 1985.)

of *Ceratophrys*, using the monophyletic sister group of *Ceratophrys*, the genus *Lepidobatrachus*, as the outgroup. Lynch's resulting phylogenetic tree (fig. 4.20a) has a consistency index of 87.5%.

The geographical distribution of the six species is shown in figure 4.20. The genus comprises two clades of three species each, one occurring from just north of the central Amazon northwards (*C. stolzmanni, C. calcarata,*

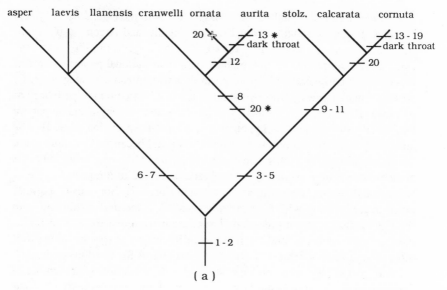

[--LEPIDOBATRACHUS --]   [--------------------- CERATOPHRYS ---------------------]

asper   laevis   llanensis cranwelli ornata   aurita   stolz. calcarata   cornuta

20

13 *
dark throat

13 - 19
dark throat

12

20

8

20 *

9 - 11

6 - 7

3 - 5

1 - 2

(a)

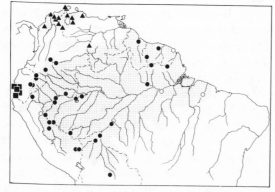

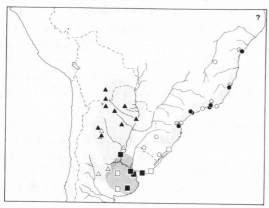

Fig. 4.20. Speciation of frogs. (*a*) Phylogenetic tree for the frog genus *Ceratophrys*, based on eighteen adult morphological characters (1–6, 8–11, 13–20), one larval morphological character (7), and one karyotypic character (12). *Names in lowercase letters* = species; *stolz.* = *C. stolzmanni.* * = homoplasious characters. Figures b and c are distribution maps for these frogs. Continuous distributions are estimated using *stippling*. *Black symbols* = specimens examined; *white symbols* = specimens in literature and/or museum records. (*b*) Frogs in the subgenus *Stombus*. *Triangles* = *C. calcarata; circles* = *C. cornuta; squares* = *C. stolzmanni.* (*c*) Frogs in the subgenus *Ceratophrys. Triangles* = *C. cranwelli; circles* = *C. aurita; squares* = *C. ornata.* ? = an unidentified taxon. (From Lynch 1982.)

C. cornuta: fig. 4.20b) and the other associated with the Paraná River system and coastal areas of southeastern Brazil, Uruguay, and Argentina (*C. cranwelli, C. ornata, C. aurita;* fig. 4.20c). The three species in the "northern" clade are all allopatric. At first glance, their distributional pattern conforms to a classical peripheral isolates scenario of a widespread central species, *C. cornuta* (black circles), with two smaller species located on the periphery of its range. However, Lynch's analysis uncovered the one phylogenetic pattern that specifically refutes this hypothesis of allopatric speciation mode II. The large central species cannot be considered ancestral because *C. cornuta* is not the sister group of the other two members of the clade, conflicting with the phylogenetic pattern predicted if speciation were due to a repeated cycle of sequential dispersal, isolation, and speciation (fig. 4.7). Nor does *C. cornuta* occur in a polytomy with *C. stolzmanni* and *C. calcarata,* conflicting with the phylogenetic pattern predicted if speciation were due to random settlements of individuals around the margins of the ancestral species' range (fig. 4.6), or to a series of microvicariance events (fig. 4.8). In addition, it is *C. cornuta,* and not the peripheral species, that is the most divergent member of the clade (i.e., it exhibits the largest number of autapomorphies). This pattern, however, does support the hypothesis that evolutionary diversification in this clade has been associated with two vicariance events (allopatric speciation mode I: fig. 4.5). The sundering of the first ancestral species resulted in the appearance of *C. stolzmanni* and the ancestor of the *C. calcarata* + *C. cornuta* clade, while the second vicariance event fragmented that ancestor, resulting in the emergence of *C. calcarata* and *C. cornuta.*

The situation is complex for the species composing the "southern" clade. The ranges of these species are relatively equal in size and overlap in two areas. Although *C. ornata* and *C. cranwelli* are parapatric, they are not sister species, so their zone of contact might initially be regarded as unimportant to speciation studies. However, this interpretation changes somewhat after a more detailed examination of the relationships depicted in figure 4.20a. The sister species of *C. cranwelli* was the ancestor (x in fig. 4.21) of the clade *C. ornata* + *C. aurita.* Both of these descendant species display autapomorphies; therefore, neither of them can immediately be identified as a persistent ancestor. However, the autapomorphy for *C. ornata* is the postulated secondary loss of character 20 (eyelid tubercles), and there are three equally parsimonious interpretations of the transformation series for this character (fig. 4.21).

Since the transformation shown in figure 4.21a eliminates the only autapomorphy postulated for *C. ornata,* we cannot eliminate the possibility that this species might, in reality, be ancestor x, based upon the information we have to date. If this possibility is realized, then the overlap between *C. ornata* and *C. cranwelli* represents a parapatric speciation event in which both spe-

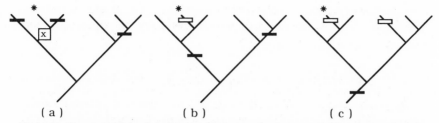

Fig. 4.21. Three equally parsimonious transformations for character 20 (the presence of eyelid tubercles) mapped onto the phylogenetic tree for the frog genus *Ceratophrys*. *White bars* = secondary loss of eyelid tubercles; *black bars* = presence of tubercles. * = the species of interest in this discussion, *C. ornata; x* = the ancestor of *C. ornata* and *C. aurita*.

cies are still extant, and thus within the scope of ongoing research. However, if future investigations uncover a solid autapomorphy for *C. ornata,* then the origins of *C. cranwelli* and the ancestor of the *C. aurita* + *C. ornata* clade are embedded deeper within the phylogenetic tree and are no longer subject to experimental investigation. *C. cranwelli* and *C. ornata* are ecologically (xeric versus grasslands) and karyotypically (both *C. ornata* and *C. aurita* are octoploid, while *C. cranwelli* is diploid) distinct. It is possible, therefore, that the ecological and/or the chromosomal change may have been associated with the parapatric speciation of *C. cranwelli* and ancestor x.

Equally exciting is the discovery that the ranges of the sister species *C. ornata* and *C. aurita* overlap in one locality. Both these species are ecologically isolated (Lynch 1982), *C. ornata* in the grasslands and *C. aurita* in the rainforests. The plesiomorphic habitat preference for the genus is hypothesized to be a xeric, nonforest environment. Thus, in both the northern and southern subgenera, there has been a movement, correlated with speciation, towards the rainforests. In the northern clade this change in habitat preference is associated with two vicariance events, and thus is not the driving force behind the initiation of speciation. In the southern clade the change in habitat preference may have been associated with two parapatric speciation events. If ecological segregation has been the motivating force behind the *C. ornata* and *C. aurita* (and *C. cranwelli*-ancestor x) differentiation, then we would expect to find evidence of ecological interactions between overlapping populations of these species. Total ecological and behavioral segregation will not provide support for the parapatric model, nor will it refute it; this type of "absence of data" can only fail to refute the hypothesis, leaving us still partially in the dark.

Overall, then, researchers interested in studying parapatric speciation should focus their attention on the two areas of overlap in the southern clade, between *C. ornata* and *C. cranwelli* and between *C. ornata* and *C. aurita.* Since several critical pieces of information are lacking, involving the status of *C. ornata* as a species independent from its ancestor (x) and the interactions, if any, between species in the areas of parapatry, the opportunities for future research are intriguing.

South American plants: Lepechinia section Parviflorae

Lepechinia is a group of small, white-flowered shrubs distributed mainly throughout tropical and subtropical Latin American highlands. Within this range, the section *Parviflorae*, comprising twelve weedy species living at altitudes of 1,700–3,900 meters in Andean shrub/forest zones, shows complicated latitudinal and altitudinal distribution patterns. Hart (1985a) analyzed the phylogenetic relationships in this group, using the sections *Speciosae* and *Salviifoliae* as the outgroups. The resultant tree (fig. 4.22a), based on twenty-four morphological characters, has a consistency index of 95.8%.

Figure 4.22b depicts the geographic distribution of these twelve species. The six oldest species in this section are allopatrically distributed with respect to the appropriate sister species. *Lepechinia graveolens,* the sister species to the rest of the section, occurs in northern Chile and southern Bolivia. *L. vesiculosa* ranges through Peru and Bolivia, while its putative sister species *L. bullata* is widely disjunct in Columbia and Venezuela. Finally, *L. heteromorpha* ranges from eastern Ecuador to southern Peru and Bolivia, while *L. radula* is distributed throughout southwestern Ecuador and northern Peru, and *L. conferta* is widespread throughout Columbia and Venezuela. Although *L. bullata* and *L. conferta* occur in moist, upper Andean forests at similar altitudes in Columbia and Venezuela, they are not sister species, so their potential geographical overlap is not pertinent to speciation studies. At the moment, the relationships in this portion of the phylogenetic tree are not strongly delineated (Hart 1985a). Further resolution of this problem is required before it will be possible to hypothesize about the type(s) of allopatric speciation mode(s) involved in these speciation events.

The remaining six species, *L. betonicaefolia, L. paniculata, L. scobina, L. dioica, L. mutica,* and *L. mollis,* from a clade united by a change in ovarian fertility (character 22; see also chapter 5). In contrast to the widespread distributions of their higher elevation, forest-dwelling relatives (five of the six preceding species in the section and all species in the outgroup), all members of this clade are locally endemic with very small ranges and are found only in lower-elevation, dry habitats (fig. 4.22a). Additionally, they all appear to be more apomorphic than their common ancestors. The distributions, habitat preferences, and hypothesized patterns of phylogenetic relationships among these six species can be explained by repeated incidents of peripheral isolates allopatric speciation by means of sequential colonization and speciation (fig. 4.7). Since this portion of the phylogenetic tree is not an unresolved poly-

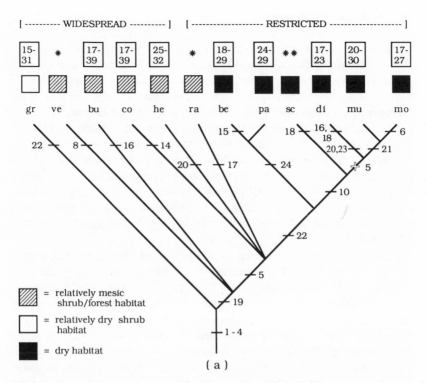

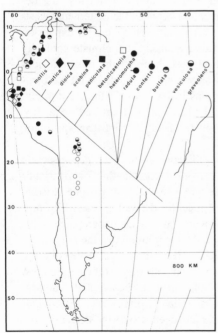

Fig. 4.22. Speciation of plants. (*a*) Phylogenetic relationships within *Lepechinia* section *Parviflorae* based on twenty-four morphological characters. Details of the characters used are provided in Hart 1985a. Habitat preferences are depicted in the bottom line of boxes above the species names. Range of inhabited altitudes, in hundreds of meters, is depicted in top line of boxes above species names, and geographical distribution is listed as either widespread or restricted. *be* = *Lepechinia betonicaefolia; bu* = *L. bullata; co* = *L. conferta; di* = *L. dioica; gr* = *L. graveolens; he* = *L. heteromorpha; mo* = *L. mollis; mu* = *L. mutica; pa* = *L. paniculata; ra* = *L. radula; sc* = *L. scobina; ve* = *L. vesiculosa*. Asterisks refer to incomplete data: * = "high" altitudes, ** = "low" altitudes. (*b*) Distribution map for twelve species of *Lepechinia* section *Parviflorae*. (From Hart 1985a.)

tomy, the microvicariant splitting of one ancestral species into several periph-
erally isolated populations (fig. 4.8) seems not to have been the initial condition for the evolutionary diversification of these plants. Hart proposed that
a series of forest expansions and contractions during the ebb and flow of the
Pleistocene glacial periods produced a repeating cycle of ancestral range expansion, range contraction accompanied by peripheral isolation, and speciation of the isolate. The members of this clade are located within the Huancabamba Deflection region of southern Ecuador and northern Peru. Unlike
the usual north-south orientation of the Andean mountain range, this area is
characterized by low mountain chains, bisected by deep, dry, east-west-
running valleys. These topographical differences, then, set the geographical
stage on which the following scenario is hypothesized to have been enacted
(Hart 1985a).

Forests in the higher mountains were reduced and the eastern and western
slopes separated by the lowering snow line during the cold, wet glacial periods (van der Hammen and Gonzales 1960; van der Hammen 1972; Geel and
van der Hammen 1973). The Pacific side of the Andes was colder and wetter,
so forests forced down the mountainsides and to the east were more likely to
find refuge than their western counterparts (Hastenrath 1971a,b; Simpson
1975). Interestingly, this scenario, based upon a variety of geological data, is
supported by the extant distributions of *L. vesiculosa* and *L. heteromorpha* in
eastern-slope forest habitats. Within the Huancabamba Deflection, the decrease in mountain size and the communication between the eastern and western slopes through valleys is postulated to have allowed a westerly expansion
of the eastern-slope forests. Following this expansion, the climate gradually
shifted towards a drier state, driving the forests back up the mountainsides
following the retreating ice fields. This movement, combined with the complex geography in the area, left populations isolated within moister refugia.
Increasing xeric conditions, small population size, and severed gene flow
eventually resulted in a series of peripheral isolates allopatric speciation
events.

At the moment, the evolutionary diversification of the *Parviflorae* section
of *Lepechinia* appears to have resulted from a combination of vicariance
events in the older members of the clade and a sequence of peripheral isola-
tions in xeric environments in the more recently derived species. The proposal that the older species appeared following large-scale geographical splitting of the ancestral range is predicated on the assumption that the polytomies
shown in the current phylogeny will be resolved when more data are collected
and examined phylogenetically. If, in fact, such data *support* the polytomies,
a new explanation, based on peripheral isolation, must be adopted. The proposal that diversification in the *L. betonicaefolia, L. paniculata, L. scobina,
L. dioica, L. mutica,* and *L. mollis* clade has been driven by sequential pe-

ripheral isolation could be examined by comparing this pattern with phylogenies for other groups in the area. Given this mode of speciation, we would expect to find congruent phylogenetic patterns only in clades containing species adapted to xeric conditions.

Although Hart's scenario is enticing, it needs to be corroborated by an independent estimate of the age of the clade, which is difficult based on the current information. For example, the Huancabamba Deflection is associated with Miocene uplifting of the Andes that resulted in the reversal of the Amazon River so that it no longer flowed into the Pacific Ocean (see also chapter 7). If, in fact, this group is older than the Pleistocene, this does not rule out a role for peripheral isolates allopatric speciation. Investigating this requires that we delve deeper into the biogeography of the area. It is possible, for example, that the Pleistocene glaciation dissected an already established flora, leaving the distributions that we see today and possibly obscuring the evidence for the speciation events that occurred prior to glaciation. This is apparently what happened with respect to the freshwater fish fauna of North America (see Wiley and Mayden 1985; Mayden 1988). Lynch (1986) discussed the origins of the high Andean herpetofauna (amphibians and reptiles), concluding that the current phylogenetic data base was inadequate to provide robust explanations, but that the available data supported an explanation that the assemblage of the Andean herpetofauna began with the origins of the Andes themselves. Because amphibians and reptiles do not exist in the absence of vegetation, we might well conclude that at least part of the flora of the Andes predates the timing of the scenario proposed by Hart. We do not know yet if *Lepechinia* section *Parviflorae* is part of the old assemblage or a relative newcomer.

The Frequencies of Different Modes of Speciation

To date, very few studies have been published that examine speciation using phylogenetic evidence. Development of this depauperate data base is vitally important to the future of speciation research because this is the only known way to assess the relative frequencies of different speciation modes based on evidence rather than on theory. Lynch (1989) has begun such an investigation, with intriguing results. He examined species ranges for members of a number of clades for which phylogenetic trees and extensive distributional data were available, estimating ancestral ranges by the sum of all descendants' ranges. Based on an analysis of sixty-six documented cases of vertebrate speciation, he suggested that 71% of the speciation events were due to vicariance (allopatric speciation mode I), 15% of the cases resulted from a combination of the three forms of peripheral isolates allopatric speciation (allopatric speciation mode II), and 6% of the evolutionary divergence

Table 4.1 Frequencies of speciation modes within seven different clades.

		Allopatric			
	N	I	II	Sympatric	Unknown[a]
Frogs					
Rana	22	17	1	1	3
Ceratophrys	6	5	1	0	0
Eleutherodactylus	8	5	1	1	1
Fishes					
Fundulus	4	4	0	0	0
Heterandria	8	5	3	0	0
Xiphophorus	13	7	4	1	1
Birds					
Poephila	5	4	0	1	0
Frequencies		71.2%	15.1%	6.1%	7.6%

[a] Unknown cases represent potential sympatric events buried deep within the phylogenetic tree. The species in question are now extinct (in the sense that ancestral species "go extinct" when they speciate), making it difficult to obtain distribution patterns and measurements of interpopulation interactions. Inclusion of these cases in the sympatric category boosts its frequency to 13.7%.

fulfilled the requirements of sympatric speciation (table 4.1). In the other 8% of the cases, Lynch discovered dichotomies buried deep within the phylogenetic trees that explained significant geographical overlap between more highly derived sister groups. Because of the age of these speciation events, this could be explained either as sympatric speciation, because of the putative widely sympatric distributions of the ancestral sister species, or as allopatric speciation followed by dispersal. If these cases represent instances of sympatric speciation, it would increase the possible frequency estimate for sympatric speciation to 14%.

Lynch's study and other examples we have discussed in this chapter (see examples in chapters 7 and 8; also Weitzmann and Fink 1983; Weitzmann and Fink 1985, using the species of *Paracheirodon* [neon tetras] and some of the Xenurobryconine fishes in South America) support the major contention of evolutionary theorists, such as Mayr (1963) and Futuyma and Mayer (1980), that sympatric speciation does not seem to occur very often. However, these studies do not support the traditional perspective that peripheral isolates allopatric speciation has been the major speciation mode on this planet; rather, it supports the view presented by Barton and Charlesworth (1984) that peripheral isolates allopatric speciation is not likely on theoretical grounds. In fact, it would seem to be as rare as sympatric speciation. Contrary to many early theoretical predictions, the predominant mode appears to be vicariant speciation, in which the roles of adaptation and speciation need not be coupled (see also Butlin 1987; Futuyma 1989). We will examine the implications of these findings in subsequent chapters.

Documenting the Influence of Microevolutionary Processes

Population biologists have documented the existence of numerous processes that might be involved in speciation, once particular sets of initial conditions have been established. Mechanisms resulting in the divergence of ecological characters correlated with speciation events are of particular importance for nonallopatric speciation models, because these models invoke a close tie between adaptation and speciation. We have already discussed some ecological factors that might be important in sympatric speciation, and we will expand upon the relationship between adaptation and evolution more fully in chapter 5. Another important aspect of microevolutionary theories about speciation concerns the role of reproductive isolation in initiating and completing the process.

Mayr (1963) emphasized the importance of postzygotic isolating mechanisms (i.e., reduction in hybrid fitness) as strong selection pressures promoting the spread of prezygotic isolating mechanisms (i.e., mate recognition) through the speciating populations. According to this scenario, the loss of fitness associated with the production of genetically and developmentally unstable hybrids precedes the appearance of behavioral isolating mechanisms; so, while reproductive isolation over a contact zone does not initiate speciation, it does complete the process. Paterson (1985) stood the problem on its head and argued that the evolution of mate recognition, driven by an individual mating advantage, precedes genetic and developmental changes; thus, such changes can be viewed as by-products rather than causes of speciation. Here, the origin of isolating mechanisms is associated with the initiation of speciation. Investigations reporting the existence of premating (e.g., studies by Ohta 1978; McLain and Rain 1986; Stratton and Uetz 1986; Moore 1987; Butlin 1989) and postmating (e.g., studies by Vigneault and Zouros 1986; Zouros 1986; Christie and MacNair 1987; Coyne and Orr 1989) isolating mechanisms appear regularly in the evolutionary biology literature, attesting to the importance of the concepts. Rarely, however, are those mechanisms examined within a phylogenetic framework.

In order for isolating mechanisms to be causally involved in the speciation event, both of these scenarios require an area of contact between two diverging populations (nonallopatric speciation modes), either following secondary contact between incompletely speciated populations (alloparapatric mode), or through parapatric or sympatric speciation. Neither of these scenarios is causally involved in allopatric speciation because in those modes **the disjunct populations speciate independently of one another.** Therefore, in order to investigate the temporal sequence of prezygotic versus postzygotic isolating mechanisms within a group of organisms, we must first differentiate between the contributions of allopatric and nonallopatric speciation events to the evolutionary diversification of the clade.

Once incidents of nonallopatric speciation have bee identified, phylogenetic analysis can be used to determine whether the evolution of mate recognition systems precedes or follows the evolution of postzygotic isolating mechanisms. Investigating this requires an explicit phylogenetic tree based on characters other than those postulated to be part of the isolating mechanisms. There are four potential patterns for the interaction between the origins of isolating mechanisms and the speciation event: Neither type of isolating mechanism is associated with speciation (fig. 4.23a); both mechanisms are associated with speciation (fig. 4.23b). This configuration does not support or refute either of the two scenarios because phylogenetic analysis cannot

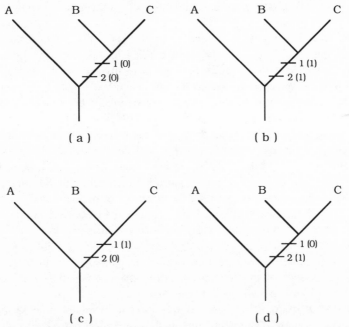

Fig. 4.23. Association between two types of isolating mechanisms and speciation. In this hypothetical example, distributional and experimental data have verified that the production of taxa B and C involved sympatric speciation. Character states: *1(0)* = postzygotic isolating mechanisms absent; *1(1)* = postzygotic isolating mechanisms present; *2(0)* = prezygotic isolating mechanisms absent; *2(1)* = prezygotic isolating mechanisms present. (*a*) Neither mechanism present. (*b*) Both mechanisms associated with speciation. (*c*) Postzygotic isolating mechanism associated with speciation. (*d*) Prezygotic isolating mechanism associated with speciation.

untangle the sequence of appearance of autapomorphic characters; postzy-gotic fitness depression is associated with speciation (i.e., it precedes the appearance of prezygotic isolating mechanisms: fig. 4.23c). This pattern sup-ports Mayr's hypothesis; prezygotic isolating mechanisms are associated with speciation (i.e., it precedes the appearance of postzygotic fitness depression: fig. 4.23d). This pattern supports Paterson's hypothesis.

Macroevolutionary Trends in Diversity: Species Number

As we discussed in chapters 1 and 3, the term "diversity" has been used in a variety of ways by biologists. One major usage is associated with the number of species in a group, the number of individuals in a population, or the relative number of individuals of different species in a given commu-nity or ecosystem. In this section we will address macroevolutionary ques-tions concerning the number of species in different groups. This question falls within the domain of the "taxic approach" to macroevolutionary studies, an approach concerned with the analysis of phylogenetic patterns resulting from processes controlling the rates of speciation and extinction (Cracraft 1985a,b). The goal of this research is to separate groups that are more species-rich from other "equivalent" groups and to distinguish between clades of "unusually high" or "unusually low" species number.

It is difficult to generalize across studies in the absence of rigorous defini-tions of "unusual" and "equivalent." Mayden (1986) suggested that two cri-teria must be satisfied before any conclusions about differences in species number, between or among groups, can be drawn. First, all groups under examination must be monophyletic. Because phylogenetic systematics stresses the recognition of monophyletic groups, it is an appropriate starting point for studies of this component of diversity. Second, the groups being compared must be of equal antiquity. A number of methods have been pro-posed for documenting the relative ages of clades. Stratigraphic and biogeo-graphic analyses (see chapter 7) attempt to use environmental parameters as independent indicators of age. The use of "genetic distances" is based on the existence of a hypothetical "internal clock" that is universally informative about evolutionary rates. Phylogenetic systematics uses yet another criterion, sister-group relationships. Remember that sister groups are the descendants of a common and unique speciation event. Hence, by definition sister groups must be of equal age.

Once these criteria have been satisfied, any disparity in diversity between sister groups may be due to unusually high or unusually low speciation or extinction rates in one of the sister clades. Delineation of these influences still leaves the underlying mechanism enclosed within a black box. The "transformational" component of macroevolution is concerned with exploring this black box in an attempt to uncover attributes associated evolutionarily with the existence of species-rich or depauperate clades. Extinctions or unusually low speciation rates will produce depauperate clades, while unusually high speciation rates will produce species-rich clades. It has been tempting to think that species-rich groups are somehow "better" or "more successful" than average, whereas groups of low species number are somehow "less successful" than average. Hence, a taxic macroevolutionary study of diversity is necessary for robust transformational macroevolutionary studies of diversity.

Unusually Low Diversity Groups

Simpson (1944) was among the first modern evolutionary biologists to consider general explanations for groups of unusually low species numbers. He considered all such groups *relicts* of one form or another and postulated that different processes could produce different kinds of relictual groups. We will be concerned with two major types of relicts. **Phylogenetic relicts** are "living fossils," members of groups that have existed for a long time without speciating very much. Such low speciation rates could result from phylogenetic or developmental constraints on phenotypic diversification, and/or from unusually pronounced ecological specialization (i.e., ecological constraints due to the effects of strong, long-term stabilizing selection from the specialized habitat). **Numerical relicts,** by contrast, are the surviving members of once more species-rich groups that have been depleted by extinction.

Brooks and Bandoni (1988) suggested that a combination of phylogenetic, biogeographical, and ecological information could be used to distinguish between phylogenetic and numerical relicts. Establishing a group's relictual status first requires methods for determining that the group is old enough to be highly diverse. There are a number of methods available for estimating the ages of clades, including molecular-clock criteria, paleontological data, and biogeographical analysis (see also chapter 7). Second, it must be established that the group is in fact unusually depauperate. As suggested above, this can be established by comparing sister groups (Mayden 1986). Brooks and Bandoni further suggested that phylogenetic relicts should be ecologically conservative, whereas numerical relicts should be ecologically diverse.

For example, ratfish (chimaeroids) are the sister group of sharks, skates, and stingrays (elasmobranchs). There are 25 species of ratfish, compared with approximately 625 species of elasmobranchs. Ratfish occur worldwide in mid-to-deep-water marine habitats, and forage on benthic invertebrates. The fossil record indicates that ratfish have been in existence for a considerable period of time but have never been highly diverse. Both the fossil evidence and the ecological homogeneity of contemporaneous species suggest that ratfish are phylogenetic relicts. Now, consider the crocodilians (Crocodilia), the sister group of the species-rich clade, the birds (Aves). Living crocodilians, numbering about 22 species, inhabit a variety of estuarine to freshwater habitats throughout the tropical and subtropical regions of the world. They prey on a wide variety of vertebrates and some invertebrates. The fossil record indicates that crocodilians were once a species-rich group, including many fully marine species; in addition, the earliest known crocodilian fossils suggest a terrestrial origin for the group. Hence, the current diversity of crocodilians represents only a fraction of the species number and ecological diversity once encompassed by the group, so we consider crocodilians to be numerical relicts.

Unusually High Diversity Groups

According to the taxic view of macroevolution, unusually species-rich groups have experienced higher speciation rates than their sister groups. However, as we have discovered in the preceding paragraphs, speciation results from a complex interaction of a variety of processes. In general, unusually high speciation rates can be attributed to three influences.

1. Cracraft (1982a,b, 1985a) suggested that the key to understanding taxic macroevolutionary patterns for any clade lay in the history of **geological change** and accompanying vicariant speciation. This suggestion is supported by three lines of evidence. First, since speciation is an irreversible phenomenon, it should be most strongly affected by irreversible environmental factors. Paleoecological studies indicate that most climatic changes are cyclical, so this would argue against a strong role for climate in the speciation process. However, geological evolution is an irreversible phenomenon and probably the only environmental process that is irreversible on time scales long enough to affect species. Second, biological diversity tends to be clumped in "hot spots" corresponding to areas with historically high rates of geological change, rather than being uniformly distributed across a given habitat or zone. For example, tropical diversity is clumped in South America and in the Indo-Malaysian region (named Wallacea, after Alfred Russell Wallace, co-discoverer of the theory of natural selection), two areas whose geological history is extremely complicated. And finally, most documented speciation patterns correspond to the predictions of vicariant (and microvicariant) speciation.

2. Certain types of **habitat** may support higher diversity than others. For example, the observation that diversity in the tropics is higher than diversity in the temperate or arctic regions is often attributed to differences in speciation rates. From this perspective, the greater energy budget in the tropics allows a finer partitioning of the environment to occur, permitting more species to evolve. Cracraft (1985a) approached the problem from a different

angle. He suggested that the critical comparison should be the rates of extinction, rather than the rates of speciation, in the different areas. From this perspective, extinction rates in temperate to arctic habitats have been higher than extinction rates in the tropics, due to historical increases in environmental harshness in the colder areas. To study the influence of habitat on speciation rates, it is necessary to filter out the effects of vicariant speciation and then search for conditions of uniformly high (or low) diversity in a given habitat.

3. A particular group of organisms might exhibit unusually high speciation rates because they possess **attributes** that allow them to invade and exploit new habitats. The extent to which this influence has affected species richness will be reflected in the amount of peripheral isolates, parapatric, or sympatric speciation that has occurred in a clade. Many authors have therefore sought "key innovations" or "key adaptations" whose origin in an ancestor increased the likelihood of nonvicariant modes of speciation (see also chapter 5). Vrba (1980, 1983) recognized two classes of key innovations that could affect speciation and extinction rates. **Species selection** involves studies of traits that emerge in an evolving lineage (i.e., in an ancestral species) that directly affect the way organisms interact with their environments. By contrast, **effect macroevolution** involves studies of traits that emerge in an evolving lineage that have an effect on speciation and extinction rates regardless of the environment (for further discussions, see Mayr 1963; Jackson 1974; Jablonski 1982; Hansen 1983; Valentine and Jablonski 1983).

As an example of the difference between these two classes of key innovations, let us consider the evolutionary diversification of the passeriform birds, a monophyletic group that is commonly perceived as unusually species-rich. One explanation of this diversity is that songbirds, as a group, possess some "key adaptation" that has allowed them to become "more successful" evolutionarily than other birds (but see Raikow 1986). Kochmer and Wagner (1988) suggested that the small size of passeriforms relative to other birds was the key adaptation. They argued that small organisms should speciate more often because of their greater success at carving up environmental resources. If this is true, then passeriforms should be ecological specialists. Fitzpatrick (1988) also suggested that small size was a key adaptation, but attributed the evolutionary success of songbirds to their role of ecological generalists. Although this conflicted with Kochmer's and Wagner's prediction, both of these explanations fall within the domain of "species selection" because the success of the key innovation is attributed to an enhanced interaction with the environment (foraging success; see also Vermeij 1988 for a discussion of diversity in male courtship songs).

From a historical ecological perspective, the question of whether passeriforms are specialists or generalists is secondary to the question, Are their foraging habits apomorphic or plesiomorphic with respect to their sister group, which is not so species-rich? Two steps are required for the clarification of this problem: first, a refinement of the data base for foraging habits within the passeriforms and, minimally, their sister group and, second, a phylogenetic analysis of the relationships within the passeriforms and, minimally, their sister group. From this, we can determine whether the foraging mode displayed by the songbirds is a plesiomorphic (ancestral) trait or a derived trait. The discovery that passeriforms display a derived feeding mode will support the hypothesis that a shift in foraging strategies associated with a change in size represented a "key adaptation" in these birds. On the other hand, the discovery that songbirds display the ancestral foraging strategy does not support a species-selection hypothesis of the relationship between body size and foraging success. For example, suppose the passeriforms are ecological generalists. If the sister group of the passeriforms includes species that are also generalists (at least primitively within the group), then the presence of that foraging mode among songbirds is not a unique attribute of those birds. Rather, the propensity to be an ecological generalist has been inherited from some common ancestor that gave rise to groups some of which are, and others of which are not, unusually species-rich.

An alternative perspective, more similar to effect macroevolution than to species selection, also assumes that "being small" is the key innovation. Regardless of the environmental context, we might expect that species of small organisms would (1) have higher reproductive rates (and hence higher probabilities of producing new variants) than their larger relatives; (2) occupy less geographic territory, and hence be more likely to speciate as a result of relatively small-scale geographic subdivision, than populations of larger organisms (see "geological change," above); and (3) require less energy to survive than species of large organisms, giving them a greater probability of survival during periods of environmental stress (see "habitat," above; see also Schmidt-Nielsen 1984 for a more detailed discussion of possible evolutionary advantages in being small). Maurer (1989) produced a mathematical model demonstrating the potential significance to speciation and extinction rates of different persistent ancestral population dynamics, regardless of the environment in which those population dynamics are manifested.

Summary

Researchers studying macroevolutionary patterns of species number have become increasingly aware of the effects of geological activity (vicariance) on speciation rates. However, what is often overlooked is that vicariant speciation also affects extinction rates because every vicariant speciation event results in the "extinction" of the ancestor giving rise to the sister species. If Lynch's (1989) estimates of the frequencies of speciation modes is representative of diversity as a whole, then as many as 71% of extinctions are due to vicariant speciation and not to the irreplaceable loss of an evolving lineage. This is an important point, because it emphasizes the dangers of letting our perceptual biases color our evolutionary explanations. "Extinction" is generally associated with destructive influences; it eliminates biological entities. "Speciation," on the other hand, is generally associated with productive influences; it creates new biological entities. Because of the special, and probably subliminal, connotations of these terms, we have overlooked the dual nature of vicariant speciation. And this, in turn, is rather a catastrophic prospect, because if roughly three-quarters of past extinctions have simply been a by-product of biological creation through speciation, where does that leave a global ecosystem in which most of the current extinctions are not accompanied by such productive processes?

5 Adaptation

The search for a functional (adaptive) fit between an organism and its environment is one of the dominant themes in evolutionary biology (see, e.g., Lewontin 1978; Dunbar 1982; Coddington 1988). There are three components to adaptation: origin, diversification, and maintenance of characters. Microevolutionary studies concentrate on the maintenance of traits in current environments where the processes shaping the interactions between the organism and its environment can be observed and measured directly. Having untangled this complicated web, these researchers then extrapolate backwards to the processes involved in the character's initial appearance in, and subsequent spread through, the ancestral species.

Historical ecology complements these studies by providing direct estimates of phylogeny, which can be used as a template for reconstructing the historical patterns of character origin and diversification. Such a template can help ecologists to focus their search for the processes underlying adaptation. For example, consider an interesting (but improbable) group of species with the following characteristics (table 5.1) and phylogenetic relationships (fig. 5.1).

In this group the relationship between a character and the environment in which it originated can be investigated for only one taxon and one character state, namely, the evolutionary increase in size (big) in species D (fig. 5.1). Researchers interested in the processes involved in character origin and adaptive success, if any, should concentrate their efforts on this species. The presence of small, blue, and square in species A, small, red, and square in species B, small, red, and round in species C, and red and round in species D are all ancestral legacies. Studies of these species will uncover the processes involved in character maintenance. Combining the results from all these studies will provide us with a more direct estimate of the relationship among the processes underlying the origin, spread, and maintenance of potentially adaptive traits. The incorporation of both the patterns of the past and the processes of the present into our framework of evolutionary explanations will thus strengthen our hypotheses of adaptation.

In this chapter we are going to examine the types of questions about adaptation that can be investigated from a historical ecological perspective. In

Table 5.1   Distribution of three characters among four members of a monophyletic group (species A + B + C + D) and the outgroup (taxon X).

Trait	Species				
	X	A	B	C	D
Color	blue	blue	red	red	red
Shape	square	square	square	round	round
Size	small	small	small	small	big

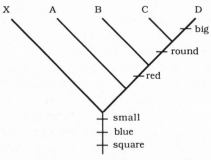

Fig. 5.1. Examining adaptationist hypotheses. Phylogenetic tree for the improbable species with the distribution of the three characters in table 5.1.

so doing, we hope to expand upon the long-standing tradition of examining adaptation within a comparative framework, advocated and illustrated by numerous researchers "ever since Darwin." Before we begin, however, it is necessary to reemphasize and expand upon some general methodological points from chapter 2.

Character Optimization: How to Interpret Characters on a Phylogenetic Tree

One cardinal rule in historical ecology is *never bias your analysis by using the ecological information you want to study to build your phylogenetic tree.* Historical ecologists must begin their explorations with two pieces of information, a phylogenetic hypothesis and the relevant ecological data. The best-supported (most-parsimonious) sequences of evolutionary transformations for the ecological characters, be they binary or multistate, can then be determined by reference to the phylogenetic tree. This is called character optimization. We will discuss optimization procedures beginning with the method originally developed by Farris (1970), which has become known as Farris optimization.

Unambiguous binary character

1. Figure 5.2 depicts a phylogenetic tree for a hypothetical group of taxa. The distribution of a binary ecological character (states a and b) is mapped at the ends of the branches.

2. "Generalizing down" the tree (fig. 5.3): Label the two nodes that are farthest from the ingroup node in the following manner: (1) label the node "a" if the two closest nodes or branches are either both a or a and a, b; (2) label the node "b" if the two closest nodes or branches are either both b or b and a, b; (3) if the closest branches/nodes have different labels (one a and the other b), then label the node "a, b." Continue working towards the ingroup node in this manner. (Does this sound familiar?) Let's start on the left side of the tree. Node 1 connects two species exhibiting character a, so we generalize

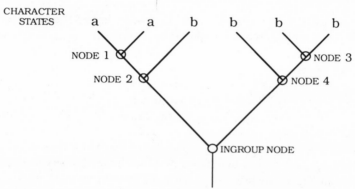

Fig. 5.2. Phylogenetic tree with the distribution of ecological character states a and b.

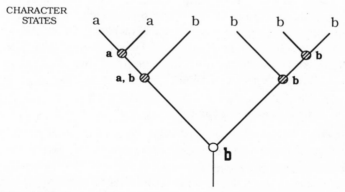

Fig. 5.3. Step 1 in Farris optimization of a binary character: generalize down the tree.

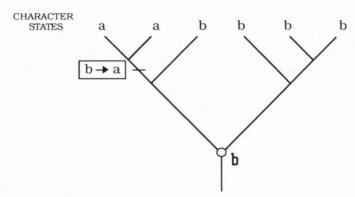

CHARACTER
STATES

Fig. 5.4. Step 2 in Farris optimization of a binary character: predict up the tree.

that the nodal (ancestral) state is a. This node, in turn, has a sister-group relationship with a taxon exhibiting state b; therefore, the state at node 2 is ambiguous for a or b. On the right side of the tree, nodes 3 and 4 are both labeled "b" because all three taxa exhibit that character state. Finally, we assign a value of b to the ingroup node because the two nodes directly above it have states a, b and b, so b "wins out" over a by majority vote (the principle of parsimony). Any ambiguity at the ingroup node may be resolved by reference to outgroups. (Starting to sound familiar yet? If not, refer back to the discussion of resolving the outgroup node, presented in chapter 2.)

3. "Predicting up" the tree (fig. 5.4): Move from the ingroup node up the tree, resolving any ambiguity by comparing the value of the ambiguous node with the value of the node directly below it. In this example only node 2, designated "a, b" in figure 5.3, is ambiguous. Since the value of the node below it is b, node 2 is reassigned state b. All nodal states have now been unambiguously resolved on the tree. Farris optimization thus provides us with the following evolutionary hypothesis for this binary character: (1) state b is a persistent ancestral condition in four of the six terminal taxa (plesiomorphy); (2) there was a change from b to a in the ancestor of the two species that now exhibit state a (apomorphy).

Unambiguous multistate character

1. Figure 5.5 depicts a phylogenetic tree for a hypothetical group of taxa. The distribution of a multistate ecological character (states x, y, and z) is mapped at the ends of the branches.

2. Generalizing down the tree (fig. 5.6): Using the same logic as that described for binary characters, node 1 is labeled with state z because both taxa connected by the node exhibit state z; node 2 is y, z because it connects a z

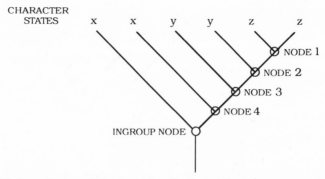

Fig. 5.5. Phylogenetic tree with the distribution of ecological character states x, y, and z.

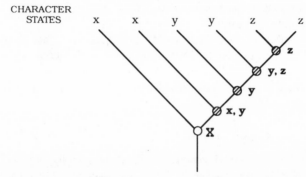

Fig. 5.6. Step 1 in Farris optimization of a multistate character: generalize down the tree.

node with a y branch (taxon); node 3 is y because it connects a y, z node with a y branch; node 4 is x, y because it connects a y node with an x branch; and finally, the ingroup node has a value of x because it connects an x, y node with an x branch.

3. Predicting up the tree (fig. 5.7): Nodes 2 and 4 are ambiguous in this example. As discussed for the binary character, resolution of these ambiguous nodes is dependent upon the character state of the node immediately below them. Accordingly, node 4 is reassigned a value of x and node 2 is labeled "y." All nodal states have now been unambiguously resolved on the tree. Farris optimization thus provides us with the following evolutionary hypothesis for this multistate character: (1) state x, occurring in the oldest two taxa, is plesiomorphic for the clade; (2) there was a change from x to y in the common ancestor of the remaining four taxa (apomorphy); (3) there was a change from y to z in the common ancestor of the terminal two taxa (apomorphy). All the ecological traits are persistent ancestral conditions in the

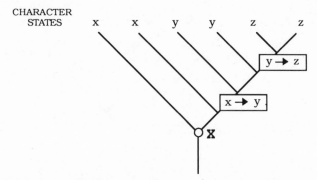

Fig. 5.7. Step 2 in Farris optimization of a multistate character: predict up the tree.

species that are extant today (and there is no reason to believe that any of them evolved more than once).

Like many aspects of the natural world, optimization does not always produce such unambiguous results. In the remainder of this book we hope to show you that the real world is a mixture of cases, some readily interpretable and others not. But first, let us consider an example that cannot be fully resolved.

Ambiguous binary character

1. Figure 5.8 depicts a phylogenetic tree for yet another hypothetical group of taxa. The distribution of a binary ecological character (states t and s) is mapped at the ends of the branches.

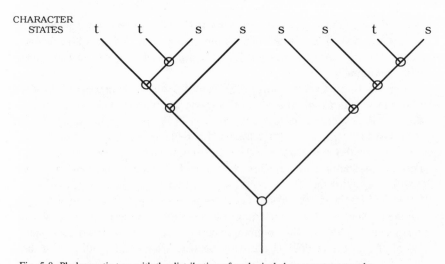

Fig. 5.8. Phylogenetic tree with the distribution of ecological character states t and s.

2. The first stage of Farris optimization proceeds as previously described (fig. 5.9).

3. The fun begins when we proceed to the second stage of the optimization procedure. Predicting up the tree produces the solution presented in figure 5.10. According to this solution, (1) s is plesiomorphic, (2) t evolved from s twice (convergence), and (3) s evolved from t once (reversal). This implies

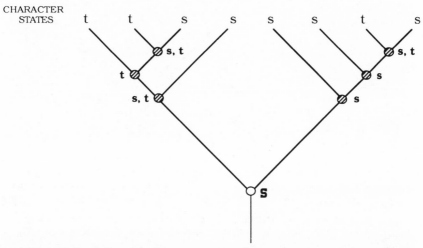

Fig. 5.9. Step 1 in Farris optimization of an ambiguous, binary character: generalize down the tree.

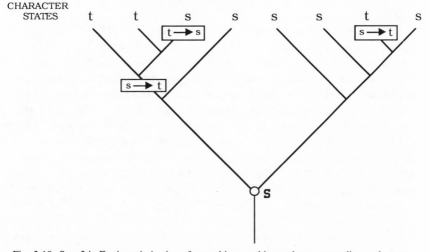

Fig. 5.10. Step 2 in Farris optimization of an ambiguous, binary character: predict up the tree.

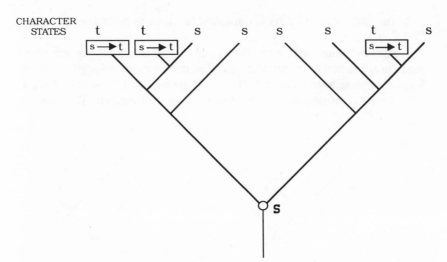

Fig. 5.11. An equally parsimonious interpretation of the evolution of the binary character in figure 5.10.

that not all the species exhibiting either t or s inherited that trait from the same ancestral source.

4. If we examine this particular example more closely, however, we realize that the result of Farris optimization is not the only hypothesis that explains the distribution of t and s among these species. An equally parsimonious interpretation of these data is that (1) s is plesiomorphic, and (2) there have been three transitions from s to t (convergence) in extant species (fig. 5.11). This implies that all species displaying s inherited that trait from the same ancestral source.

Despite the fact that phylogenetic optimization does not provide an unambiguous answer in this case, all is not lost. In some systems, additional biological information (e.g., developmental or biochemical data) can be brought to bear on the problem. In other systems, we can attempt to evaluate the conflicting hypotheses by direct experimental studies in both the field and the laboratory. For example, figure 5.11 suggests that t has arisen independently from s three times in recent species. In cases of convergent evolution, common selective pressures from the environment are immediately suspected. If the species exhibiting t share an environmental regime that is not associated with the other members of the clade, all of which exhibit s, we would have reason to support this hypothesis. If we found that the species exhibiting t did not occur in common environments, we might turn our attention to the evolutionary sequence depicted in figure 5.10.

As you can imagine, the potential for alternative explanations increases when multistate characters are considered (Swofford and Maddison 1987). Donoghue (1989) noted additional theoretical and empirical examples in which optimization failed to resolve the sequences completely. Although we would like nature to provide us with a perfect record of evolution, this does not always happen. In such cases, we expect that phylogenetic optimization will tell us exactly where the ambiguity lies, even if it cannot provide an unambiguous interpretation.

Methodological Caveats for the Historical Ecologist

First, remember that phylogenetic trees can be rotated at their nodes without changing relationships. For example, figures 5.12a and 5.12b are the same phylogenetic tree.

Second, remember that a robust phylogenetic tree is based on analysis of numerous characters, each of which was polarized using outgroup comparisons. Any of the characters used may exhibit some degree of homoplasy, in which case the distribution of the characters on the phylogenetic tree does not correspond exactly to the sequence postulated originally when the characters were polarized (the transformation series; fig. 5.13). *Thus, it is not methodologically sound to polarize one character using an outgroup and then to assume that the postulated character transformation series represents the true phylogenetic sequence of events.* This type of error results from a subtle misunderstanding of the purpose of outgroup comparisons. While it is true that polarizing character states through outgroup comparison produces a hypothesis of the sequence of evolutionary changes in that character, it is also true that this hypothesis can only be tested by comparing it to the hypothesized evolutionary sequences of other characters. In other words, a character

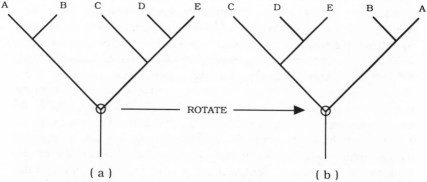

Fig. 5.12. Rotating phylogenetic trees around their nodes does not change the phylogenetic relationships among the taxa.

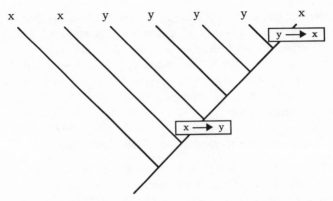

Fig. 5.13. Dangers inherent in assuming that the transformation series for one character represents the true evolutionary sequence of character change. In this example, the outgroup state for the character is x. Since both the outgroup and some members of the ingroup possess the character state x, the character transformation series is hypothesized to be x to y. However, when the character is optimized on a phylogenetic tree constructed from a number of different characters, the actual phylogenetic transformation is discovered to have been x to y to x.

transformation series does not stand on its own, independent from the underlying phylogenetic relationships of the organisms.

A final point to remember is that an adaptive explanation can be weakened if some members of the study clade are removed from the analysis. The world of phylogenetic reconstruction is fraught with enough difficulties in terms of sampling procedures and the incorporation of fossil evidence (Doyle and Donoghue 1986; Donoghue et al. 1989), without augmenting the problem by eliminating available information from the analysis. For example, consider a monophyletic group of fishes bearing the nest-building characters w, x, y, and z (fig. 5.14).

Failure to incorporate some of the taxa bearing trait w into the analysis will produce the incorrect postulate of a linear sequence of character change from w to x, x to y, and y to z (fig. 5.14a). When all the members of the group are considered, the evolutionary sequence is demonstrated to be the nonlinear w to x, w to y, and w to z. The situation depicted in figure 5.14b is slightly different. Here, although the hypothesized sequence from w to x to y to z is correct, this analysis will grossly underestimate the degree of phylogenetic constraint on the proposed evolutionary change. In this clade, 75% of the taxa bear the ancestral character state w (i.e., 25% of the distribution supports a proposal of historical lability in the trait). Eliminating taxa bearing the plesiomorphic condition from the analysis artificially inflates to 50% the proportion of taxa displaying a derived character state, thus making it appear that the character has been evolutionarily labile.

Now, on to the exciting world of adaptive changes in phylogeny. We will present this material in the following sequence: (1) formulating hypotheses of adaptation within a phylogenetic framework; (2) examining the temporal sequence of adaptive changes; (3) examining the evolutionary association of traits ("coadapted traits" within a clade); (4) examples of convergent adaptation; (5) examples of divergent adaptation; and (6) examining phylogenetic constraints on ecological diversification, including a discussion of phylogenetic approaches to documenting adaptive radiations.

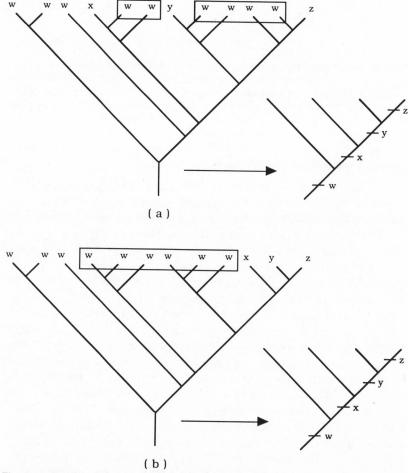

Fig. 5.14. Example demonstrating the problems that can arise when adaptive hypotheses are formulated within an incomplete phylogenetic framework. (a) A nonlinear evolutionary transformation of traits appears linear when taxa are removed. (b) The amount of phylogenetic conservatism is greatly underestimated by removing taxa from consideration.

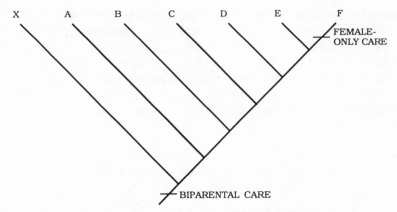

Fig. 5.15. Hypothetical example outlining the applicability of phylogenetic systematics to questions concerning the evolution of behavioral/ecological traits. The distribution of parental-care behaviors in the hypothetical fishes has been mapped onto a phylogenetic tree constructed from characters other than the parental-care trait (e.g., morphological data). The occurrence of biparental care is correlated with the phylogenetic relationships of species A, B, C, D and E. The appearance of female-only care is an autapomorphy for species F.

Formulating the Question

Ecologists frequently strive to understand "why" the members of one species act the way they do. For example, suppose you were interested in studying a hypothetical species of fish, C for short, in which both the male and the female guard their offspring. Question: Why does C show biparental care? Traditionally, you would seek the answer to this question by performing a series of sophisticated cost/benefit analyses on populations of C. Now, let us consider the question within the larger framework of phylogeny. When the distribution of parental care behaviors is mapped onto the phylogenetic tree for the genus to which C belongs, the pattern shown in figure 5.15 emerges.

Biparental care occurs in the outgroup (X) and taxa A, B, C, D, and E. Since its distribution is correlated with the phylogenetic relationships of these taxa, its presence in species C represents the persistence of an ancestral trait in that species. The answer to the question, Why does C show biparental care? is thus, because its ancestor did. Although cost/benefit analyses will uncover valuable information concerning the maintenance of biparental care (**plesiomorphic** trait) under **current** environmental conditions, such analysis does not address why that character was originally successful. In order to explain this, we need two additional pieces of information: an expanded phylogenetic analysis incorporating additional genera of fishes to establish the evolutionary "point of origin" of the biparental trait and an assessment of that *ancestor's* biology in *its* environment. But this is a different question altogether, and takes us on an excursion into the past, both genealogical and environmental. From an adaptationist perspective, the pertinent question for this clade is, Why does species F show female care only? Since female care represents the evolution of a **derived** character state in conjunction with **current** environmental conditions, cost/benefit analyses in this case would address the question of character origin, the more relevant component of the "why" of evolution to the adaptationist program.

Why are there one- and two-horned species of rhinos?

The preceding hypothetical example demonstrates the dangers of addressing evolutionary "why" questions within the context of a single species. However, omitting the historical component from evolutionary explanations will generate similar problems in formulating the relevant evolutionary hypothesis, regardless of the number of species investigated (Ridley 1983; Wanntorp 1983; Brooks 1985; Brooks and Wiley 1988). Coddington (1988) provided an excellent example of this in his reanalysis of the adaptationist explanation for the presence of one-horned and two-horned species of rhinos. Lewontin (1978) postulated that the one-horned condition and the two-horned condition represent independently achieved, equally functional, adaptive peaks in the evolution of antipredator defense structures. This scenario implies that both conditions arose *de novo* from a hornless condition. Coddington investigated this hypothesis by mapping horn number onto a phylogenetic tree for the rhinoceratids (fig. 5.16).

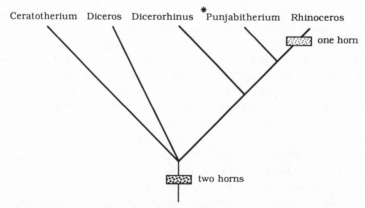

Fig. 5.16. Phylogenetic tree of rhinoceratids based upon dental and osteological characters, mapping distribution of horn number. * = an extinct genus. (Based on Groves 1983; redrawn from Coddington 1988.)

The distribution of horn numbers on the phylogenetic tree suggests that the two-horned condition is plesiomorphic for the group, including extinct species. The single-horned condition is derived from the two-horned condition through the loss of the frontal horn. Since the two conditions were not achieved independently from a nonhorned condition, and thus do not represent two alternative adaptive peaks, it is difficult to postulate an equally adaptive, antipredator function for each character state. Nor is it possible to state that two horns are better than one, because the plesiomorphic condition is two horns. However, as in the preceding example, placing the problem within a phylogenetic framework generated at least two new questions. The "whys" of two horns must be examined within the context of the ancestor and the environment in which the trait originated. Although difficult, this call for a detailed knowledge of past environments should excite students of paleoecology (a discipline also called "historical ecology" [Rymer 1979] or "zooarchaeology" by archaeologists). A more tractable question concerns the reduction in horn number from two to one in the genus *Rhinoceros*. If we cannot unequivocally postulate that horn number represents adaptive antipredator responses, perhaps we should consider other possibilities. Ample opportunities exist for students of behavior and ecology to collect data, map them onto the phylogeny, and investigate associations between changes in, for example, sexual behavior, habitat preference, or predation pressures, and changes in horn number.

The Temporal Sequence of Evolutionary Change

Lorenz (1941) cautioned, "The similarity of a series of forms even if the series structure arises ever so clearly from a separation according to characters, must not be considered as establishing a series of developmental stages." In his opinion, without reference to phylogenetic relationships the criterion of similarity was, of itself, a dangerously misleading evolutionary marker. The last two decades have witnessed a rise of a new methodology based, in direct contrast to Lorenz's warning, upon arranging characters as a "plausible series of adaptational changes that could easily follow one after the other" (Alcock 1984:432). Although intuitively pleasing, this method relies heavily on subjective, a priori assumptions concerning the temporal sequence of evolutionary modifications and dissociates character evolution from underlying phylogenetic relationships. The reintroduction of "history" into our evolutionary perspective has prompted researchers to seek alternate methods for uncovering the direction and sequence of character change (for an excellent discussion of this area, complete with additional examples, see Donoghue 1989).

The evolution of Dioecism in Lepechinia *section* Parviflorae

Hypotheses concerning the evolution of breeding systems in plants are currently a hot topic among botanists (see references in Hart 1985b). Various adaptive pathways have been proposed, based upon mathematical models of changes in gene frequencies in populations and studies at either the population level or across broad taxonomic categories. Hart (1985b) pointed out that, since these hypotheses were generally not formulated within a phylogenetic context, questions concerning the evolutionary transitions from one breeding system to another were difficult to address. He focussed his attention on one particular area of the debate: the evolution of dioecism (separate sexes) in the *Lepechinia* section *Parviflorae* (Lamiaceae). Among members of *Lepechinia* some species are monoecious, some are gynodioecious, and some are dioecious. Originally gynodioecism was viewed as a stable strategy that would not give rise to dioecism (Darwin 1877; Lewis 1942). This view was expanded by Ross (1978) to include two forms of gynodioecism, "stable" and "unstable," in which only the latter form allowed the possibility of dioecism evolving. Within the Lamiaceae, the stable form was thought to predominate. Hart investigated this problem by mapping the character "breeding system type" onto a phylogenetic tree depicting the relationships within three Lamiaceae sections (fig. 5.17). He was particularly interested in the relationships within the section *Parviflorae,* a group of small, white-flowered shrubs inhabiting the Andean regions of South America, which we have already discussed in chapter 4.

The phylogenetic relationships within this group of plants indicate that gynodioecism is the plesiomorphic condition for members of the monophyletic section *Parviflorae*. Within this section, dioecism is a derived state that has arisen independently at least twice. Hart cautioned that robust conclusions concerning the number of times dioecism has arisen in *Parviflorae* cannot be drawn based upon his phylogenetic tree, since characters involved in this type of breeding system were incorporated in the original systematic analysis. It is important to note, however, that (1) apomorphic characters independent of breeding-system type confirm the monophyletic status of this section and (2) outgroup comparisons have demonstrated that gynodioecism is plesiomorphic for the clade. Given this, the phylogenetic hypothesis of an evolutionary sequence from gynodioecism to dioecism is a valid one based on the available data. Hart's caution involves the resolution of the number of times the transition occurred, not the sequence in which it occurred. This result weakens the hypothesis that gynodioecism is "stable" within this section, and highlights some potentially interesting areas for further research. Hart noted that the gynodioecious outgroups occurred in mesic upland sites

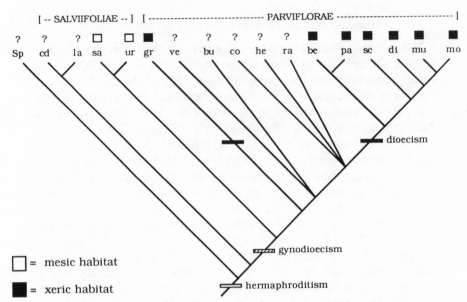

Fig. 5.17. Distribution of breeding-system type mapped onto a phylogenetic tree for three *Le-pechinia* sections. *Speciosae, Salviifoliae,* and *Parviflorae. be = Lepechinia betonicaefolia; bu = L. bullata; cd = L. codon; co = L. conferta; di = L. dioica; gr = L. graveolens; he = L. heteromorpha; la = L. lancifolia; mo = L. mollis; mu = L. mutica; pa = L. paniculata; ra = L. radula; sa = L. salviifolia; sc = L. scobina; Sp = Speciosae; ur = L. urbanii; ve = L. vesiculosa.* The *boxes* above the phylogenetic tree represent habitat preference for each species where known. (Redrawn and modified from Hart 1985b.)

in the Andes, while the dioecious species preferred xeric habitats and demonstrate higher degrees of weediness. If further analysis of the ecology of the gynodioecious species within the *Parviflorae* demonstrates that preference for mesic habitats is the plesiomorphic condition in this section, then the hypothesized correlation between breeding-system shift and habitat shift would represent a correlation between apomorphic traits about which investigations of adaptation could be profitable.

In the preceding example, we used a phylogenetic analysis to establish the sequence of evolutionary changes for one character. We can also use this method to evaluate the evolutionary relationships between two or more different traits (see also Resh, Morse, and Wallace 1976; Morse 1977; Morse and White 1979). There are three general hypotheses concerning such relationships: (1) The traits neither co-originate nor co-occur in any members of the study group (fig. 5.18a). (2) The traits do not co-originate, but they do co-occur in some members of the study group. In this case the researcher is primarily interested in examining the sequence of origins for the different

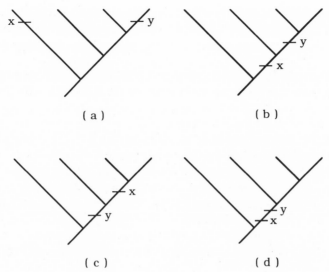

Fig. 5.18. Examining the phylogenetic relationships between two different characters. (*a*) Traits x and y arose independently in different species; the traits neither co-originate nor co-occur. (*b*) Trait x arose before trait y; the traits do not co-originate, but they do co-occur in some members of the group. (*c*) Trait y arose before trait x; the traits do not co-originate, but they do co-occur in some members of the group. (*d*) Both characters originated in the same ancestor; the traits co-originate and co-occur.

traits. For example, the hypothesized sequence "trait x arose first, followed by trait y," is refuted by the phylogenetic patterns depicted in figures 5.18a and 5.18c and supported by the pattern depicted in figure 5.18b. Interestingly, the relationships outlined in figure 5.18d offer no resolution of the problem. These associations can only fail to refute hypothesis 2 because phylogenetic analysis cannot determine the sequence of trait origins within a species. (3) The traits co-originate and co-occur in members of the study group (fig. 5.18d).

The evolution of gregariousness in aposematic butterfly larvae

Although most butterflies are solitary ovipositors, there are species that lay their eggs in clusters. Once hatched, larvae from these egg aggregations tend to remain clustered. This observation requires an explanation because groups of small, relatively immobile but tasty prey items should be highly susceptible to predation, and thus be at a strong selective disadvantage. Closer examination reveals that (1) the larvae of many butterfly species are unabashedly aposematic, displaying vivid combinations of black, red, orange, yellow, or white against the brown-and-green backgrounds of the surrounding

vegetation; (2) there is a correlation between larval clustering (gregariousness) and aposematism; and (3) there is a correlation between larval distaste-
fulness and aposematism. Based on these observations, Fisher (1930) proposed that the development of prey unpalatability required the involvement
of kin selection. Turner (1971) and Harvey et al. (1982) extended the kin-
selection hypothesis to include the development of aposematic color patterns.
The idea was straightforward; some bad-tasting mutants with warning coloration would be eaten, predators would associate bad taste with the mutant's
color pattern, and those predators would avoid their last meal's similarly colored siblings. Under this hypothesis, gregariousness should either (1) evolve
first, providing a context in which kin selection could work, followed by the
evolution of distastefulness/aposematic coloration, or (2) appear at the same
time as warning coloration/distastefulness. The former evolutionary sequence
provides the strongest corroboration for the hypothesis of a causal link, provided by kin selection, between gregariousness and the appearance of "taste
bad/look bad" traits. Sillen-Tullberg (1988) examined this hypothesis by
referring to phylogenies (and, in cases where phylogenies were not available, through reevaluation of taxonomic information by optimization techniques) for a number of butterfly lineages. She compiled data from the literature concerning larval aggregation habits (solitary versus gregarious) and
color (cryptic versus aposematic). Her focus was on the order of events:
Did gregariousness evolve before, with, or after warning coloration?
Her results were startling. For example, consider the tribe Papilionini
(fig. 5.19).

In this particular clade, the ancestral condition was determined by outgroup
analysis to be "larvae solitary and larvae cryptic." Diversification within the
clade involved the evolution of gregariousness in three separate lineages and
the transition from crypsis to warning coloration in two lineages. Although
all four possible combinations of the grouping and color traits exist in these
butterflies, in no case did the evolution of gregariousness precede the evolution of warning coloration. This example is representative of the results found
for all the butterfly groups examined phylogenetically by Sillen-Tullberg: of
the twenty-three independent appearances of gregarious larval interactions
during the course of butterfly evolution, three evolved in conjunction with
the appearance of warning coloration, fifteen evolved after warning coloration, and five evolved without the subsequent development of aposematism.
Gregariousness did not evolve before warning coloration in any lineage.
Based upon the temporal sequence of trait appearance, Sillen-Tullberg concluded that "kin selection is of minor importance for the evolution of both
unpalatability and warning coloration" and suggested that researchers turn
their attention to hypotheses of individual selection for explanations of the
evolution of these characters.

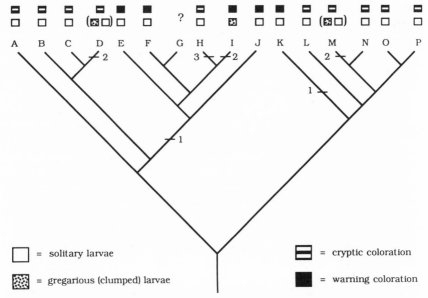

Fig. 5.19. Temporal sequence of the evolution of larval clustering and warning coloration mapped onto a phylogenetic tree for the butterfly tribe Papilionini. *A* = *Pterourus* (5 species groups); *B* = *Heraclides thoas; C* = *H. torquatus; D* = *H. anchisiades; E* = *Chilasa elwesi; F* = *C. clytia; G* = *C. veiovis; H* = *C. agestor; I* = *C. laglaizei; J* = *Eleppone anactus; K* = *Papilio machaon; L* = *Princeps xuthus; M* = *P. demolion; N* = *Princeps* (9 species groups); *O* = *Princeps* (6 species groups); *P* = *Princeps* (10 species groups). The distribution of the two larval-aggregation and color characters is mapped above the taxa and optimized onto the tree. *1* = warning coloration present; *2* = gregariousness present; *3* = reversal to crypsis.

Evolutionary Associations of Traits: Coadapted Trait Complexes within a Clade

Observations of correlations among traits in extant species have been used as the basis for adaptive hypotheses concerning the evolutionary relationship between the traits. For example, consider the distribution of two behavioral characters, number of mates per male and male nuptial plumage, among a taxonomically "closely related" yet hypothetical group of birds (table 5.2).

The nonhistorical explanation for this distribution might be that the characters are "travelling together" evolutionarily, acting as coadapted trait complexes (Mayr 1963). Advocates of this hypothesis would turn their attention to the special case, species C, in an attempt to discover the current environmental pressures that have led to a change in the relationship between the traits. Now consider an alternate method for examining the relationships be-

Table 5.2 Distribution of nuptial plumage and polygyny in a hypothetical, but monophyletic, group of birds.

	Characters	
Taxon	1 Nuptial Plumage	2 Polygyny
A	absent	absent
B	absent	absent
C	present	absent
D	present	present
E	present	present
F	present	present

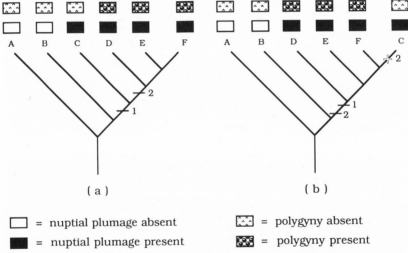

☐ = nuptial plumage absent   ☒ = polygyny absent

■ = nuptial plumage present   ▩ = polygyny present

Fig. 5.20. Phylogenetic trees, differing in the placement of species C, for the hypothetical birds (table 5.2). Derived states of the two characters are mapped onto each tree. *1* = nuptial plumage present; *2* = polygyny present.

tween the two traits. Figure 5.20 depicts two phylogenetic trees for this clade of birds.

Both trees indicate that the correlation of "nuptial plumage absent" with "polygyny absent" and "nuptial plumage present" with "polygyny present" is the result of common phylogenetic history. While this in itself is interesting, it does not help us to distinguish between a coadaptive and a coincidental phylogenetic association between the two characters. The critical piece of evidence is provided by examining species C *within the phylogenetic frame-*

work. The relationships depicted in figure 5.20a reveal that, given the evolutionary "placement" of C, the presence of "polygyny absent" and "nuptial plumage present" in that species indicates different rates of evolutionary divergence in the two characters. This provides support for a hypothesis of relative evolutionary *independence* of male plumage development and number of mates in this group. The relationships depicted in figure 5.20b corroborate the hypothesis that traits 1 and 2 are acting as a coadapted unit. This tree further suggests that species C seems so anomalous because it has undergone an evolutionary reversal to the plesiomorphic mating system, monogamy, while retaining the derived trait of "nuptial plumage present." In this case, as suggested by a nonphylogenetic approach, the change in the relationships between male plumage and number of mates could be examined within the context of current environmental selective pressures.

The association between ritualization of behavior and increasing levels of aggression in stickleback fishes

Tinbergen (1953) predicted that the evolution of increasing levels of aggression in any group should be accompanied by increased ritualization of agonistic behaviors. He proposed that the evolutionary ritualization of aggression was adaptive because threat and intense, but noncontact, fighting would permit the establishment of dominance/territorial relationships with a minimum amount of physical injury to the combatants. McLennan, Brooks, and McPhail (1988) examined this prediction using a fascinating and widespread family of north temperate fishes, the Gasterosteidae, or sticklebacks. Subjective observations have placed *Spinachia spinachia* (the sea stickleback) and *Gasterosteus aculeatus* (the famous three-spined stickleback) at opposite ends of a continuum of "aggressiveness," with *G. aculeatus* being the most pugnacious of the sticklebacks (Wootton 1976). Bearing this in mind, McLennan et al. mapped aggressive and submissive behaviors onto a phylogenetic tree for the Gasterosteidae (fig. 5.21). Since the analysis was not based on any a priori coupling of aggression and ritualization, examination of the relationship between the two on a macroevolutionary level provided a rigorous test of Tinbergen's hypothesis.

Inspection of figure 5.21 reveals that the directional change in levels of "aggressiveness" is correlated with both the phylogenetic diversification of aggressive behaviors through the predicted sequence of "chase/bite" to "threat" to "circle fight" and the simultaneous appearance of aggressive and appeasement behaviors (broadside and head-down "threat" coupled with "head up" submission; "circle fight" coupled with "dorsal roll"). This analysis thus provides corroboration for Tinbergen's hypothesis from the perspective of the evolutionary origins of agonistic characters, their associations dur-

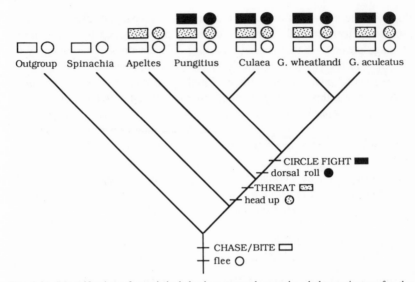

Fig. 5.21. Diversification of agonistic behaviors mapped onto the phylogenetic tree for the stickleback fishes, family Gasterosteidae. *G* = *Gasterosteus*. Aggressive behaviors are in *uppercase letters,* submissive behaviors in *lowercase letters.* The figures above the tree are a visual representation of the character changes. *Rectangles* = aggressive behaviors; *circles* = submissive behaviors. The evolutionary diversification of the sticklebacks was accompanied by the simultaneous *addition* of aggressive/submissive behaviors to the repertoire of the new species.

ing their subsequent diversification, and the sequence of the diversification of the traits. The analysis also highlights an interesting and unexpected result. Experiments have demonstrated that *Pungitius pungitius* is substantially less aggressive than *G. aculeatus* (Wilz 1971; Wootton 1976); however, according to the phylogenetic hypothesis presented in figure 5.21, there has been no further diversification (ritualization) of agonistic behavior in this clade. The simplest explanation for this is that *Pungitius* + *Culaea* + *G. wheatlandi* retain the ancestral level of aggression (and the associated ancestral behaviors), and the increased levels detected in the three-spined stickleback (*G. aculeatus*) are derived. If this is so, then explanations for the increase in aggression without the simultaneous increase in the ritualization of agonistic behaviors must be sought for *G. aculeatus.* If all the members of this monophyletic group demonstrate some increase in aggressive levels, the interpretation would be more complex. Further resolution of the relationship between aggressiveness and ritualization awaits a more detailed analysis of levels of aggression in these fishes. In such studies, aggressiveness should be measured by using the plesiomorphic indicator of aggression in this family: the number of bites directed towards a test subject.

The relationship between preferred environmental temperature and optimum sprinting temperatures in skinks

Skinks are hyperactive, predominantly terrestrial lizards. They generally eat insects and other arthropods, but large skinks will sometimes eat baby mice or birds, or small bird eggs (Conant 1975). As is the case with many middle-level predators, skinks are themselves tasty morsels for larger animals. They often sprint to avoid predators, and this running speed is, in turn, strongly influenced by environmental temperature because lizards are poikilotherms. Since predator avoidance is such a highly desirable outcome in the life of a lizard, the evolution of a species' habitat/body-temperature preference should be tightly coupled with the evolution of the temperature that produces its optimum sprinting speed. Huey and Bennett (1987) examined this expectation within both comparative/nonhistorical and comparative/phylogenetic frameworks. Statistical analysis of the relationship between thermal preference and optimum running temperature revealed a significant association between the two traits; however, this resulted from a very strong association between the characters in species with high thermal preferences and a weak association in species with low thermal preferences. The nonphylogenetic approach thus uncovered some flaws in the original expectation that thermal preferences and thermal sprinting optima would be tightly coupled in all species of skinks. Huey and Bennett then examined the relationship between thermal preference and optimum body temperature for sprinting speed within a phylogenetic framework. This was a difficult proposition because the states of the two characters are quantitative, with each species having a range of variation. Huey and Bennett based their analysis on the mean value of the character states for each species. Figure 5.22 depicts a simplified version of their results.

With an eye to both the direction and magnitude of evolutionary change, the authors delved further into the question of coadaptation. Phylogenetic examination corroborated the statistical conclusion that coadaptation was stronger in genera with high thermal preferences (*Egernia* and *Ctenotus*: same rate and direction of character changes) than in genera with low thermal preferences (*Leiolopisma, Sphenomorphus, Eremiascincus,* and *Hemiergis*). Skinks of the genera *Leiolopisma* and *Sphenomorphus* display moderate levels of partial coadaptation based upon similar directions in character change coupled with slightly different rates of evolution. The two weakest associations were displayed by *Hemiergis* and *Eremiascincus,* which, interestingly, have greatly reduced limbs. Huey and Bennett proposed that coadaptation was weak in *Hemiergis* because, although the direction of the change was the same for both traits, the lowering of the thermal preference occurred at a

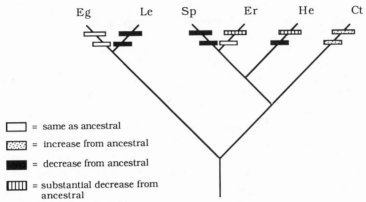

Fig. 5.22. Phylogenetic tree for Australian skinks, mapping thermal preference (*top line of bars*) and optimum sprinting temperature (*bottom line of bars*). *Eg* = *Egernia; Le* = *Leiolopisma; Sp* = *Sphenomorphus; Er* = *Eremiascincus; He* = *Hemiergis; Ct* = *Ctenotus*. The traits have been simplified to four categories based upon general trends in both the direction and magnitude of evolutionary change compared with the ancestral condition. Presumptive ancestral conditions were reconstructed based on outgroup comparison and a method called "the minimum-evolution method" (Huey and Bennett 1987).

substantially faster rate than the lowering of optimal sprint temperature. Within *Eremiascincus,* thermal preferences also underwent a major decrease from the ancestral preference; however, sprinting temperature remained at the ancestral high point. In this instance, both the rate and the direction of evolutionary change varied between the traits. So, rather than operating as a tightly coadapted unit in all cases, it appears that the relationship between thermal preference and optimal sprinting temperature can vary independently to some extent.

Convergent Adaptation

Convergent evolution of similar traits in different lineages is considered to be one of the strongest types of evidence for adaptation; however, convergence is often asserted without its demonstration by phylogenetic analysis. As we noted in chapter 2, phylogenetic systematics provides a strong test of homoplasy because the homoplasious characters are highlighted against a background of presumed evolutionary homology. Putative convergences are identified a posteriori from phylogenetic analyses based on a set of characters for which no postulate of convergence was proposed a priori. Convergences in quantitative or qualitative traits can be identified by optimizing such characters onto phylogenetic trees constructed using other data. Once convergence has been identified, adaptive hypotheses can be constructed by looking for similarities in environments inhabited by taxa exhibiting the convergent traits.

Run-jump behavior in Richardson's ground squirrels and black-tailed prairie dogs

Dobson (1985) documented the presence of an interesting "run-jump" flight behavior in Richardson's ground squirrels (*Spermophilus richardsonii*) and the black-tailed prairie dog (*Cynomys ludovicianus*), in which fleeing individuals "push the anterior part of the body off the ground with the forelegs, while continuing to run with the hind legs." Mapping the trait onto a phylogenetic tree reconstructed from the current classification scheme produces a disjunct and apparently convergent distribution of run-jumping (fig. 5.23a).

When the distribution is reanalyzed on a second tree, reconstructed from a different classification scheme, the appearance of run-jumping in both species is now hypothesized to be the result of common history (fig. 5.23b). Clarification of the problem resides in the resolution of the phylogenetic relation-

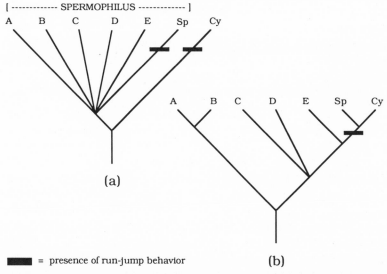

■■■■ = presence of run-jump behavior   (b)

Fig. 5.23. The presence of the run-jump trait mapped onto two hypothetical phylogenetic trees for ground squirrels and prairie dogs. (*a*) According to this tree, the genus *Spermophilus* includes the subgenera A, B, C, D, E, and Sp. (*b*) Under the second tree, subgenera *Spermophilus* (Sp) and *Cynomys* (Cy) are each other's closest relatives; therefore, if this is the correct phylogeny for the group, the genus *Spermophilus* is paraphyletic (i.e., it does not include all the species descended from the common ancestor).

ships within the possibly paraphyletic genus *Spermophilus*. Dobson's analysis reiterated the importance of examining character evolution on phylogenetic trees reconstructed by phylogenetic systematic methodology, rather than equating taxonomy with phylogeny. It also emphasizes a critical aspect of the importance of phylogeny in the testing of adaptationist hypotheses. Consider the following scenario for the origin of run-jumping. Prairie dogs and ground squirrels live in tall grasses, which makes it difficult to detect terrestrial predators. When run-jumping appeared in each species, it became fixed because it increased an individual's detection of predators in that environment. Now consider an elaborate and lengthy series of field experiments that ultimately do not produce supportive data for the "predator perception" hypothesis. Result: rejection of the hypothesis. However, if figure 5.23b represents the true phylogenetic relationships within this group of rodents, then run-jump originated in the common ancestor of Richardson's ground squirrels and black-tailed prairie dogs. Since the presence of run-jumping in extant species is due to phylogenetic constraints on its evolution, there is no a priori reason to assume that the benefits conveyed upon the ancestor in which the trait originated will still be in effect in the descendants of that ancestor. It is possible that run-jumping was advantageous to the ancestor of *C. ludovicianus* and *S. richardsonii* in terms of predator perception, and that the environment has changed sufficiently to decrease or obscure that benefit, but not sufficiently enough to select against the trait. This study underscores the dangers of formulating hypotheses about convergence based upon studies of paraphyletic groups.

Daisy trees: convergent adaptations to living in cloud forests?

The Compositae genus *Montanoa,* or "daisy trees," comprises some thirty taxa living throughout Central America, extending as far north as central Mexico and as far south as central Colombia. All species are woody, possessing flowers with white to cream "petals" (rays) surrounding a central yellow to grey-green or black disc floret. Twenty-one of these species are shrubs, five are trees reaching twenty meters in height, and four share the vine "habit." Funk (1982) performed a phylogenetic analysis of the genus based on morphological characters, then mapped the habits of each species on the phylogenetic tree (fig. 5.24).

She discovered that the shrublike habit is plesiomorphic in daisy trees. If the common ancestor of the genus was a shrub, the presence of shrub forms in 70% of the known taxa indicates a rather ancient origin for, and considerable phylogenetic conservatism in, the diversification of the shrub habit. Therefore, studies involving the convergence of two or more species on the shrub habit are inappropriate for this genus. Seven out of forty-seven

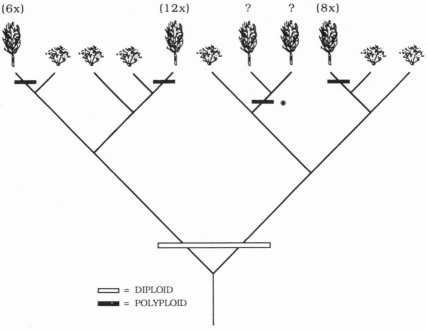

(6x)   (12x)   ?   ?   (8x)

⊏⊐ = DIPLOID
■·■ = POLYPLOID

Fig. 5.24. Simplified phylogenetic tree for members of the genus *Montanoa*, showing the distribution of the shrub and tree habits. Ploidy levels of the trees are listed above the trees. * = a case of hypothesized, but as of yet unconfirmed, polyploidy.

branches on the phylogenetic tree (about 15%) demonstrate an evolutionary change in habit. These evolutionary events have given rise to the tree forms four times and to vines three times. It is the tree and vine forms for which studies of convergent adaptation in particular species of *Montanoa* might be appropriate.

Now let us consider the tree forms in particular. One species grows in cloud forests at higher elevations than other species in the genus in each of the following locations: Guerrero, Mexico; Chiapas, Mexico, and northern Guatemala; Costa Rica; Venezuela and Colombia; and the Santa Marta mountains, Colombia. All five species have a number of similar morphological and anatomical characters that allow them to survive in cloud forests, and none has ever been found at lower elevations. They are members of four different clades, their sister species in each case being shrubs living at adjacent lower elevations. They represent cases of convergent adaptation because they exhibit a convergent trait correlated with the same environmental variable. One could speculate that in each case natural selection had favored the evolution of the same kind of strategy for surviving in cloud forests, but we do not know if this convergent adaptation was environmentally driven. In fact, an understanding of the development of tree forms within a phylogenetic context leads to a different interpretation.

Examination of the distribution of the *Montanoa* habit and ploidy level on the cladogram reveals a repeated pattern of association between the appearance of the tree form and increased ploidy levels (fig. 5.24). Although diploidy is the ancestral condition in *Montanoa,* all the tree species, and only the tree species, are high-level polyploids (Funk and Raven 1980; Funk 1982; Funk and Brooks 1990). Developmental studies have demonstrated that it is common for polyploids to be larger than diploids in the Compositae; hence, the distinction "tree versus shrub" may simply be a by-product of ploidy level. It is also common for diploids to produce polyploid seeds in this group, so the parallel and convergent appearance of polyploids is not a surprising phylogenetic pattern. Based on these data, it would appear that the repeated convergent evolution of treelike species is a result of developmental plasticity in polyploid production in this genus (something common among composites). What needs to be explained is the restriction of tree forms to cloud forests and the absence of shrub and vine forms from cloud forests. For this, we need to examine the adaptive **divergence** of each tree species from its shrubby ancestor and sister species.

Divergent Adaptation

Phylogenetic analysis of ecological diversification can be used to investigate hypotheses about mechanisms involved in evolutionary diversification within a clade. In contrast to convergent adaptation, where we looked for evidence of distantly related species inhabiting similar habitats and exhibiting convergent traits associated with the habitat, the evidence for divergent adaptation is the occurrence of apomorphic changes in ecological traits associated with the appearance of descendant species inhabiting new habitats.

Why don't tree and shrub forms of daisy trees occur together?

The Compositae genus *Montanoa* offers an example of the way in which this can be done. As noted previously, daisy trees, contrary to the name, exist in vine, shrub, and, fortunately, tree forms. The convergent evolution of the trees was hypothesized to be a result of the propensity for the diploid sister species to produce polyploid seeds, with concomitant changes in shape due to the increase in ploidy level. While this explains the convergence of tree forms, it does not explain the strict habitat divergence between each tree species and its sister shrub. The trees all grow in cloud forests at higher elevations than the other species in the genus, while their sister shrubs live at lower elevations. Prior to the phylogenetic analysis (Funk 1982), all the trees

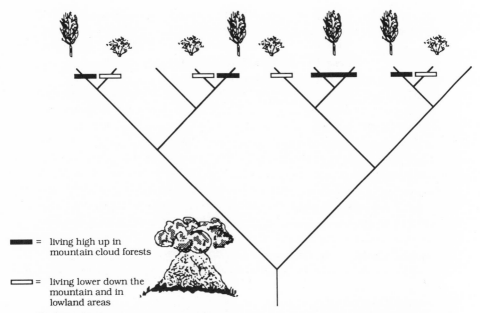

Fig. 5.25. Habitat distributions of the tree species of the genus *Montanoa* and their shrubby sister groups. Tree forms only live high in the mountain cloud forests, while the shrub forms are located lower down the mountain sides and into lowland areas.

were classified together in a single taxon, and their disjunct distribution was explained as either the remnants of a once widespread ancestral distribution or as the result of long-distance dispersal from mountaintop to mountaintop. The phylogenetic relationships within the group do not corroborate either of these hypotheses, because the tree forms do not form a monophyletic group. Rather, they suggest a possible alternative evolutionary explanation. The absence of sympatry between the tree forms and the shrub forms could have arisen repeatedly through interspecific competition between the derived tree forms and ancestral shrub forms, leading to divergent adaptation as the competition was resolved when the tree forms moved into cloud forests (fig. 5.25).

As we have noted previously, assumptions about particular evolutionary mechanisms are not intrinsic to a phylogenetic analysis. Such analyses are most useful in differentiating between those possibilities that are supported and those that are not supported by the phylogenetic evidence about macroevolutionary structuring of character appearance and divergence. In this case, it is possible that the adaptive divergence of the tree forms from shrub forms involved competitive exclusion. This hypothesis, in turn, can be examined within the larger framework of potential constraints on divergence. In general, this entails examining the system from several levels. For example, in the case of trees versus shrubs, developmental and physiological data bring an interesting new perspective to the question. Physiologically, composites are relatively inefficient water conductors, so it is unlikely that the large tree forms could survive outside of high-moisture habitats like cloud forests. On the other hand, greenhouse experiments (Funk, pers. comm.) indicate that the diploid seeds of shrublike species of *Montanoa* cannot survive under conditions of high moisture. The developmental constraints on size (ploidy), moisture requirements, and moisture tolerance lead to three conclusions: any polyploid seed that is produced too far from a cloud forest to disperse into a high-moisture zone will never germinate successfully; no diploid seed that disperses into a cloud forest will survive in the high-moisture conditions; and no polyploid seed that disperses outside a cloud forest will germinate successfully in the low-moisture conditions. The habitat segregation between trees and shrubs results from developmental constraints on moisture requirements/tolerances in the clade leading to strong selection against diploids in cloud forests and against polyploids in more mesic environments. No matter how many times the treelike polyploid forms evolve, they will never occur sympatrically with any diploid species, so the necessary initial conditions for interspecific competition will never be established.

Why don't all the snakes on Hispaniola eat the same things?

On the zoological side, snakes on Hispaniola exhibit a variety of feeding strategies, preying on local frogs and lizards. Some are generalists and some are specialists; some are active foragers and others are sit-and-wait ambushers. Traditionally, the differences in the foraging and feeding behaviors of each species have been explained as adaptive responses to interspecific competition. Henderson et al. (1988) provided ecological and phylogenetic data for some members of *Hypsirhynchus, Uromacer,* and *Alsophis,* using *Antillophis, Ialtris,* and *Darlingtonia* as outgroups (fig. 5.26).

Their analysis suggests that the active-foraging, generalist feeding mode employed by *Alsophis cantherigerus* is plesiomorphic in these snakes. The change to a sit-and-wait ambushing strategy occurred in the common ancestor of *Uromacer* and *Hypsirhynchus* and has been retained in all its descendants. These relationships emphasize the importance of historical constraints on foraging behaviors in these groups. *U. catesbyi* is reported to show both active and sit-and-wait behaviors. This raises some interesting points about the control of these two behaviors, which have been treated as alternative character states in the analysis. The most parsimonious phylogenetic explanation is that active foraging has reevolved in *U. catesbyi.* However, when this problem is examined in light of the snake's diet, a new explanation presents itself. All

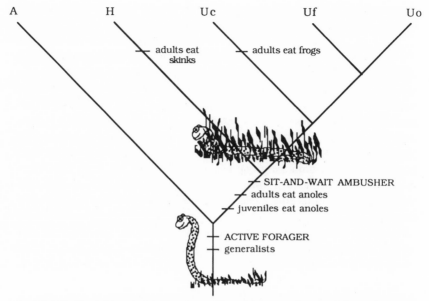

Fig. 5.26. Feeding habits of colubrid snakes on Hispaniola, mapped onto a phylogenetic tree for the group. *A = Alsophis; H = Hypsirhynchus; Uc = Uromacer catesbyi; Uf = U. frenatus; Uo = U. oxyrhynchus.*

the sit-and-wait species on the island feed predominantly on some type of anoline lizard, except for *U. catesbyi*, which eats frogs. Perhaps the appearance of active foraging is associated with this shift to such a novel prey item. Observations recorded in the literature also suggest that *U. catesbyi* is the worst sit-and-wait predator in the genus *Uromacer*, and that it is frequently seen chasing its prey, presumably after missing them by the first strategy! Further field observations are required to discover whether *U. catesbyi* does indeed display both foraging modes (which requires quite a complex phylogenetic explanation) or whether its chasing behavior is simply a by-product of its inability to capture frogs very efficiently. Additional resolution of the relationships among all the colubrid snakes in this area is also required before this problem can be adequately formulated. For example, three species of colubrids were not included in the study group because information on their feeding habits and behavior is lacking. Of these, all are active foragers; one feeds on frogs almost exclusively, one feeds on anoles primarily and frogs secondarily, and the feeding habits of the third are unknown. Interestingly, the most plesiomorphic member of the study group, *Alsophis,* is an active forager feeding on frogs among other things. Although the recurring association between active foraging and feeding on frogs is a tantalizing one, without further phylogenetic analysis such a relationship is, for the moment, speculative.

Although foraging mode is phylogenetically conservative among these colubrid snakes, there has been moderate evolutionary divergence in the group with respect to the principal prey item. All species retain the plesiomorphic juvenile feeding preference for anoles, and *U. oxyrhynchus* and *U. frenatus* carry this preference into adulthood. *Hypsirhynchus ferox* becomes more specialized on teiid lizards (*Ameiva*) and *U. catesbyi* switches to frogs (*Osteopilus*). These species pinpoint profitable foci in the search for adaptationist explanations of ecological diversity in this group of snakes. Although there has been a trend within the group towards an increase in dietary specialization, only three of the eight branches on the phylogenetic tree are characterized by any shift in feeding mode. It is therefore unlikely that such diversification has been driven solely by competitive interactions favoring niche partitioning. Instead, phylogenetic analysis suggests that the observed ecological diversity is a reflection of more than one factor, only one of which might be competition.

Discovering Constraint: Is the Study Finished?

The belief that the evolution of ecological traits for a given species can be explained by studying only that species implies that ecological diversification occurs at a higher rate than speciation. In fact, many evolutionary biologists operate under the implicit assumption that new species form as a result of ecological diversification. Ross (1972a,b) was among the first of recent authors who have questioned this assumption. Now even more authors have begun to examine the question of overall degree of ecological diversity within a phylogenetic framework, with some interesting results.

The evolution of mating-behavior repertoires in stickleback fishes

Many biologists have voiced concern over the structural and temporal plasticity of behavior, arguing that it is far more environmentally sensitive than morphology and thus difficult to characterize (Parsons 1972; McClearn and DeFries 1973; Dunford and Davis 1975). On the surface, this concern seems valid. Verbs are intuitively more labile than nouns. These perceptions, however, are based on an assumption that variability within a species automatically disqualifies a character from examination of relationships among species. Consider the following example: All male three-spined sticklebacks court females with an elegant, horizontal zigzag dance. The frequency and duration of the dance is highly variable both within and among individuals. However, no male *Gasterosteus aculeatus* ever executes the semihorizontal zigzag of *G. wheatlandi* or the vertical zigzag dance typical of the related

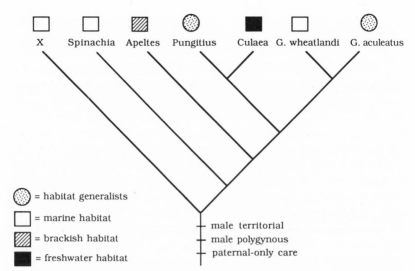

Fig. 5.27. Distribution of general mating-system characteristics, mapped onto the phylogenetic tree for the sticklebacks. Habitat preferences are listed above each species. *G.* = *Gasterosteus.*

species *Pungitius pungitius.* So, although some aspects of the character "male courtship dance" show high intraspecific variability (i.e., frequency, duration), other aspects (orientation of the zigzag) are autapomorphic for the species, and still other components (presence of a zigzag dance) are synapomorphic for a larger clade and therefore are useful systematic tools at those levels (see also fig. 9.4). If this is true, then behavior, like any other biological system, should be a product of the interplay between constraints from the past and adaptations to the present.

McLennan, Brooks, and McPhail (1988) examined this prediction by performing a phylogenetic analysis for the stickleback family Gasterosteidae based solely upon twenty-seven behavioral characters. Traits were polarized using the sister group of the Gasterosteidae, the Aulorhynchidae, or tubesnouts, as the outgroup. One tree was produced from this analysis (fig. 5.27) with a very high consistency index (90.3%), indicating a high overall level of historical constraint and a low degree of evolutionary plasticity for the behavioral traits used to construct the tree. That tree, in turn, was congruent with trees based on morphological and genetical traits, and with a higher consistency index!

Having discovered that behavioral characters contain phylogenetic information, let us turn our attention towards an investigation of what phylogeny can tell us about the evolution of behavior. In a paper about mating systems and ecology, Vehrencamp and Bradbury (1984) stated, "One expects mating behaviour to be a prime focus of selection for all sexual organisms. . . . One of the early important insights about mating system evolution was the recognition that the form of mating systems is more closely correlated with environmental contexts than it is with phylogenetic heritage." Is this statement corroborated by the distribution of mating behaviors on the gasterosteid tree?

Within the Gasterosteidae, *Pungitius* and *Gasterosteus* are habitat generalists. *G. aculeatus,* for example, exists as marine, estuarine, anadromous, and freshwater populations. Among the freshwater populations, habitats range from ephemeral, weed-choked ditches to large, oligotrophic lakes. At the other end of the spectrum, the fifteen- and five-spined sticklebacks are habitat specialists: *Spinachia spinachia* is restricted to marine habitats and *Culaea inconstans* to freshwater ones. *Apeltes quadracus* falls somewhere between these extremes, preferring brackish habitats but venturing into freshwater areas on occasion. Although they live in vastly different environments, all gasterosteids exhibit a male territorial, polygynous, paternal-care mating system. The presence of these behavioral patterns is thus a reflection of tight phylogenetic constraints on mating-system evolution within this family of fishes. So examination of the relationship between the form of the mating system and the environment on this coarse level of analysis reveals that this system is more closely tied to phylogeny than to the environment. However, the discovery of this phylogenetic influence is only the first step in developing a comprehensive evolutionary picture. We might next ask, What factors were responsible for the initial success of that mating system, and what factors are responsible for its current maintenance throughout the entire family? Answering this requires an analysis of the "fitness components" of the mating system for each species (i.e., female fecundity, adult survival rates, female encounter rate; see Vehrencamp and Bradbury 1984 for a detailed discussion). Once these have been optimized on a phylogenetic tree, we can investigate, for example: (1) whether there are any macroevolutionary correlations between the appearance of this mating system and a change in one (or more) of the fitness components; (2) which components appear to have been evolutionarily fixed throughout the family, and which are highly variable; and (3) the flexibility of the mating system (how many components can change and still maintain a male territorial, polygynous, paternal-care system). The results of such an analysis will provide a more robust estimate of the relative roles for the effects of *both* "phylogenetic heritage" *and* environmental factors in the evolution of this mating system.

Sexual dimorphism and social systems in lizards

Carothers (1984) investigated the relationship between sexual dimorphism and social organization in nine species of herbivorous iguanid lizards. Based

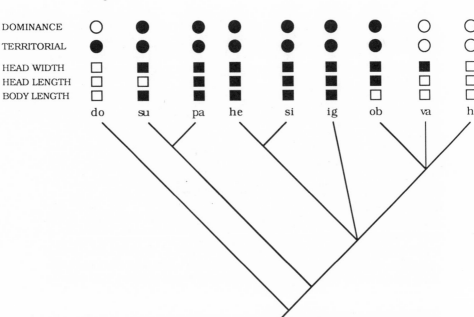

Fig. 5.28. Distribution of morphological and behavioral traits for eight species of herbivorous iguanid lizards. *Boxes* = sexually dimorphic morphological characters; *circles* = male behavioral traits; *white symbols* = the trait is absent in that species; *black symbols* = the trait is present. *do = Dipsosaurus dorsalis; su = Conolophus subcristatus; pa = Conolophus pallidus; he = Ctenosaura hemilopha; si = Ctenosaura similis; ig = Iguana iguana; ob = Sauromalus obesus; va = S. varius; hi = S. hispidus.*

on extensive experimental work and a search through the current literature, Carothers compiled data concerning sexual differentiation in three morphological characteristics: body length, head length, and head width. He also compiled data pertaining to two behavioral traits involved in the structuring of lizard social systems: the presence or absence of male territoriality and male dominance hierarchies. He examined the distribution of those five traits among the species in the context of their phylogenetic relationships (fig. 5.28).

Carothers did not designate outgroups for his study, so character polarizations could not be determined by outgroup comparison. Instead, he used *Dipsosaurus dorsalis* as the functional outgroup because, according to the phylogenetic relationships among the members of the ingroup, it is the sister group of the remaining eight lizard species considered. Based upon this, the characters displayed in figure 5.28 can be optimized onto the phylogenetic tree in the following manner (fig. 5.29).

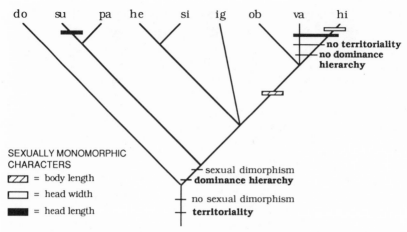

do   su   pa   he   si   ig   ob   va   hi

no territoriality
no dominance
hierarchy

SEXUALLY MONOMORPHIC
CHARACTERS

⬚⬚ = body length
⬚ = head width
■ = head length

sexual dimorphism
dominance hierarchy

no sexual dimorphism
territoriality

Fig. 5.29. Optimization of characters onto the phylogenetic tree for eight species of herbivorous lizards. Three morphological characters involved in sexual dimorphism (head length, head width, and body length) and two behavioral characters involved in mating-system "type" (territoriality and dominance hierarchy) were polarized using *Dipsosaurus dorsalis* as the "outgroup." *do* = *Dipsosaurus dorsalis; su = Conolophus subcristatus; pa = Conolophus pallidus; he = Ctenosaura hemilopha; si = Ctenosaura similis; ig = Iguana iguana; ob = Sauromalus obesus; va = S. varius; hi = S. hispidus.*

This analysis demonstrates that the morphological/behavioral attributes of these species represent a mixture of past (historical constraints) and present (selection in current environments). The presence of male territoriality is primitive in this group, has been retained in the majority of species, and has been lost either once or twice within the chuckwallas, depending upon the phylogenetic relationships between *Sauromalus hispidus* and *S. varius*. In contrast to this, the consistency index for the three morphological characters on the phylogenetic tree is 37.5% (3/8 steps), demonstrating that these traits are highly labile (in this case easily reversed). However, trait vagility does not necessarily translate into lack of phylogenetic constraints. For example, four of the eight species in the ingroup, *Conolophus pallidus, Iguana iguana* (the green iguana), *Ctenosaura hemilopha,* and *Ctenosaura similis,* express the derived state for all three characters as a legacy of the change that occurred in their common ancestor. Additionally, the loss of sexual dimorphism in body length occurred in the common ancestor of the genus *Sauromalus* and has been retained in all its descendants. The evolution of male dominance hierarchies is more tightly constrained, having appeared once, been retained in six of eight species, and, like territoriality, lost either once or twice within the chuckwallas.

The association between sexual dimorphism and social structure is a fascinating one, according to this preliminary phylogenetic hypothesis. Male territoriality has not played a direct role in the appearance of sexual dimorphism, since it appeared before the change in morphology. On the other hand, the co-occurrence of the appearance of sexual dimorphism in all three morphological characters and the appearance of male dominance hierarchies suggests a possible link among the four characters. This relationship is maintained even when one of the characters reverses to a sexually monomorphic state—loss of head-length dimorphism in *Conolophus suberistatus* and loss of body-length dimorphism in *S. obesus*—suggesting that the characters are not perfectly coupled. However, the association cannot withstand the reversal of more than one character, and the loss of head-length dimorphism, in addition to the ancestral loss of body-size dimorphism in *S. hispidus* and *S. varius*, is associated with the loss of both male territoriality and dominance structure. Overall, then, the association between sexual dimorphism and dominance hierarchies is an old one maintained in five of the nine species considered. Studies involving the relationship between these characters should concentrate on comparing species that exhibit this ancestral association with sister species that exhibit evolutionary changes in the association. For example, questions about the relationship between head length and dominance structure might focus on comparisons between *Conolophus pallidus* (dimorphism present, dominance present) and *C. subcristatus* (dimorphism absent, dominance present). Investigations of the suite of correlated morphological and behavioral reversals demonstrated in *S. hispidus* and *S. varius* might also prove rewarding. If, as the distributions of characters on the phylogenetic tree suggests, these two chuckwallas are sister species, then these changes occurred in a relatively recent ancestor. If they are not sister species, then questions about convergent evolution would be exciting. Until the characters are analyzed by outgroup comparison, however, these conclusions are still in the preliminary stage. The hypothesized origin of male territoriality prior to the evolutionary divergence of these lizards will be retained regardless of the state in the sister group. Resolution by outgroup analysis will only tell us whether territoriality originated in the ancestor of this group (synapomorphic for the group) or at some point before that speciation event (plesiomorphic for the group). The conclusions regarding the evolutionary origins of sexual dimorphism and male dominance hierarchies may change depending upon the state of these characters in the outgroup. Further phylogenetic analysis, based upon adequate outgroups and including the species belonging in this group of iguanids that were not considered by Carothers (e.g., the Galapagos marine iguana; see Estes and Pregill 1988 for a current estimate of lizard phylogeny), are necessary before any robust hypotheses concerning the evolutionary relationships between sexual dimorphism and social structure can be drawn.

Adaptive Radiations

When we examine populations within individual species, we generally discover extensive evidence of changes in gene frequencies correlated with environmental changes over relatively short time scales and relatively small spatial scales. The extrapolationist view of macroevolution is based upon the assumption that this degree of intraspecific adaptive plasticity can be smoothly extended to explain the degree of diversity among species within a clade. Futuyma (1986:32) defined adaptive radiation as "a term used to describe diversification into different ecological niches by species derived from a common ancestor." This concept has played an important role in evolutionary biology, as an explanation for differences in species richness among groups. Such differences are postulated to result from unusually high speciation rates in the more speciose group. Some authors have suggested that there should be an adaptive explanation for all speciation events (Stanley 1979; Stanley et al. 1981). Simpson (1953) believed that adaptive radiations resulted from diversification accelerated by ecological opportunity, such as dispersal into new territory (see peripheral isolates allopatric speciation in chapter 4), extinction of competitors, or adoption of a new way of life (i.e., an adaptive change in ecology or behavior). Other factors, including the adoption of a specialist foraging mode (Eldredge 1976; Eldredge and Cracraft 1980; Vrba 1980, 1984a,b; Cracraft 1984; Novacek 1984; Mitter, Farrel, and Wiegemann 1988), sexual selection and population structure (Spieth 1974; Wilson et al. 1975; Carson and Kaneshiro 1976; Ringo 1977; Templeton 1979; Gilinsky 1981; West-Eberhard 1983; Barton and Charlesworth 1984; Carson and Templeton 1984), or the origin of key ecological innovations in an ancestral species (Cracraft 1982a; Mishler and Churchill 1984; Brooks, O'Grady, and Glen 1985a), have also been postulated to have a positive effect on speciation rates. The consensus view of adaptive radiations today remains one with emphasis on "adaptive" (Futuyma 1986:356):

> a lineage may enter an adaptive zone and proliferate either because it was pre-adapted for niches that became available, or because it evolves "key innovations" enabling it to use resources from which it was previously barred.

Nevertheless, some authors have argued that differential rates could emerge naturally from a stochastic model of speciation (Raup et al. 1973; Raup and Gould 1974; Gould et al. 1977).

Radiations, whether "adaptive" or not, can only be distinguished on the macroevolutionary level of analysis. At the moment, however, there have been relatively few discussions about the patterns that should identify such radiations. Because of this, it is difficult to objectively examine the potential influences of various adaptive, geological, or stochastic factors on changes in speciation rates. Historical ecology, with its emphasis on macroevolutionary patterns, offers a solution to this problem. Armed with a set of explicit criteria for recognizing an adaptive change, we can search for historical coupling between high speciation rates and one or more of these criteria. Such a discovery would corroborate the hypothesis that a given radiation was, in fact, adaptive.

Adaptive Radiations in Ecological Preferences

The search for rigorous criteria by which we can document such radiations is a relatively new one. Ross (1972a) used the following ecological criteria when discussing the adaptive radiation of a variety of insect groups: (1) geographic dispersal from the primitive climatic zone to a derived one, and (2) shifts from the plesiomorphic condition to any apomorphic state in ecological life-history traits, behavior, and host preference. Based on the discovery that only approximately one out of every thirty speciation events in these groups was correlated with some form of ecological diversification, Ross concluded that adaptive changes (in any of the above characters) were consistent with, but much less frequent than, phylogenetic diversification. Furthermore, he felt that there were no predictable patterns explaining the shifts that did occur and suggested that ecological diversification in evolution comprised a biological "uncertainty principle." This interpretation was certainly at odds with hypotheses of ecologically driven phylogenetic change.

Where, and on whom, do parasitic wasps prefer to lay their eggs?

The Labeninae are a monophyletic group of ichneumonid hymenopterans (parasitic wasps) residing mainly in Australian and neotropical regions. Based upon a phylogenetic analysis of adult morphological characters, Gauld (1983) recognized four monophyletic groups as tribes within the subfamily. He then examined the distribution of two ecological characters on the phylogenetic tree (fig. 5.30).

Members of the tribe Labenini display the plesiomorphic condition for both characters, oviposition through lignified tissue and development of the larvae on coleopteran hosts. The Groteini retain the plesiomorphic oviposition-site preference, but the host chosen by all members has switched from beetles to bees (one group on ground-nesting bees, another on stem nesters). Divergence within the Poecilocryptini has involved switches in both host and oviposition-site preferences. *Poecilocryptus* retains the plesiomorphic oviposition site but attacks gall-forming insects, while species in the anomalous genus A retain the plesiomorphic attachment to beetle larvae, but have changed their oviposition site from wood to seeds. Finally, members of the Brachycyrtini display derived conditions in both characters. All species in

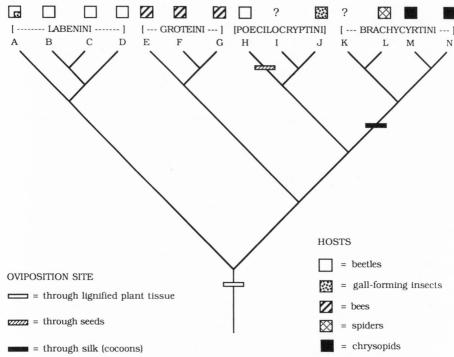

Fig. 5.30. Phylogenetic tree for parasitic wasps of the subfamily Labeninae, with the distribution of oviposition-site-preference and host-preference characters. *A* = *Labena; B* = *Asperellus; C* = *Certonotus; D* = *Apechoneura; E* = *Labium; F* = *Macrogrotea; G* = *Grotea; H* = genus A; *I* = genus U; *J* = *Poecilocryptus; K* = *Pedunculus; L* = *Adelphion; M* = *Habryllia; N* = *Brachycyrtus.* The general types of preferred host, represented in boxes atop the taxa names, have not been optimized onto the tree because data are missing for several genera. (Redrawn and modified from Gauld 1983.)

this tribe oviposit through silk cocoons, with the choice of host varying according to the genus of wasp. *Habryllia* and *Brachycyrtus* are united in their preference for chrysopid cocoons, while some species within the genus *Adelphion* develop in spider egg sacs. The evolutionary divergence of these parasitic wasps thus appears to involve a combination of (1) phylogenetic constraints, indicated by conservative changes in oviposition-site preferences, as only two changes, from lignified tissue to seeds and from lignified tissue to silk, have been documented in the entire subfamily, and (2) potential adaptive radiation of the genera due to switches in host preference. These changes, although quite extensive, are congruent with the phylogeny proposed for the wasps and demonstrate some historical influence on a general "host type" level. For example, all members of the Labenini develop on some type of coleopteran larvae, a pattern that is paralleled in the Groteini (bee larvae) and the *Habryllia* + *Brachycyrtus* clade (chrysopid cocoons).

The preceding study corroborates Ross's observations of a high degree of phylogenetic concordance and conservatism in ecological diversification within, at least, the Insecta. Andersen (1982) reported similar findings in a study of the Gerromorpha, a large group of semiaquatic hemipteran insects. Erwin (1985) extended the line of thinking represented by the above investigations to include yet another group of insects, the incredibly species-rich carabid beetles. Coupling the observed phylogenetic conservatism in ecological traits with a hypothesized high number of peripheral isolates allopatric speciation events in that group, Erwin proposed a macroevolutionary model called the "taxon pulse" to explain their adaptive radiation. Under this model, a group of beetles begins with an ancestral species displaying a certain ecological propensity. As time progresses, the ancestor and its descendants spread over a larger and larger geographical area, with descendant species fulfilling the same or very similar ecological roles in different locations. Subsequent to this first wave of dispersal, a new ecological trait arises in one of the descendant species in one of the localities. The species bearing this novel trait then undergoes widespread dissemination and a new "pulse" of diversity occurs, producing a new set of descendant species, all performing similar functions in different locations. Diverse and highly structured communities of carabid beetles could be formed in many different areas in this manner, with every community containing a member of each of the "pulses." According to this model, the number of "occupied niches" within a community would correspond to the number of pulses represented by the beetles present (see Roughgarden and Pacala 1989 for a similar example using anoline lizard communities on Caribbean islands).

Adaptive Radiations in Life Cycle Patterns

Brooks, O'Grady, and Glen (1985a) modified Ross's macroevolutionary criteria of adaptive radiations to include the diversification of life cycle patterns. According to this proposal, any phylogenetic diversification in ecological or behavioral traits, or in developmental characters relevant to the successful completion of the life cycle, were considered evidence of adaptive change. In other words, studies of adaptive radiations should focus on assessing the degree of diversification in ecological and reproductive strategies. A similar proposal was made by Duellman in a study of the diversification of reproductive modes among frogs.

Adaptive radiation in a free-living group:
reproductive modes in frogs

Duellman (1985) examined frog breeding systems within a phylogenetic context. Based on a combination of oviposition site, parental care, and developmental characters, he identified twenty-nine reproductive modes among

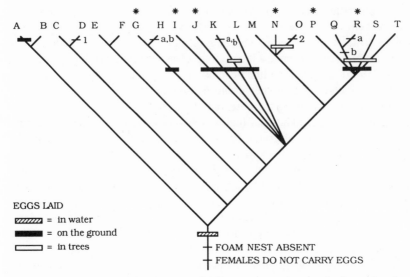

Fig. 5.31. Phylogenetic tree of major anuran family groups, mapping reproductive modes. A = Leiopelmatidae; B = Discoglossidae; C = Rhinophrynidae; D = Pipidae; E = Pelobatidae; F = Pelodytidae; G = Sooglossidae; H = Myobatrachidae; I = Brachycephalidae; J = Rhinodermatidae; K = Leptodactylidae; L = Bufonidae; M = Pseudidae; N = Centrolenidae; O = Hylidae; P = Dendrobatidae; Q = Hyperoliidae; R = Rhacophoridae; S = Microhylidae; T = Ranidae. Two general characters are represented. Type of female egg carrying: 1 = aquatic; 2 = terrestrial. Type of foam nest: a = aquatic; b = terrestrial. * = no members lay their eggs in water. The plesiomorphic conditions for these characters are mapped on the stem of the tree.

twenty anuran family groups (fig. 5.31). Six of the families are characterized by the possession of only one mode. The rhinophrynids (C), pelodytids (F), and pseudids (M) possess the primitive strategy "deposit eggs free in ponds/ feeding tadpoles develop in ponds." The remaining three families each exhibit a different derived mode associated with the deposition of terrestrial eggs. Eggs of the brachycephalids (I) develop directly into froglets; dendrobatid (P) adults carry their newly hatched, feeding tadpoles to water; and the tadpoles of centrolenids (N) fall out of their arboreal hatching place into (with luck) an underlying pond or stream. The other fourteen families comprise species exhibiting more than one mode.

When the twenty-nine strategies are optimized onto the phylogenetic tree for the frog families (Duellman and Trueb 1986), a consistency index of only 36% is achieved. This indicates a great deal of convergent evolution of reproductive strategies at the family level. However, this convergence is not totally random; it has been played out against a phylogenetic background. As mentioned previously, three of the twenty families display only the plesiomorphic reproductive mode. Interestingly, none of the suprafamilial (nonterminal) branches on the phylogenetic tree are characterized by any evolutionarily derived reproductive strategy, suggesting that much of the early radiation of frogs occurred in the context of "eggs and tadpoles in ponds/no parental care." This suggestion is corroborated by the fact that all members of the Pseudidae, one of the most recently derived groups, retain this plesiomorphic reproductive mode, while thirteen of the twenty families contain members displaying this strategy. Subsequent to the *radiation* of the ancestors of the frog families bearing this initial reproductive strategy, there has been an evolutionary "drive" in several families towards placing the eggs out of water; in fact, six of the families have severed the association between egg laying and water altogether (marked with an asterisk in fig. 5.31). This movement onto land has been coupled with a great amount of divergent and convergent adaptive change. This *adaptive radiation* is particularly pronounced in the myobatrachids (H) and in seven of the twelve most recently derived frog families, the leptodactylids (K), bufonids (L), hylids (O), hyperoliids (Q), rhacophorids (R), microhylids (S) and ranids (T). However, just to reinforce the background of phylogenetic constraints, the majority of the species in all these families except the Leptodactylidae exhibit the plesiomorphic reproductive mode. Within the leptodactylids, the majority of species outside the genus *Eleutherodactylus* exhibit the plesiomorphic reproductive mode, while within *Eleutherodactylus* there are a variety of derived reproductive modes. Because *Eleutherodactylus* is the most species-rich vertebrate genus, comprising more than half the members of the family Leptodactylidae, we cannot say that most leptodactylids exhibit the plesiomorphic reproductive mode.

Divergent adaptive change is suggested by the appearance of eleven reproductive modes, which are each restricted to members of a single family. Seven of these changes involve the appearance of some form of parental care: two modes in the pipids (involving eggs embedded in the dorsum of the female), three modes in the hylids (the eggs carried by the female), and one mode each in the myobatrachids (the eggs swallowed by the female) and the discoglossids (the eggs carried by the male). The remaining four divergences involve changes in the aquatic oviposition site from ponds to either small basins (hylids) or to water in tree holes or aerial plants (microhylids), the appearance of foam-nest–building behavior in pools (myobatrachids), and the appearance of viviparity (bufonids). Convergent adaptive change is suggested by the independent appearance of twelve reproductive modes in several families. In contrast to the predominance of parental care behaviors within the divergent radiation category, this category is primarily associated with changes from aquatic to terrestrial oviposition site and the concomitant modifications in development. Of these twelve convergences, one (ovoviviparity) is shared between two families; six (various changes in oviposition or foam-nest site/development) are scattered among three different families; one (carrying tadpoles to the water) is exhibited by four families; two (changes in oviposition site/development) are found in five different families; one (direct development of terrestrial eggs) is shared among eight families; and one (eggs and tadpoles in streams rather than ponds) is scattered among nine families.

In summary, the distribution of these reproductive modes across the families of frogs supports the interpretation of an interplay between phylogenetic constraints and adaptive radiation in these animals. Six families display only one reproductive mode and have not undergone widespread speciation. At the other end of the spectrum, the eight most species-rich families are characterized by either widespread convergent adaptation (leptodactylids, hyperoliids, rhacophorids, and ranids) or a combination of widespread convergence and the appearance of novel parental-care behaviors (myobatrachids, bufonids, hylids, and microhylids).

Adaptive radiation in a parasitic group: life cycle patterns in digeneans

Every student who has ever taken a course in parasitology has been left reeling by a seemingly never-ending procession of life cycle descriptions. It seems that every single species of parasitic organism has evolved a unique life cycle pattern that is specifically designed to enhance the chances that the parasite and its offspring will be able to bedevil hosts and students for all of eternity. In the following discussion, we will provide some evidence that the evolution of life cycle patterns in the parasitic flatworms conforms to the major postulates of this chapter. First, these life cycle patterns have been assembled in a historically coherent sequence. Second, the transformations in life cycle patterns, like other aspects of ecological evolution, are more conservative phylogenetically than is morphological diversification. Third, because the life cycles have been assembled piecemeal, rather than arising *de novo,* different components in the evolution of the life cycle patterns are detectable only at particular phylogenetic (temporal) scales. And fourth, in the absence of outgroup comparisons and explicit phylogenetic hypotheses, it is possible to make mistakes about evolutionary transformations (e.g., "simple" does not always mean "primitive"). We also hope to convince you that host-parasite systems are widely represented in this book because they are good general models for studies in evolutionary biology (not because DRB is fixated on life cycles).

Price (1980) stated that no groups of free-living organisms exceed parasitic organisms in the extent of their adaptive radiation. If this is true, parasitic taxa should be good model systems for studying adaptive evolution. Brooks, O'Grady, and Glen (1985a) and Brooks, Bandoni, Macdonald, and O'Grady (1989) presented phylogenetic systematic studies of sixty-three major family groups of the most handsome of flatworms, the digenetic trematodes, or flukes. Besides being an integral part of every first-hear biology laboratory (recall *Fasciola* and *Clonorchis,* the liver flukes), digeneans are one of the most extensively studied parasite groups in existence. Second only in abundance and distribution to the species-rich nematodes, they inhabit a wide range of vertebrates, preferring sites such as the intestine, liver, lungs, or circulatory system of their hosts. The morphological data base for the phylogenetic analysis of the group comprised 180 characters, which supported a phylogenetic tree with a consistency index of 75%. The variety of life cycle patterns exhibited by digeneans has been the focus of discussions about the adaptive radiation of the group. It was generally believed that this diversification, as a reflection of adaptive responses, should be more closely related to the ecology of individual species than to their phylogeny. Brooks, O'Grady, and Glen (1985a) investigated this prediction within a phylogenetic context. They considered five classes of life cycle attributes: developmental changes in invasive larvae that increased the numbers of such colonizing stages; changes in preferences for the first intermediate host, the second intermediate host, and the final host; and changes in the mode of infection of the second intermediate host (a reflection of juvenile colonization ability). The outcome of this investigation is depicted in figure 5.32.

In contrast with the anuran example, derived changes in life cycle characters for the digeneans are not concentrated within family groups. Rather, they appear interspersed throughout the more basal (suprafamilial) branches of the phylogenetic tree. This conforms more closely to Ross's (1972a,b) observations, indicating that diversification in life cycle patterns occurred relatively sporadically during the early evolution of the digeneans. However, there is once again a strong phylogenetic background to this evolutionary change. Only a small proportion (about 28%) of the branchings on the phylogenetic tree are correlated with any diversification in these life cycle characters. The modification of life cycle patterns is therefore more conservative than phylogenetic diversification at the family level. Studies at the genus and species levels indicate that the conservatism in the evolution of life cycle patterns is even more pronounced than the family-level analysis would suggest (Brooks and Overstreet 1978; Brooks 1980b; Brooks and Macdonald 1986; Macdonald and Brooks 1989).

Furthermore, the diversification of digenean life cycle patterns is more closely correlated with phylogeny than were the reproductive modes for frogs (see fig. 5.33). There is a historically coherent sequence of elaboration of life cycles at this level of phylogenetic analysis that explains much of the diversity of digenean life cycle patterns. Interestingly, most of the diversification appears to have been initiated by evolutionary changes in the cercarial stage. The cercaria is a juvenile stage that develops in the molluscan intermediate

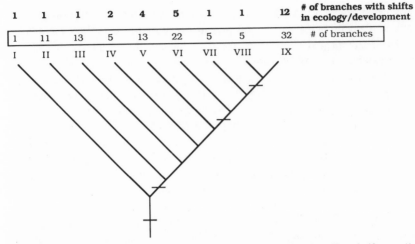

Fig. 5.32. Phylogenetic tree for nine orders of digenetic trematodes. *I* = Heronimiformes (1 family); *II* = Paramphistomatiformes (6 families); *III* = Echinostomatiformes (7 families); *IV* = Haploporiformes (3 families); *V* = Hemiuriformes (7 families); *VI* = Strigeiformes (12 families); *VII* = Opisthorchiiformes (3 families); *VIII* = Lepocreadiiformes (3 families); *IX* = Plagiorchiiformes (23 families). *Numbers* above each order indicate the number of evolutionary changes in five classes of life cycle traits, discussed in text, and the total number of terminal (family) and nonterminal branches within each order. *Slash marks* on nonterminal branches of the phylogenetic tree indicate additional points of diversification in life cycle traits for this group. Of the 115 branches on the phylogenetic tree, 32 (28%) are associated with some form of diversification in life cycle patterns. (Brooks, Bandoni, Macdonald, and O'Grady 1989.)

host and becomes infective to the second intermediate host (or sometimes to the final, definitive host). When cercariae began emerging from their molluscan hosts and encysting on vegetation and animal exteriors (most often on the exoskeletons of aquatic arthropods), the range of potential vertebrate hosts was enlarged greatly. No longer would trematodes be restricted to molluscivores. When the cercariae began penetrating particular intermediate hosts and encysting within them, a high degree of specificity in type of second intermediate host emerged, possibly enhancing adaptive modes of speciation. The evolution of cercarial emergence and the evolution of cercarial encystment and penetration therefore had significant adaptive consequences. The general evolutionary trend in the case of the digeneans appears to have been from relatively simple to relatively complex life cycles. Departures from congruence are due primarily to the reappearance of plesiomorphic life cycle patterns in relatively derived groups, generally involving the secondary loss of a host (see Brooks, O'Grady, and Glen 1985a). Overall, the diversification of life cycle patterns for this group of parasitic organisms appears to have been much slower than the evolution of individual species. We can gain additional insights into the evolutionary assemblage of life cycle patterns by examining the larger clade within which the digeneans are nestled. Three plesiomorphic life cycle traits are listed at the base of the phylogenetic tree for the flukes (fig. 5.33): a mollusc first intermediate host, a vertebrate final host, and endoparasitic, rather than ectoparasitic, adults. Where did these traits originate?

The digeneans belong to a larger group, called the Cercomeria (Brooks, 1982, 1989a,b; Brooks, O'Grady, and Glen 1985b), which encompasses the major groups of parasitic flatworms (including flukes and tapeworms). Within this group, the simplest life cycle is displayed by the monogeneans. Adult monogeneans live on the exterior of their vertebrate hosts and transmit their offspring directly to another vertebrate, where they mature and begin the cycle again. Among parasitologists, it has often been considered axiomatic that simple life cycles are more primitive than complex life cycles (cf. fig. 5.33). By that reasoning the one-host/direct-transmission pattern dis-

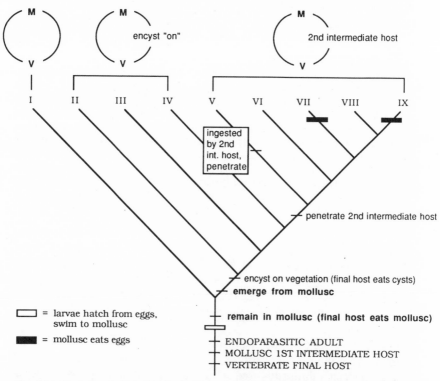

Fig. 5.33. Phylogenetic diversification in life cycle patterns for digenetic trematodes. Summary of the degree that can be explained by ordinal- and supraordinal-level phylogenetic relationships. Orders of digeneans: *I* = Heronimiformes; *II* = Paramphistomatiformes; *III* = Echinostomatiformes; *IV* = Haploporiformes; *V* = Hemiuriformes; *VI* = Strigeiformes; *VII* = Opisthorchiiformes; *VIII* = Lepocreadiiformes; *IX* = Plagiorchiiformes.

ENDOPARASITIC
VERTEBRATE
ARTHROPOD

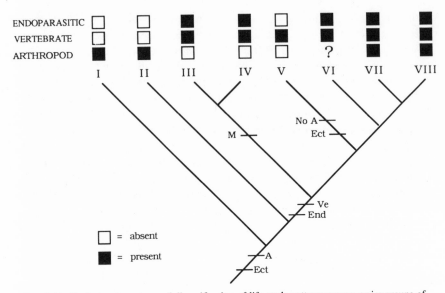

Fig. 5.34. Phylogenetic patterns of diversification of life cycle patterns among major groups of cercomerian platyhelminths. *I* = Temnocephalidea; *II* = Udonellidea; *III* = Aspidobothrea; *IV* = Digenea; *V* = Monogenea; *VI* = Gyrocotylidea; *VII* = Amphilinidea; *VIII* = Eucestoda. *Boxes* above taxa indicate distribution of traits for three components of life cycle patterns. Top row: *white boxes* = adults ectoparasitic; *black boxes* = adults endoparasitic. Middle row: *white boxes* = no vertebrate host; *black boxes* = vertebrate host. Bottom row: *white boxes* = arthropod host; *black boxes* = molluscan host; *?* = presence or absence, and type, of invertebrate host unknown. Slash marks on tree, and accompanying abbreviations, summarize phylogenetically the data presented in the boxes. *A* = arthropod host acquired (primitive one-host ectoparasitic life cycle); *V* = vertebrate host acquired (primitive two-host endoparasitic life cycle); *M* = molluscan host acquired in exchange for anthropod host (derived two-host life cycle); *No A* = arthropod host lost (derived one-host life cycle); *Ect* = ectoparasitic adult; *End* = endoparasitic adult.

played by monogeneans would be considered plesiomorphic among flatworms parasitizing vertebrates. According to that hypothesis, the addition of intermediate hosts, and the appearance of endoparasitic modes of life, have been independently derived by the digeneans and the tapeworms from this primitive pattern.

Figure 5.34 suggests an alternative interpretation. To begin with, the plesiomorphic life cycle pattern for all the parasitic flatworms appears to be one in which an arthropod is used as the only host by an ectoparasitic species (see the Temnocephalidea [I] and Udonellidea [II] in fig. 5.34). The pattern became more complicated in the ancestor of the trematodes + cercomeromorphs, as a vertebrate hose was added and the adult parasites became endoparasitic. At this level, then, the basal life cycle pattern involves an arthropod intermediate host plus a vertebrate final host, with the adult parasite living endoparasitically in the vertebrate. The current information on life cycle patterns in the gyrocotylids + amphilinids + tapeworms (a group called the Cestodaria) suggests that most of them have *retained* this primitive life cycle pattern. The trematodes (III + IV) display one variation on this central life cycle theme: in their ancestor, a molluscan host was substituted for an arthropod host. Finally, we come back to the monogeneans. Figure 5.34 suggests that the monogeneans [V] have a highly derived life cycle pattern, in which both the arthropod intermediate host and the endoparasitic life style have been lost. Virtually every discussion of the evolution of life cycle patterns in parasitic flatworms has assumed that the life cycle pattern displayed by monogeneans is primitively simple. The discovery that these flatworms display a *secondarily* simplified life cycle pattern forces parasitologists to rethink long-established assumptions and evolutionary scenarios.

Adaptive Radiations and Species Richness

We might also consider another perspective on adaptive radiations. It is possible that adaptive changes early in the ancestry of a group might constitute a productive theme that led to unusually high speciation rates and survival (i.e., low extinction rates) in descendant species. The result would be a species-rich clade whose members share a plesiomorphic trait that could explain the group's success. Studies along these lines would concentrate on finding "key innovations" that arose in one lineage and are correlated with an unusually high diversity in that lineage compared with its sister group (Cracraft 1982a,b). This line of research would also help provide a bridge between the taxic and transformational views of macroevolution.

Larson et al. (1981) proposed that speciation rates in clades could be regulated by the appearance of particular key innovations or evolutionary novelties in ancestors that give the descendant species in the clade an advantage over competitors. Traditionally, a key innovation was considered to be any novel feature that characterizes a clade (i.e., any synapomorphy) that is proposed to be correlated with the adaptive radiation of the clade (Mayr 1960; Liem 1973). The possession of this novelty alone was thought to be necessary and sufficient to explain the adaptive radiation of the clade. Liem (1973), for example, suggested that the extensive diversification of cichlid fishes in the African Rift lakes was due to the origin of a lower pharyngeal jaw suspended in a muscular sling in their common ancestor. Lauder (1981; see also Liem and Wake 1985 and Stiassny and Jensen 1987) discussed several reasons for caution in applying this concept for explanations of adaptive radiations. First, each ancestral branch on a phylogenetic tree may be characterized by more than one apomorphic trait. Hence, there is no a priori way to determine which of those traits would be "the" key innovation (perhaps even a combination of traits could be the innovation). Second, there is rarely strong evidence about the manner in which the trait considered to be the key innovation enhances speciation rates (or reduces the likelihood of extinction) in the clade. And finally, if the key innovation arose only once in evolution, how are we to test whether the putative key innovation confers a competitive advantage in all cases? As we have noted previously, our hypotheses of adaptation are strongest when we can compare convergent acquisition of traits under similar environmental conditions. Once again, historical ecological methods can help set the stage for more detailed studies.

Why are there so many species of digeneans and so few species of aspidobothreans?

As we discussed above, the digeneans are the sister group of the aspidobothreans. Both groups share an ancestral life cycle pattern involving a molluscan and a vertebrate host (fig. 5.34). As sister groups, both lineages are the same age, and as monophyletic groups, both are evolutionary units. However, there are fewer than five hundred described species of aspidobothreans and more than five thousand species of digeneans. Why the disparity in species richness? We have already discovered that the radiation of the digeneans was more strongly associated with developmental, rather than ecological or behavioral, diversification (figs. 5.32 and 5.33). In the aspidobothreans, larvae hatch from eggs and develop directly into juveniles in the molluscan host, are then ingested by a mollusc-eating vertebrate, and develop to the adult state. Hence, each embryo potentially can give rise to a single adult. Digeneans are characterized by a series of complex developmental stages in the molluscan host, at least one (and usually two) of which produce a large number of cloned larvae or juveniles (depending on the species and the stage). The reproductive potential of a single digenean embryo may exceed ten thousand adults, in contrast to the single individual produced by each aspidobothrean embryo. Although the changes in the ancestral digenean that are correlated with the high success of digeneans relative to their sister group were developmental rather than ecological (effect macroevolution rather than species selection), they were certainly adaptive and resulted in a markedly greater diversity for the digeneans. Significantly, this radiation was enhanced by further developmental changes within the digeneans themselves (see fig. 5.33).

A Last Look at Adaptive Radiations

It is clear that the term "adaptive radiation" means different things to different people. To some, it has been virtually synonymous with speciation. To others, it involves an association between overall diversification and adaptive changes in ecological and behavioral characters, as well as a high degree of homoplasious phenotypic change. Coddington (1988) expanded the criteria for assessing adaptive evolution to incorporate functional morphological as well as ecological information. His study added the following general criteria to the list of "ways to recognize an adaptive radiation": (1) the appearance of homoplasy correlated with functional change on a phylogenetic tree, (2) the appearance of predicted homoplasy correlated with predicted functional changes, and (3) the appearance of particular structural change correlated with particular functional change, regardless of homoplasy. Lauder and Liem (1989) suggested an experimental approach to testing hypotheses of this form of adaptive radiation. Their approach examines patterns of structural diversification throughout particular clades, thus equating adaptive radiation more with degree and extent of structural diversification than with speciation rates per se. They also emphasize the importance of having a causal model that predicts what the relationship should be between possession of an innovation and the pattern of structural diversification in a clade, in order to recognize particular synapomorphies as key innovations.

"Adaptive radiation" may also mean different things to different groups of organisms. For example, consider the insects, the frogs, and the digeneans presented in this chapter. In each case, a unique combination of phylogenetic conservatism and adaptive innovation has contributed to the contemporaneous diversity of these organisms. Much of early frog evolution occurred in species bearing the initial reproductive strategy "deposit your fertilized eggs in ponds/leave the tadpoles to fend for themselves." Diversification in reproductive modes apparently arose subsequent to this *radiation* by the ancestors of the frog families. By contrast, ecological changes appear to have occurred relatively early in the diversification of many insect groups. The evolution of digenean life cycles illustrates yet another pattern, in which episodes of ecological change are sprinkled throughout the phylogenetic tree. Given the new evolutionary insights uncovered from just three studies, what treasure troves are in store for us, buried within the phylogenetic histories of the cichlid fishes in African Rift lakes, the gammarid amphipods in Lake Baikal, the Hawaiian honeycreepers, or the Galapagos finches?

A Comment on Transformational Aspects of Macroevolution

The examples we have presented indicate that ecological and behavioral diversification within clades lags behind phylogenetic diversification of clades. Interestingly, this mirrors the information provided in the fossil record (Bakker 1983).

The fossil record of mammals shows that stasis for 10^5 or 10^6 generations is the rule, not the exception, for species . . . and that in many cases chronic stasis preserved phenotypes probably very far from the optimal compromise possible within the given habitat. Ever since Darwin, most evolutionary theory has concentrated on providing explanations of why populations should evolve. The fossil record demands more emphasis on explaining why populations do not evolve. . . . Perhaps, as a complement to our Society for the Study of Evolution, we need a Society for the Study of the Prevention of Evolution, to explore explanations of the apparent rarity of major adaptive change.

So, in contrast to the expectations of the extrapolationist view of macroevolution, it would appear that most of the adaptive plasticity exhibited by any given species is not translated into adaptive diversity among species within a clade. This supports the perspective of some researchers that macroevolution is more than just microevolution "writ large." However, it is important to recognize that this interpretation is derived from a limited phylogenetic data base, so at the moment it is simply an observation and not a theory of diversification per se. The generality of this observation can be tested by examining different groups of organisms using the methods described in this chapter. Once a trait has been identified as an "adaptation" at the macroevolutionary level using these methods, there is still work to be done at the microevolutionary level to strengthen the hypothesis that the trait is adaptive. This includes studies of the selective environment, and the character's function and its fitness with respect to that function. These areas of research, beyond the purview of historical ecology, highlight the need for a closer collaboration between micro- and macroevolutionary research programs.

If it is generally true that ecological and behavioral diversification is a conservative feature of **macroevolution,** we should not expect adaptive radiations to be manifested as a one-to-one mapping of species and divergent ecologies on phylogenetic trees. If, as Coddington has suggested and we have tried to show, adaptive radiations result from developmental and functional morphological, as well as ecological, changes within a phylogenetic context, we will need to map information from a variety of sources in order to provide robust explanations for the adaptive radiation of any group. We may also have to rethink our theories about the causal basis of such radiations. Cracraft's perspective on the importance of taxic macroevolutionary phenomena in shaping macroevolutionary patterns of diversity becomes even more important if the transformational aspects of macroevolution discussed in this chapter are cohesive, rather than diversifying, influences in evolution.

Summary

Biological diversity has been shaped on this planet by countless years of interactions between the evolutionary processes of speciation and adaptation. It should be clear by now that no one process can be assigned the dominant role in evolution; every clade is a unique combination of historical and environmental influences. Because each species is a unique evolutionary lineage, we cannot predict the exact pathway that clades will travel in the future. However, because each species carries with it the burden of history and, as such, is constrained by the past, we may be able to determine pathways that will not be available for that journey. These constraints can be investigated from both the macroevolutionary perspective of historical ecology (i.e., analysis of origins, elaborations, and associations) and from the microevolutionary perspective of population ecology (i.e., analysis of character maintenance and the effects of constraints on current population structure). Combining the information from both these research programs will produce a more robust theory of evolutionary ecology, and this, in turn, will allow us to make more informed decisions in our attempts to understand and preserve our ecosystem.

PART THREE

Phylogeny and the Evolution of
Ecological Associations

6 Preamble to Cospeciation and Coadaptation

In the preceding two chapters, the evolutionary diversification of ecological and behavioral characters was discussed as if it were a within-clade phenomenon. Obviously, the situation is more complex in nature, where each organism experiences a wide range of interactions with both closely and distantly related species; or, in the language of historical ecology, the evolutionary diversification of any given clade occurs in association with members of other clades. The evolution of ecological associations involves combinations of speciation and adaptation events within both common and independent phylogenetic contexts. This evolutionary interchange, in turn, produces four major classes of associations: the associated species and their interactions have evolved in a phylogenetic context (I in table 6.1); the associated species have evolved in a phylogenetic context, but their interactions show phylogenetic divergence (II in table 6.1); the associated species have not evolved in a phylogenetic context, but their interactions reflect ancestral attributes (III in table 6.1); and the associated species have not evolved in a phylogenetic context, and their interactions indicate phylogenetic divergence (IV in table 6.1). We expect biotas to be influenced evolutionarily by combinations of these four general classes of processes.

We are going to simplify this discussion by adopting neutral terms for a variety of biological associations. In the discussion of coevolution, one member of each association will be designated as the "host" taxon, and we will refer to the other members of the association as "associated with" or "inhabiting" the host taxon. This will relieve us, and the reader, of the burden of distinguishing among an array of ecological associations (e.g., competitive, host-parasite, insect-plant, herbivore-crop, predator-prey, mutualistic, commensal), while attempting to understand the historical ecological perspective on coevolution. When we discuss multispecies associations, we will use the general terms "biota," "ecological association," and "community" interchangeably. At the moment, the data base is not large enough for us to differentiate among these terms within a macroevolutionary context.

Table 6.1 Heuristic table depicting the four types of associations that result from the interaction between speciation and adaptation within a phylogenetic framework.

	Interactions	
Associations	Phylogenetic	Nonphylogenetic
Phylogenetic	I	II
Nonphylogenetic	III	IV

A Broad-based Coevolutionary Paradigm

In the past, ecological associations have been "diagnosed" by their geographical location, by their species composition, and by the interactions among the associated species. We believe that there is a need for a more general theory of ecological associations that includes an explicit historical component. Furthermore, we suggest that such a general theory should be a broad-based "coevolutionary" paradigm.

The term "coevolution" is generally associated with the pioneering paper by Ehrlich and Raven (1964), although the theoretical and mathematical basis of their perspective, as well as the term itself, was developed by Mode (1958). Ehrlich and Raven's perspective was population biological and emphasized the possibilities of short-term mutual adaptive interactions, although they did briefly address the macroevolutionary patterns that might result from the microevolutionary dynamic they proposed. This paper established the direction of current coevolution studies. As a consequence, the term coevolution has come to be used in a relatively restricted sense within the evolutionary framework, referring only to cases of reciprocal adaptive responses between ecologically interacting species (see, e.g., the excellent review by Futuyma and Slatkin 1983). Our sense of "coevolution" is broader, encompassing both the *degree of mutual phylogenetic association* (what we will call cospeciation) and the *degree of mutual modification* (what we will call coadaptation). As with our discussion of speciation and adaptation, we do not think there is a simple linear relationship between cospeciation and coadaptation; both aspects of ecological associations must be studied in order to produce robust explanations of their evolutionary origins.

The historical record of studies in coevolution actually begins with von Ihering's (1891) observations about the close similarities between some flatworm parasites inhabiting crayfish in New Zealand and those inhabiting crayfish in the mountains of Argentina. He postulated that the species in the two disjunct areas were derived from ancestral crayfish and flatworms that were themselves associated. Hence, von Ihering argued, South America and New Zealand must have at one time been connected by fresh water. In 1902 he presented a study of the origins of South American mammals, concluding that North and South America were not connected until the Pliocene, because the South American mammal fauna included both an autochthonous (endemic) element, which had originated in South America, and an allochthonous (invasive) element derived from North America, comprising taxa that were not known from South America prior to the Pliocene.

Von Ihering's proposal that close ecological associations between parasites and hosts could be used as indicators of phylogeny and ancient geographical configurations was supported by other systematists. Fahrenholz (1913) examined the relationships between catarrhine (Old World monkeys) and hominoid (great apes) primates based on their blood-sucking lice, with an eye on the possibility that the associations might indicate something about phylogenetic relationships. He postulated that the information derived from the occurrence of related lice on different primates demonstrated that the catarrhine primates were more closely related to hominoids than to any other primate group, a conclusion that remains the consensus view today. Additionally, Kellogg (1896, 1913), who studied birds and their associated biting lice, Harrison (1914, 1915a,b, 1916, 1922, 1924, 1926, 1928a,b, 1929) and his associate Johnston (1912, 1914, 1916), who studied a variety of vertebrate and associated parasite systems, and Metcalf (1920, 1922, 1923a,b, 1929, 1940), who studied the opalinid protists inhabiting frogs, all thought that contemporaneous host-parasite associations held keys to answering evolutionary questions. Metcalf, in fact, recognized that all the preceding studies had a common basis. He called studies using the interactions between species and their geographic locations or ecologically associated species to infer phylogenetic relationships and paleontological conditions the "von Ihering method."

The von Ihering approach, using biogeographical and ecological association data, and the Fahrenholz approach, using only data from ecological associations, inspired two generations of parasitologists to formulate abundant evolutionary "rules" about the evolution of hosts and parasites (see references in Brooks 1979b, 1985). The studies underlying these rules had two things in common: because the systems under investigation showed rather obvious patterns of phylogenetic association, they were based in systematics rather than in ecology; and they were plagued by the lack of a rigorous analytical method for documenting their hypotheses. The early development of this systematic perspective was paralleled by an ecologically based research paradigm, based on efforts to understand patterns of host utilization by phytophagous insects (e.g., Verschaffelt 1910; Brues 1920). Because the associations often showed no clear phylogenetic component, this tradition tended to develop without the influence of systematic information. Instead, researchers were interested in uncovering the ecological ties between organisms, that is, the particular cues being used by insects to locate their host plants. The Ehrlich and Raven (1964) paper was the ultimate synthesis of this line of research.

And so today we have two very different perspectives on "coevolution," each of which has a long history and has developed independent sets of models, explanatory hypotheses, and research methods. Needless to say, coevolution, like any evolutionary change, represents the complex outcome of numerous processes; therefore, we believe that a robust theory of coevolution must incorporate both perspectives. Furthermore, we think that the methodological rigor provided by phylogenetic systematics now makes it possible to begin integrating the two approaches. In fact, if the inclusion of both systematic and ecological information in recent texts on coevolution is a barometer of change, then the integration has already begun (Futuyma and Slatkin 1983; Nitecki 1983; Wheeler and Blackwell 1984; Kim 1985).

The coevolutionary paradigm that will emerge from this integration is based on attempts to answer three questions: (1) How did the species come to be in this area? This is the central question in historical or comparative biogeography. (2) How did the species come to be in this association? Within any group of geographically associated species there will be nonrandom ecological interactions, ranging from casual to obligatory, among some of those species. The problem lies in determining which components, if any, of the associations among extant organisms can be traced through a history of association between the ancestors of those organisms. (3) How are the members of an association interacting with one another? This question focuses our attention on the details of specific interactions between organisms by examining information about the origins and the modifications of these interactions within a phylogenetic framework.

Cospeciation

The production of both microevolutionary and macroevolutionary patterns in biological associations, in analogy with our discussion of diversification within and among clades, relies on two evolutionary processes, speciation and adaptation. Questions 1 and 2 above are concerned with speciation patterns and thus address biological associations at the level of the interacting clades. This is the focus of chapter 7. Most researchers approach this problem from a macroevolutionary perspective; however, microevolutionary theorists have recently included implications for cospeciation in their models (e.g., Kiester, Lande, and Schemske 1984).

Studies of cospeciation attempt to uncover the patterns of geographical or ecological associations among clades. There are two components to these patterns. First, two or more species may be ecologically associated today because their ancestors were associated with each other in the past. In this

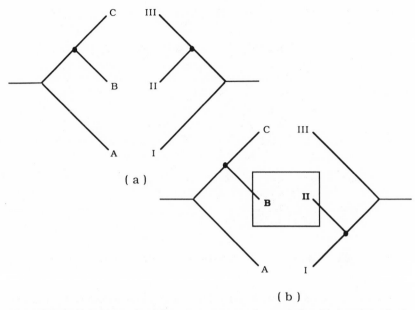

Fig. 6.1. Hypothetical example of the two components of cospeciation. (*a*) Association by descent: all species associations, A with I, B with II, and C with III, are the result of common history; that is, the ancestor of A + B + C was associated with the ancestor of I + II + III and the ancestor of B + C was associated with the ancestor of II + III. (*b*) Association by colonization: in this case, species II is hypothesized to have colonized species B. In this example, A–C and I–III may be two different groups of species or a group of species and a group of areas inhabited by the species.

case the contemporaneous relationship is a persistent ancestral component of the biotic structure within which the interacting species reside. This is referred to as **association by descent** because each of the species has "inherited" the association. Second, two or more species may be associated because at least one of the species originated in some other context and subsequently became involved in the interaction by colonization of a host or dispersal into a geographical area. This is referred to as **association by colonization** (Brooks 1979b; Mitter and Brooks 1983; Brooks and Mitter 1984). In the first case the associated species share a common phylogenetic history (fig. 6.1a); in the case of association by colonization, they do not (fig. 6.1b). Both association by descent and association by colonization may be manifested in geographical patterns or in purely ecological patterns of species co-occurrence. Of course, any species that originally appears as a colonizing influence in an ecological association can become phylogenetically associated in the subsequent evolutionary history of that association (fig. 6.2).

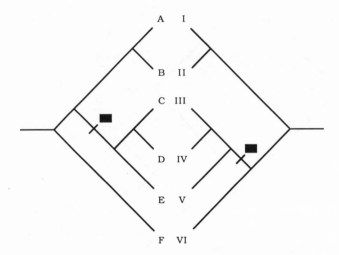

■ = association due to colonization of the ancestor of the C + D + E
clade by the ancestor of the III + IV + V clade

Fig. 6.2. Hypothetical example of the interrelationships between association by descent and association by colonization. Taxa A, B, and F are historically associated with taxa I, II, and VI, respectively. However, the ancestor of the III + IV + V clade colonized the ancestor of the C + D + E clade. Subsequent to that colonization, the interaction between the two ancestors was carried through two speciation events: (1) an association between taxa E and V and between the ancestor of C + D and the ancestor of III + IV, followed by (2) an association between C and III and between D and IV. This particular pattern of association through shared speciation demonstrates the existence of a historical interaction growing out of an original colonization event.

Cospeciation in a Geographical Context: How Did the Species Come to Be in This Area?

One of the fundamental advances of the past twenty years in evolutionary biology has been the formal articulation of two different perspectives on the general manner by which species achieve their geographical distributions. These perspectives may be categorized loosely as "island biogeography" (MacArthur and Wilson 1967), which calls attention to the propensity for organisms to move about, and "vicariance biogeography" (Croizat, Nelson, and Rosen 1974; Rosen 1975, 1985; Platnick and Nelson 1978), which reminds us that those movements may not be unconstrained (see Cain 1944, Camp 1947, and Wulff 1950 for early examples of this type of research). Since both research programs have contributed valuable insights into the problem of species co-occurrence, we are advocating the following "fusion" perspective: a species may be represented in a particular area because it evolved elsewhere and dispersed into that area, or it may occur in the area because it evolved there. In the first instance, the dispersal may have occurred a long time ago or relatively recently; in any event, there is no reason to expect the species' history to be congruent with the history of the area into which it dispersed. In the second instance, the species' history must be congruent with the history of the area (allopatric speciation mode I). It is likely that many, if not most, ecological associations contain both vicariant and dispersalist elements; therefore, it is important to have a method that elucidates the relative roles that geological changes and colonization have played in determining the patterns of species origin and geographic occurrence.

The first methods developed in this area were designed to provide qualitative documentation of general biogeographic distribution patterns based upon the sister-group relationships of members of different clades. Rosen (1975) presented an approach using "reduced area cladograms" in which distributional elements not common to all clades were eliminated from the data base, resulting in a simplified area cladogram depicting the "general pattern" (fig. 6.3).

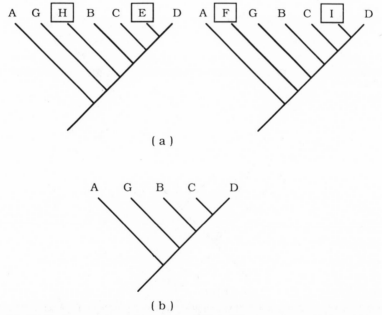

Fig. 6.3. "Reduced area cladogram" approach to historical biogeography. (a) Phylogenetic trees for two different clades, with geographic distributions of member species (*letters*) noted, indicate that some areas (A, G, B, C, D) contain members of both clades. (b) When the taxa inhabiting other areas (E, F, H, I) are removed from the two phylogenetic trees, reducing the number of taxa being considered, the biogeographical relationships for each clade appear identical.

The Platnick and Nelson (1978) and Nelson and Platnick (1981) approach, called "component analysis" (see also Humphries and Parenti 1986), relied on the use of "consensus trees" (Adams 1972; Nelson 1979, 1983) to summarize the common biogeographic elements (for a discussion of the limitations of consensus trees see Miyamoto 1985 and Wiley et al. 1990). In this method, elements that depart from the general pattern are depicted as ambiguities (fig. 6.4). The possibility that these ambiguous elements actually represent instances agreeing with the general pattern is then investigated by invoking one of two assumptions.

Both the reduced-area-cladogram and the component-analysis approaches have been criticized by ecologists and by systematists. Simberloff (1987, 1988) and Page (1987, 1988) pointed out that removal of ambiguous or conflicting data might make it appear that there is more evidence for general patterns than the data actually support. These authors also objected to the lack of statistical significance tests (but see Simberloff 1987) to determine the probability that apparent general patterns are due to a common cause (vicar-

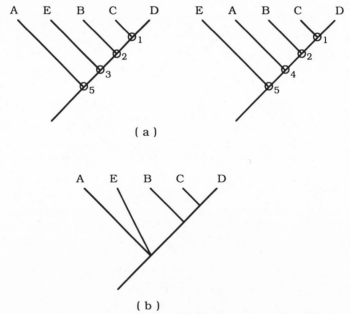

(a)

(b)

Fig. 6.4. "Component analysis" approach to historical biogeography. (*a*) Phylogenetic trees for two different clades, with geographic distributions of member species (*letters*), are compared by enumerating nodes that link different combinations of areas. There are five different nodes: (1) CD, (2) BCD, (3) EBCD, (4) ABCD, and (5) AEBCD. Nodes 1, 2, and 5 are common to both phylogenetic trees. (*b*) The common biogeographic relationships for both clades are those supported by those three nodes indicated in the consensus tree.

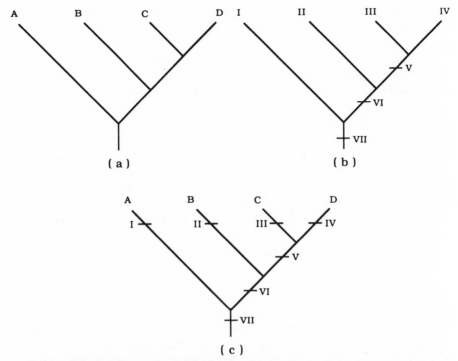

Fig. 6.5. Hypothetical example of optimizing the phylogeny for the study group onto the area cladogram for the geographical region under investigation. (*a*) Area cladogram. (*b*) Phylogeny. (*c*) Phylogeny optimized on area cladogram.

iance: allopatric speciation mode I) or to a series of unrelated parallel speciation events (Endler 1982). Wiley (1988a,b) and Zandee and Roos (1987) also criticized component analysis for obscuring evolutionarily relevant aspects of biogeographic patterns. They called for methodological changes that would allow an integration of the exceptions with the more general patterns. The methodology we will present in chapter 7 is based on the changes proposed by Wiley and Zandee and Roos and further modified by Brooks (1990). This approach is not designed to ask, either qualitatively or quantitatively, if there is a single general distribution pattern that might explain all of the data. Rather, it is based on the assumption that any given biota is likely to be a combination of species that (1) evolved where they now occur and (2) evolved elsewhere and dispersed into the area where they now occur.

If nonbiological data can be used to reconstruct the geological history of an area, we can use phylogenetic optimization methods to map the phylogeny of any clade onto this "area cladogram" (fig. 6.5). Since the geological history serves as an independent variable, optimizing the phylogenetic data on the area cladogram is analogous to regressing the evolutionary history of the taxa onto the geography. Given this, we can potentially identify the species in any biota that co-occur due to a common history of allopatric speciation, and those that are present due to dispersal from another area.

At the moment, actual geological data for specific studies may exist only in a historical ecologist's utopia. If these data are not available, a secondary source of information can be extracted from the phylogenetic relationships of co-occurring groups of organisms. If we have phylogenies for two or more clades whose members are associated geographically, we may ask if any of the taxa share a common history of speciation. The assumption here is that points of congruence among the phylogenies of geographically associated species identify episodes in which the clades experienced common speciation events. Such studies assess (1) historical geological relationships among areas if the speciation events took place as a result of geographic isolation (allopatric speciation mode I) or (2) the degree to which biotas share common histories if speciation resulted from colonization over common routes (allopatric speciation mode II; see reviews by Wiley 1988a,b).

It is important to determine which components of an ecological association are linked with either vicariance or dispersal events, in order to assess the context in which particular interactions emerged. However, when discussing "cospeciation in a geographical context," the possibility of shared phylogenetic histories among geographically co-occurring species is investigated without making assumptions about the type or extent of their ecological interactions. In fact, beyond occupying the same general area, the study species need not interact at all. In a simple sense, we are interested in asking to what extent speciation has been promoted by the active movements of organisms and to what extent it has been promoted by the active movements of the earth. The robustness of such studies depends on whether we have an explicit independent "evolutionary history" for the areas and how many clades we are able to analyze phylogenetically.

Cospeciation in an Ecological Context: How Did the Species Come to Be in This Association?

Speaking of "cospeciation in an ecological context" does not imply that historical biogeography is not an ecological phenomenon. We use this nomenclature to refer to cospeciation analyses in which the geographic context of the ecological associations is not considered; only the associations themselves are investigated. This is conceptually and analytically analogous to biogeographic studies, except that in this case one taxon, designated as the "host," replaces the concept of "area" in a historical biogeographic analysis.

This type of research is the "genealogical descendant" of the von Ihering method. If the "host" phylogeny is articulated, it serves as an independent variable onto which the phylogenetic relationships of the taxa associated with the host can be optimized (cf. the area cladogram reconstructed from geological evidence). If the host phylogeny is unknown, the history of the associations can be inferred by substituting the names of the hosts for the names of the associates on the phylogenetic tree of the associates. The latter method provides the weakest hypothesis of historical association, because data from the associated taxa are used to reconstruct host relationships and are then mapped back onto those relationships. Because of the inherent circularity in such an approach, it should only be used in a preliminary study (or as a last resort when no other information is available). And finally, if we have phylogenetic information about more than one clade associated with the same "host" group, we may ask questions about common histories of association between members of the host group and members of the associated groups.

We will use a single analytical method for both types of cospeciation analysis (see chapter 7). This application is based on the assumption that, on the macroevolutionary scale of cospeciation, areas and "hosts" can be equated to "taxa" in the phylogenetic analysis and the actual taxa, in turn, can be equated to characters for the areas and "hosts." Cracraft (1988) delineated two problems that may arise when microevolutionary phenomena are replaced with their macroevolutionary counterparts in phylogenetic analyses (see also Sober 1988). The first possible problem surrounds the use of hosts or areas as taxa and stems from the possibility that species inhabiting areas or in an association may represent multiple historical origins. By contrast, "real" taxa have only a singular origin. Now, in a phylogenetic systematic analysis the unique history of a taxon is demonstrated by the homologous characters (shared derived characters). The explanation for homology, common history, is the same for phylogenetic analyses at all levels of biological organization. Therefore, the same methodological principles should be applicable to the history of clades or to the history of associated clades. By contrast, the explanations for homoplasy invoke different processes at different levels of analysis. When we speak of homoplasious traits for individual taxa, we postulate multiple origins of traits mediated by the effects of natural selection and/or common developmental pathways. In cospeciation analysis, homoplasies indicate episodes of colonization, called "dispersal" in biogeography and "resource switching" in coevolutionary studies (when the homoplasies are parallelisms or convergences), or of extinction (when the homoplasies are reversals). So long as we are aware of the level at which we are analyzing evolution, we should be able to invoke appropriate mechanisms to explain departures from common history. Hence, the analytical method produces patterns of phylogenetic congruence for those elements sharing a common history and phylogenetic incongruence for those having independent histories, calling the elements in the analysis "characters" or "taxa" notwithstanding. The question, then, is one of interpretation and not of analysis.

The second problem arises from a perception that this approach forces data into a single model of evolutionary process, and as such is unrealistically reductionist and may obscure the influence of independent lineages in the evolutionary assemblage of biotas and other ecological associations. Kluge (1988) suggested that the vicariance explanation should be the null hypothesis for studies of cospeciation in geographically associated species. He advocated this explanation because it is based on the assumption that speciation has occurred as a result of active geological changes and this, in turn, does not require a supposition about the involvement of any particular biological process. Like Kluge, we believe that what we call cospeciation analysis uses common phylogenetic history as a null hypothesis: not in an a priori attempt to favor one explanation over another, but as the background against which the influences of independent elements can be most efficiently highlighted.

Macroevolutionary patterns of species co-occurrence result from the interactions of the geographical and ecological components of biological associations. The current data base supports an interpretation that there is a substantial phylogenetic component in some geographical and ecological associations. However, there is also evidence of associated taxa "escaping" from this underlying phylogenetic structure. Cospeciation analyses that include both historical and nonhistorical influences on geographic distributions and ecological associations will thus be richer than analyses that concentrate on either one component or the other. In chapter 7 we will discuss how this approach can be used to ferret out information about the assembly of biotas with complex evolutionary histories.

Coadaptation

How Are the Members of an Association Interacting with One Another?

Since cospeciation analysis does not incorporate assumptions concerning particular adaptive processes, such analysis will not provide any information about the ways in which associated lineages are influencing, or have influenced, each other's evolution. This falls into the domain of coadaptation studies, which are designed to uncover the adaptive components within the macroevolutionary patterns of association between and among clades (Mitter and Brooks 1983; Brooks and Mitter 1984). Brooks (1979b) referred to this as the degree of **coaccommodation** among ecological associates, but coaccommodation refers more to an evolved state than to a particular evolutionary process. *Investigations of this putative coevolutionary process attempt to establish whether there is evidence for coadaptation in an association and, given the presence of such evidence, to discern whether this is a legacy of history or a result of current interactions between the associated species.* Although the microevolutionary and macroevolutionary levels of coevolution are closely linked by the process of coadaptation, surprisingly few phylogenetic studies have examined this problem. The studies that are available will be discussed in chapter 8.

Studies of coadaptation must begin with a cospeciation analysis to provide the phylogenetic background against which episodes of mutual modification can be highlighted. Without this analysis, it is impossible to objectively differentiate scenarios based on the assumption that current associations reflect historical associations from other scenarios that presuppose little or no history of interaction. We will examine coadaptation from the perspectives of (1) pairs of very closely interacting clades and (2) larger multiclade associations. Type 1 associations are more commonly investigated by students of coevolution, whereas type 2 associations pertain to studies of the evolution of communities.

The Evolution of Closely Interacting Clades

Most theoretical information about coadaptational influences on coevolution comes from the studies of population biologists who stress the degree of mutual modification, or reciprocal adaptation, of the population ecology or population genetics of ecologically associated species. Three general classes of ecological "models" have emerged as a result of this research. **Allopatric cospeciation** is the "null" coadaptation model, because it assumes that the evolutionary associations between the taxa are due to cospeciation events that do not necessarily reflect any mutual interaction or evolutionary modification (reciprocal adaptation). Being the null model, allopatric cospeciation is, by itself, a weak hypothesis of coadaptation. For example, discovering that current associations are due to allopatric cospeciation rules out coevolutionary models based on an assumption of resource switching, but does not allow us to ascertain whether the association is merely a historical correlation or is due to some mutual interaction that maintains or promotes the association and its diversification. This model predicts congruence between the phylogenies of associated taxa. The second class is **Resource-tracking,** or **colonization,** models, in which diversification of each associated taxon occurs independently. For example, plant-eating insects may have colonized new host plants many times during their evolution (Mitter, Farrel, and Wiegemann 1988). In each case the colonization is hypothesized to have been the result of the evolution of insects that cued in on a particular biotic resource that was present in at least one plant species. Since this host switching involves the "tracking" of a resource that is shared among a group of hosts that do not form a clade, this model predicts no congruence between the phylogenies of the interacting taxa. Finally, there is the **evolutionary arms-race,** or **exclusion,** class of models. Briefly summarized, this classical model of coevolution proposes the following sequence of evolutionary interactions: Members of taxon A (say, spiders) reduce the fitness of members of taxon B (say, butterflies). Butterflies that, by chance, acquire traits (defense mechanisms) that increase their resistance to spiders, increase their fitness, so the new defense mechanism will spread throughout the population. However, some mutant spiders will, in their turn, overcome the new defense mechanism and be able to feed on the previously protected butterfly group. These spiders increase their fitness because they will avoid competition from other spiders; therefore, the ability to overcome the new defense mechanism will spread throughout the spider population, and this population will feed on the previously protected butterfly taxon. This model predicts some degree of congruence between the associated taxa's phylogenies, with "gaps" in the historical association due to the time lag between the evolution of "defense" and "counterdefense" traits. In chapter 8 we will show that distinguishing among these coevolutionary models is more complex than originally thought.

These models, developed by population biologists interested in microevolutionary processes, are more useful in explaining particular interactions than in investigating the coevolutionary dynamics of entire clades. On a macroevolutionary time scale the dynamics of many, if not all, clades of ecological associates may include contributions from one, two, or even all three of the above processes. We will provide some evidence of this in chapter 7, when we document the degree of congruence and incongruence between "host" and "associate" phylogenies, and in chapter 8, when we discuss the coevolution models in detail.

The Evolution of Interacting Biotas

Because of the complexity of the systems, there has been a tendency in community ecology to base models of community species composition and structure on one particular process (reviewed in Brown 1981; McIntosh 1987). Many of these processes, including interspecific competition, random colonization, predation, and disturbance, have been widely studied, so a wealth of information is now available. Given this, perhaps the time has come to return to the viewpoint, articulated so succinctly by Hutchinson (1957), that most biotas represent a complex interaction among a variety of different influences. In other words (Brown 1981), success in understanding complex systems usually comes from . . . taking them apart from the top down, inducing the processes underlying their organization from patterns in the relationships of the components to each other. The alternative approach of trying to recreate the entire system by assembling the components rarely works because if the system is really complex, there is an overwhelming number of possibilities.

In chapter 8 we will use this top-down approach to the study of community evolution. Starting from the macroevolutionary patterns of evolution, we will attempt to distinguish the relative contributions of the following influences on species composition and species interaction in a community: (1) **Phylogeny:** the species occurs in the association because of cospeciation, and the traits involved in interactions amongst members of the association are plesiomorphic. This represents a potential cohesive influence in the evolution of biotas. (2) **Colonization by "preadapted" species:** the species occurs in the association because of colonization, but still exhibits plesiomorphic interaction traits (i.e., these traits were brought into the new association during the colonization event and have been retained during the evolutionary diversification, if any, of the interacting clades). (3) **Colonization by competitive exclusion:** the species occurs in the association because it colonized the area and successfully competed with the resident species. In this case, either the colonizer or the competing resident(s), or both (all), exhibit evolutionarily derived (apomorphic) interaction traits relevant to the competition (i.e., the plesiomorphic condition in both species was originally the same trait, and now one or both species exhibit a different trait). (4) **Stochastic ("nonequilibrium") phenomena:** the species is a resident (associated by descent) but exhibits an apomorphic trait that is not due to competitive "displacement" by a colonizer. This indicates a certain degree of evolutionary "wandering" by resident species, which has been "allowed" by the overall structure of the association.

Summary

Evolution results from a dynamic interaction of several processes on many different levels. In chapters 4 and 5 we discussed some of the interesting questions that can be asked by examining the effects of two of these processes, speciation and adaptation, at the level of evolving clades. In chapters 7 and 8 we will allow our gaze to sweep across a broader evolutionary landscape to encompass the influences of speciation and adaptation at the level of evolving ecological associations. There is a continuum of interactions among species composing a biota. In some cases, the interactions are so predictable and pronounced that they play a role not only in the maintenance of associations but also in their evolution. Ecological associations do not evolve in the same way that species evolve; rather, they are "assembled" evolutionarily, and some of the "assembly rules" are phylogenetically based. Information from the studies discussed in the following chapters can be combined to produce robust explanations for the evolutionary "assemblage" of multispecies ecological associations.

7 Cospeciation

There are a variety of ways in which phylogenetic trees can be used in studying the evolution of biological associations. Many of these applications involve investigating the degree of congruence between the history of one group and the history of the areas in which its members reside, or the histories of other groups with which it is ecologically associated. These comparisons, in turn, provide a way to distinguish among hypotheses concerning the observed species composition of the associations. For example, a species may occur in a certain geographic area because its ancestor lived in that area and the descendant evolved "*in situ.*" You will recognize this as the result of vicariant speciation (allopatric speciation mode I). Alternatively, the species may have evolved elsewhere and dispersed into the area where it now resides, or it may have evolved as a result of dispersing into the area where it now resides (allopatric speciation mode II). In the first case we would expect the history of the species to coincide with (to be **congruent** with) the history of the area, whereas in the second case we would not. Similar reasoning can be applied to studies of interspecific associations regardless of their geographic context. Two or more species may be associated ecologically today because their ancestors were associated, or they (or their ancestors) may have evolved in association with other species and subsequently "switched allegiances." Such allegiance switching (or the common, but more restrictive, term "resource/host switching") in ecological associations is equivalent to dispersal in biogeographical associations. In the first case we would expect the histories of the taxa involved in the association to coincide with each other, whereas in the second case we would not expect to find such congruence. Taxa that show historical association either with geographical areas or with other taxa exhibit **cospeciation** patterns. Phylogenetic systematic methods can help to distinguish components of associations that are due to history (**association by descent**) from those that are due to dispersal or resource/host switching (**association by colonization**). Differentiating between these two components of any association in both geographical and ecological contexts will be the focus of this chapter.

Investigating the macroevolutionary components of ecological associations requires some advanced applications of phylogenetic systematic methods. First, new terminology: Up to this point, we have equated "branching diagram" with **"phylogenetic tree."** In this chapter we will discuss methods for comparing the amount of ecological association between different clades throughout their evolutionary history. Such comparisons are summarized in branching diagrams derived by using the phylogenetic relationships of the taxa as "characters" and the ecological associations as "taxa." We will refer to these diagrams as **cladograms,** which literally means "branching diagram": that is, "area cladograms" when phylogenies are being compared with respect to common geographic distributions, or "host cladograms" when phylogenies are being compared with respect to common ecological associations. The term "phylogenetic tree" will be reserved for estimates of historical relationships among taxa based on characters intrinsic to the taxa. We realize that the term "cladogram" is commonly used synonymously with "phylogenetic tree"; however, we believe that it is important to distinguish between a genealogical reconstruction and an ecological reconstruction in evolutionary biology. We hope this distinction will be more helpful than confusing.

Cospeciation in a Geographic Context: How Did the Species Come to Be in the Same Geographical Area?

Basic Methodology

We will begin our discussion of this section with an example drawn from a group of rare flatworms, the Amphilinidea. Amphilinids, the sister group of the species-rich true tapeworms, are a small (eight known species) but widespread group of parasites that live in the body cavities of freshwater and estuarine ray-finned fishes and in one species of freshwater turtle.

1. The first step in the search for possible historical components in the association between the amphilinids and their geographical distributions is *the reconstruction of the phylogenetic relationships of the organisms.* Phylogenetic systematic analysis of the eight species of amphilinids, based on forty-six morphological characters, produced a single tree with a consistency index of 87.5% (Bandoni and Brooks 1987a). Figure 7.1 depicts these relationships for five of the eight species (we will include the other three later).

2. The next step is to *designate the areas in which the species occur as if they were taxa.* Geological evidence (e.g., Dietz and Holden 1966) is then used to produce an area cladogram showing the historical connections among the study areas (fig. 7.2).

3. We then prepare a list *placing the species* of amphilinid flatworms *with the areas in which they occur* (table 7.1).

4. The phylogenetic relationships of the five amphilinid species, previously

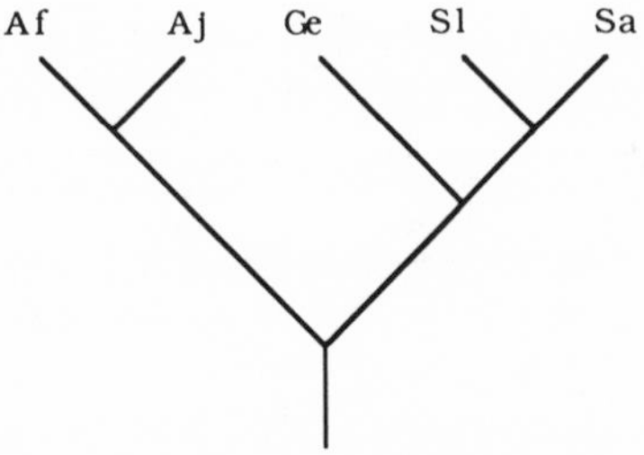

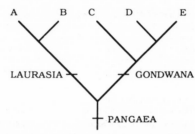

Fig. 7.1. Phylogenetic tree for five species of amphilinid flatworms. *Af* = *Amphilina foliacea; Aj* = *A. japonica; Ge* = *Gigantolina elongata; Sl* = *Schizochoerus liguloideus; Sa* = *S. africanus.*

Fig. 7.2. Area cladogram for the major continents on this planet, based on historical geological data. *A* = Eurasia; *B* = North America; *C* = Australia; *D* = South America; *E* = Africa.

Table 7.1 List of geographical areas and species of amphilinid flatworms that inhabit them.

Area		Taxon	Taxon Name
A	Eurasia	1	*Amphilina foliacea*
B	North America	2	*Amphilina japonica*
C	Australia	3	*Gigantolina elongata*
D	South America	4	*Schizochoerus liguloideus*
E	Africa	5	*Schizochoerus africanus*

reconstructed using morphological data (fig. 7.1), can now be treated *as if they were a completely polarized multistate transformation series,* in which each taxon and each internal branch of the tree is numbered (fig. 7.3). The sequence of numbering is arbitrary, but each internal branch of the tree must have a number.

5. Each species of amphilinid now has a "code" that indicates both its identity *and* its common ancestry. For example, the code for *Amphilina japonica* is (2, 6, 9) and the code for *Schizochoerus africanus* is (5, 7, 8, 9).

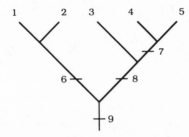

Fig. 7.3. Phylogenetic tree for five species of amphilinid flatworms, with internal branches numbered for cospeciation analysis. *1 = Amphilina foliacea; 2 = A. japonica; 3 = Gigantolina elongata; 4 = Schizochoerus liguloideus; 5 = S. africanus.*

Table 7.2 Matrix listing binary codes that represent the phylogenetic relationships among five species of amphilinid flatworms.

Taxon	Binary Code
1 *Amphilina foliacea*	100001001
2 *Amphilina japonica*	010001001
3 *Gigantolina elongata*	001000011
4 *Schizochoerus liguloideus*	000100111
5 *Schizochoerus africanus*	000010111

These codes, in turn, can be represented in a data matrix in which the presence of a number in the species code is listed as one and the absence of a number in the species code is listed as zero (table 7.2).

6. You should recognize this as an application of **additive binary coding** (see chapter 2). The phylogenetic relationships of the study group are now represented by the binary codes. This can be confirmed by performing a phylogenetic systematic analysis for species 1–5 using the binary codes from table 7.2 (fig. 7.4). If all is correct, this will reproduce the tree shown in figure 7.3.

7. Now we *replace the species names* in table 7.2 *with their geographic distributions* (table 7.3).

Table 7.3 Matrix listing binary codes indicating phylogenetic relationships among five species of amphilinid flatworms inhabiting five geographic areas.

Area	Binary Code
A Eurasia	100001001
B North America	010001001
C Australia	001000011
D South America	000100111
E Africa	000010111

8. Finally, we construct a new area cladogram based, this time, on the phylogenetic relationships of the species (fig. 7.5). This produces a "picture" of the historical involvement of areas in the evolution of the species.

In this example the area cladogram based upon geological evidence (fig. 7.2) and the area cladogram reconstructed from the phylogenetic relationships of the taxa occurring in each region (fig. 7.5) are identical. In addition, the consistency index for the area cladogram is 100%, indicating that all the speciation events postulated by the phylogenetic tree are congruent with the area cladogram. Therefore, we can hypothesize that the occurrence of the study species in the study areas is a result of a long history of association between amphilinids and the areas in which they now occur.

Dispersal of organisms is a common phenomenon in nature, so real data sets will generally show less than the 100% congruence depicted in the preceding example. Let us return to the amphilinids and complicate the picture somewhat by including the remaining three members of the group in the analysis.

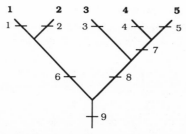

Fig. 7.4. Cladogram for five species of amphilinid flatworms, based on the additive binary matrix representing the phylogenetic tree for those species. *Numbers accompanying slash marks* indicate codes for species and their relationships from table 7.2.

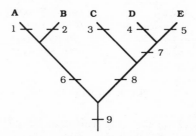

Fig. 7.5. Area cladogram for five areas, based on the phylogenetic relationships of five species of amphilinid flatworms that inhabit those areas. *A* = Eurasia; *B* = North America; *C* = Australia; *D* = South America; *E* = Africa. *Numbers accompanying slash marks* indicate codes for species and their relationships from table 7.2.

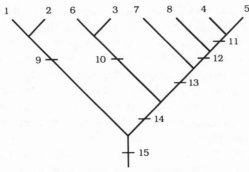

Fig. 7.6. Phylogenetic tree for eight species of amphilinid flatworms, with internal branches numbered for cospeciation analysis. *1 = Amphilina foliacea; 2 = A. japonica; 3 = Gigantolina elongata; 4 = Schizochoerus liguloideus; 5 = S. africanus; 6 = G. magna; 7 = S. paragonopora; 8 = S. janickii.*

1. The complete phylogenetic tree for the Amphilinidea is shown in figure 7.6, with internal branches numbered for additive binary coding. The "new" flatworms are *Gigantolina magna* (taxon 6), *Schizochoerus paragonopora* (taxon 7), and *S. janickii* (taxon 8).

2. The three additional species of amphilinids inhabit South America (*S. janickii*) and Indo-Malaysia (*G. magna* and *S. paragonopora*). Species codes (from fig. 7.6) are converted to binary codes and listed for each area in table 7.4 (when more than one species occurs in an area, the codes are combined— we will discuss this more fully later).

Table 7.4 Matrix listing binary codes for species of amphilinid flatworms inhabiting six geographic areas.

	Area	Binary Code
A	Eurasia	100000001000001
B	North America	010000001000001
C	Australia	001000000100011
D	South America	000100010011111
E	Africa	000010000011111
F	Indo-Malaysia	000001100100111

3. The area cladogram reconstructed from that data matrix is depicted in figure 7.7.

The consistency index for this area cladogram is 93.75%. Note that "10" appears twice on the tree. This indicates that the common ancestor of species 3 and species 6 (taxon 10) occurred in both area C and area F. Its occurrence in area C coincides with the geological history of the areas, so we explain this by saying that species 3 evolved in the same place (area C) as its ancestor (taxon 10). On the other hand, the occurrence of 10 in area F does not coincide with the geological history of the areas. We explain this by hypothesizing that at least some members of ancestor 10 dispersed to area F, where the population evolved into species 6. Hence, the occurrence of species 7 in area F is due to common history, whereas the occurence of species 6 in area F is due to dispersal of its ancestor into that area. If this is true, then what we have called area F is, from a historical perspective, two different areas for species 6 and 7.

We can test this possibility and further examine the question of ancestor 10's dispersal by recoding the data matrix in table 7.4, listing species 6 and species 7 in different subsections of area F (F$_1$ and F$_2$; table 7.5). When we

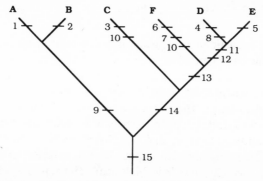

Fig. 7.7. Area cladogram based on phylogenetic relationships among eight species of amphilinid flatworms. *A* = Eurasia; *B* = North America; *C* = Australia; *D* = South America; *E* = Africa; *F* = Indo-Malaysia. "Characters," represented by numbers accompanying slash marks, are species. *1* = *Amphilina foliacea; 2* = *A. japonica; 3* = *Gigantolina elongata; 4* = *Schizochoerus liguloideus; 5* = *S. africanus; 6* = *G. magna; 7* = *S. paragonopora; 8* = *S. janickii.* Numbers *9–15* = ancestral species (see fig. 7.6).

Table 7.5   Matrix listing binary codes for species of amphilinid flatworms inhabiting six geographic areas.

Area[a]		Binary Code
A	Eurasia	100000001000001
B	North America	010000001000001
C	Australia	001000000100011
D	South America	000100010011111
E	Africa	000010000011111
F$_1$	Indo-Malaysia	000001000100011
F$_2$	Indo-Malaysia	000000100000111

[a]Indo-Malaysia is listed once for species 6 (F$_1$) and once for species 7 (F$_2$).

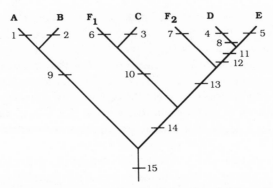

Fig. 7.8. Area cladogram based on phylogenetic relationships of eight species of amphilinid flatworms, listing Indo-Malaysia (*F*) as two separate areas. *A* = Eurasia; *B* = North America; *C* = Australia; *D* = South America; *E* = Africa.

perform a phylogenetic analysis using this new matrix, we obtain the area cladogram depicted in figure 7.8.

We now find areas F_1 and F_2 in different parts of the geographic cladogram, with F_1 connected to area C (Australia) and F_2 associated with areas D (South America) and E (Africa). The placement of F_2 is in accordance with the patterns of continental drift, but the placement of F_1 is not. The "misplacement" of area F_1, according to the original area cladogram based on geological evidence (fig. 7.2), strengthens our hypothesis that ancestor 10 did some dispersing (from Australia into Indo-Malaysia). The separation of F_1 and F_2 strengthens our suspicions that they are not the same areas historically, and this conclusion is reinforced by the observations that F_1 encompasses estuarine Indo-Malaysian habitats, while F_2 represents freshwater, nuclear Indian subcontinent habitat.

Special Applications

More than one member of the clade in the same area

The preceding example highlights some of the analytical problems that can occur when the patterns of species distribution include incidents of colonization. In the case of the amphilinids, this movement resulted in more than one member of the group occurring in the "same" area. When this happens, the binary code for the area is a composite of the codes from all the taxa in that area. For example, the codes for taxa 6 and 7 were combined to give a composite binary code for area F, while the codes for taxa 4 and 8 were combined to give a composite binary code for area D. This procedure is called "inclusive ORing" (Cressey, Collette, and Russo 1983). In this example, the patterns of dispersal did not override the historical patterns in the analysis, so the resulting area cladogram coincided with the geological history of the areas. The dispersal episode appeared as a homoplasy, and the two speciation events within one area (producing species 4 and 8) appeared as an autapomorphy for the area. The ambiguity resulting from the occurrence of species 6 and 7 in Indo-Malaysia was resolved by assuming that "Indo-Malaysia" for species 6 was different from "Indo-Malaysia" for species 7. No ambiguity resulted from the "extra" speciation event in South America.

Dispersal in the Amphilinidea and the subsequent problems generated for researchers (although presumably not for the flatworms) demonstrated one kind of ambiguity that can arise from this procedure, and one way in which the source of the ambiguity could be discovered. A more serious analytical problem resulting from using the inclusive-ORing method can arise *when a large enough number of relatively derived taxa disperse into relatively primitive areas.*

1. Figure 7.9 depicts an area cladogram for hypothetical areas A–D based on geological evidence.

2. Figure 7.10 depicts the phylogenetic tree for hypothetical species 1–6, based on, say, morphological characters.

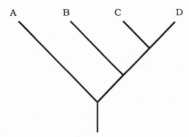

Fig. 7.9. Area cladogram for hypothetical areas A–D.

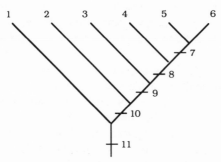

Fig. 7.10. Phylogenetic tree for hypothetical species 1–6, with internal branches numbered for cospeciation analysis.

Table 7.6   Matrix listing the geographic distribution of hypothetical species 1–6 among hypothetical areas A–D, along with the binary codes representing the phylogenetic relationships among species 1–6.

Area	Taxon	Binary Code
A	1	10000000001
B	2, 5, 6	01001111111
C	3	00100000111
D	4	00010001111

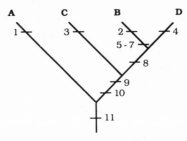

Fig. 7.11. Area cladogram for hypothetical areas A–D, based on phylogenetic relationships of hypothetical species 1–6.

3. The matrix depicting the geographic distributions of species 1–6 among areas A–D is shown in table 7.6.

4. The new area cladogram constructed from the binary codes is shown in figure 7.11. This cladogram has a consistency index of 100%; however, the positions of areas B and C are reversed. So, although we have perfect congruence between the relationships among the taxa and the **new** area cladogram, the new cladogram is not congruent with the area cladogram based on the geological history of the areas (fig. 7.9). This occurs because area B contains two highly derived members of the clade, species 5 and 6; therefore, area B is assigned a highly derived status when taxon codes are combined.

Since three species currently occur in area B, we will redo the analysis, treating area B as three areas, B_1 for taxon 2, B_2 for taxon 5, and B_3 for taxon 6.

1. This treatment produces a new data matrix (table 7.7).

2. Phylogenetic analysis of the new data matrix produces the area cladogram depicted in figure 7.12. According to the historical geological associations of the areas, B_1 is in the correct location on the new area cladogram, while areas B_2 and B_3 are misplaced. This leads us to hypothesize that the relatively derived taxa 5 and 6 are currently found in area B because their ancestor (species 7), which links areas B_2 and B_3, dispersed from area D into area B and subsequently speciated, producing species 5 and 6.

Table 7.7 Matrix listing the geographic distribution of hypothetical species 1–6 among hypothetical areas A–D, along with the binary codes representing the phylogenetic relationships among species 1–6.

Area[a]	Taxon	Binary Code
A	1	1000000001
B_1	2	0100000011
C	3	0010000111
D	4	0001000111
B_2	5	0000101111
B_3	6	0000011111

Note: [a]Area B is listed as three separate areas, one each for species 2, 5, and 6.

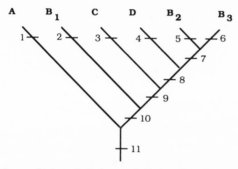

Fig. 7.12. Area cladogram for hypothetical areas A–D, based on phylogenetic relationships of hypothetical species 1–6 and treating area B as if it were three separate areas historically.

"Widespread taxa"

We now turn to the ambiguity that may result from the occurrence of a widespread taxon in the data set. *Species that occur in more than one of the areas being studied may occur there because they have dispersed from their area of origin into the other areas, or because they have failed to speciate in response to vicariance events.* Phylogenetic analysis will tend to treat widespread taxa as if their presence is plesiomorphic for all the areas inhabited. This may produce two types of ambiguity, relationships supported by the area cladogram that are inconsistent with the original estimates of phylogeny used as characters, and postulates of secondary loss (extinction) of the widespread taxon if it does not occur in all of the areas that are linked historically.

1. Figure 7.13 depicts our geologically based area cladogram for hypothetical areas A–D.

2. Figure 7.14 depicts a phylogenetic tree for hypothetical species 1–4 (not to be confused with hypothetical species 1–6 from the previous example).

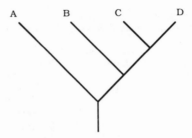

Fig. 7.13. Area cladogram for hypothetical areas A–D.

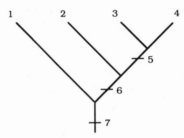

Fig. 7.14. Phylogenetic tree for hypothetical species 1–4, with internal branches numbered for cospeciation analysis.

Table 7.8   Matrix listing the geographic distribution of hypothetical species 1–4 among hypothetical areas A–D, along with the binary codes representing the phylogenetic relationships among species 1–4.

Area	Taxon	Binary Code
A	1	1000001
B	1, 2	1100011
C	1, 3	1010111
D	4	0001111

3. Table 7.8 lists the data matrix for the areas and the species (plus the codes for their phylogenetic relationships) that inhabit them.

4. Phylogenetic analysis of this data matrix produces the area cladogram depicted in figure 7.15. In this instance, the area cladogram derived from biological data is congruent with the area cladogram derived from geological data (fig. 7.13). Nevertheless, something is still amiss, because interpreting the absence of species 1 in area D as a reversal, or extinction event, in that area requires placing species 1 in a position ancestral to species 2, 3, and 4. This conflicts with the phylogenetic tree for the clade, which places species 1 as the sister group to the remaining taxa (fig. 7.14). Hence, while the most

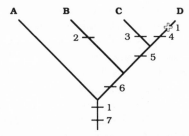

Fig. 7.15. Area cladogram for hypothetical areas A–D, based on phylogenetic relationships of hypothetical species 1–4. *Cross* = putative extinction of species 1 in area D.

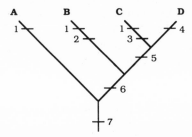

Fig. 7.16. Area cladogram for hypothetical areas A–D, based on phylogenetic relationships among hypothetical species 1–4 and using an optimization rule that allows no reversals. In this case, species 1 is hypothesized to have colonized areas B and C subsequent to its origin in area A.

parsimonious interpretation based on the occurrence of species in particular areas supports the postulate of extinction in area D, other evidence rules against it.

There are at least two general strategies available for dealing with this problem. The first is to perform a phylogenetic analysis using an optimization rule that allows no reversals. For the preceding example, this produces an area cladogram showing two episodes of colonization by taxon 1 (fig. 7.16). However, there might be cases in which extinction is a better explanation than colonization, so this option should be invoked with great care. This leads us to the second general strategy: expand the scope of a biogeographic study to include more than one group of species at a time.

Multiple groups and "missing taxa"

The composition of species within particular biotas is not always similar among biotas. For example, species may be present in some areas and not in others because of different rates or degrees of dispersal. We have discussed methodological protocols for distinguishing species that have been *added to* (dispersed into) an area from those that have evolved in situ. There are a number of reasons why certain clades might be *absent from* an area. In many cases, especially in the tropics, the observation of absence is simply an artifact of inadequate sampling. However, suppose we have sampled an area thoroughly and still can find no evidence of the species in question. The question now becomes, Are we dealing with a primitive absence of the species in this area or a secondary loss (extinction)? To answer this, we must examine more than one clade. Fortunately, the methods applied in a single clade analysis can be used to compare the degree of congruence between geographical history and phylogeny for more than one group at a time. In general, nothing will change: we will continue to treat areas as taxa and the phylogenetic tree for each clade as a separate transformation series, then perform a multicharacter phylogenetic analysis. Consider the hypothetical example of two clades inhabiting the same areas.

1. Figure 7.17 is the area cladogram based on geological evidence.

2. Table 7.9 lists the occurrence of species representing two hypothetical clades in the five areas.

3. Figure 7.18 depicts the phylogenetic trees for the clades containing the species 1–5 and species 10–14, with internal branches numbered for cospeciation analysis.

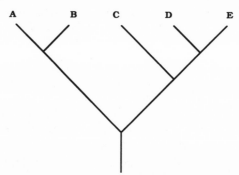

Fig. 7.17. Area cladogram for hypothetical areas A–E, based on geological evidence.

Table 7.9 Occurrence of species representing two hypothetical clades (1–5 and 10–14) in five hypothetical areas.

Area	Taxon 1	Taxon 2
A	1	10
B	2	11
C	3	12
D	4	13
E	5	14

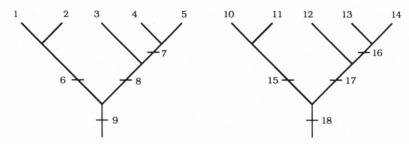

Fig. 7.18. Phylogenetic trees for the clades containing species 1–5 and species 10–14, with internal branches numbered for cospeciation analysis.

Table 7.10   Matrix listing binary codes for members of clades 1–5 and 10–14 inhabiting areas A–E.

Area	Binary Codes
A	100001001100001001
B	010001001010001001
C	001000011001000011
D	000100111000100111
E	000010111000010111

4. Table 7.10 lists the binary codes for members of each clade for each area.

5. Figure 7.19 portrays the area cladogram that results from phylogenetic analysis of this data matrix.

In this hypothetical example, the consistency index for the area cladogram based on the covarying phylogenies of two different clades is 100%. The new area cladogram, in turn, is congruent with the geological history of the areas (fig. 7.17). Therefore, the evolutionary history of the clades represents an example of two co-occurring groups that have speciated in response to the same episodes of geological disruption of gene flow (allopatric speciation mode I). In the next example, two groups of species are not equally represented throughout the areas under investigation. Specifically, the members of one clade occur in five areas (A–E) and the members of the other clade occur in only four areas (A–D).

1. Figure 7.20 depicts the phylogenetic trees for the clades containing species 1–5 and species 10–13.

2. Table 7.11 lists the binary codes for the members of clades 1–5 and 10–13 in each area.

3. Two equally parsimonious area cladograms are produced based on phylogenetic analysis of the data matrix. One of these (fig. 7.21a) is congruent

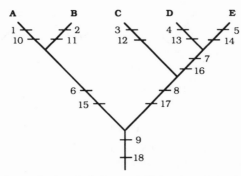

Fig. 7.19. Area cladogram for hypothetical areas A–E, based on phylogenetic relationships of species representing clades 1–5 and 10–14.

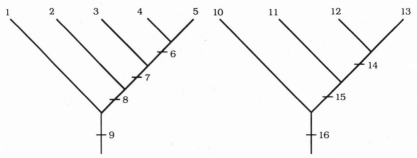

Fig. 7.20. Phylogenetic trees for hypothetical clades 1–5 and 10–13, with internal branches numbered for cospeciation analysis.

Table 7.11   Matrix listing hypothetical areas, the species that inhabit them, and the binary codes representing those species and their phylogenetic relationships.

Area	Taxon	Binary Code
A	1, 10	1000000011000001
B	2, 11	0100000110100011
C	3, 12	0010001110010111
D	4, 13	0001011110001111
E	5	0000111110000000

with the historical relationships among the areas, while the other (fig. 7.21b) places area E at the base of the cladogram rather than with area D. The reversals in characters 14–16 on the first area cladogram implies that clade 10–13 went extinct in area E. In contrast, the second area cladogram postulates three cases of parallel dispersal by ancestral taxa 8, 7, and 6, the latter producing species 5 in area E.

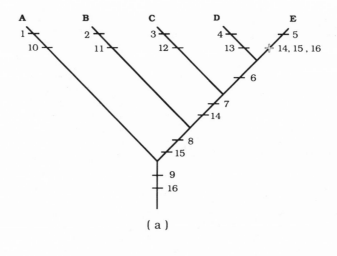

(a)

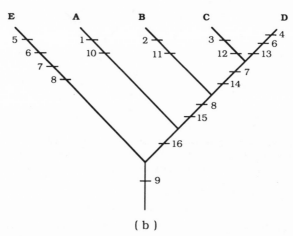

(b)

Fig. 7.21. Two equally parsimonious area cladograms based on phylogenetic relationships of members of clades 1–5 and 10–13.

Secondary loss (extinction) appears as a series of reversals in a phylogenetic analysis because taxa that are absent in an area are coded with a zero, which is equivalent to saying that they were primitively absent from the area. Wiley (1988a,b) suggested that absent taxa should be treated as missing data for the relevant area (fortunately, the newest computer programs for phylogenetic analysis have such an option). So, let us reanalyze the preceding example coding missing taxa with a question mark (?).

1. Table. 7.12 is the matrix produced by coding absent taxa as question marks.

Table 7.12 Matrix listing hypothetical areas, the species that inhabit them, and the binary codes representing those species and their phylogenetic relationships.

Area	Taxon	Binary Code[a]
A	1, 10	1000000011000001
B	2, 11	0100000110100011
C	3, 12	0010001110010111
D	4, 13	0001011110001111
E	5	000011111???????

[a]? = taxa missing from an area.

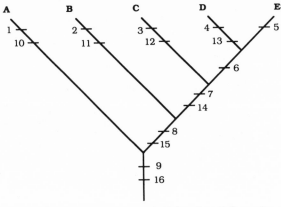

Fig. 7.22. Area cladogram for hypothetical areas A–E, based on phylogenetic analysis of clades 1–5 and 10–13 using "missing data" coding option (note absence of notation for clade 10–13 on branch connecting area E to the rest of the area cladogram).

2. Only one area cladogram, congruent with figure 7.21a, is produced by a phylogenetic analysis of this new data matrix (fig. 7.22). In this case, the absence of clade 10–13 from area E is not considered informative in constructing the area cladogram (i.e., relationships cannot be reconstructed from question marks, or *it is dangerous to group on the basis of character absence*). Note that there is no hypothesis concerning either the presence or absence of taxon 10–13 in area E. This is an important point to remember because some explanation must be offered a posteriori. For example, we might propose either that no members of taxon 10–13 ever inhabited area E or that some member of the clade reached it and then became extinct. The first proposal, in turn, implies that species 5 occurs in area E because of some form of allopatric speciation involving dispersal (allopatric speciation mode II). In the second case we would be postulating that species 5 occurs in area E because of vicariant speciation (allopatric speciation mode I).

A special case of this situation results in the creation of "pseudomissing taxa." Specifically, if you are working with multiple clades, one of which has two or more members occurring in the same area, duplicating the areas containing more than one clade member in the manner discussed above will produce areas with missing taxa. So, given that you now have multiple new areas, in which ones do you allocate the clades that have only single representatives in the area? The answer is straightforward: place them in the areas that contain other species with which they show historical congruence. *This is equivalent to invoking Hennig's auxiliary principle: assume homology unless forced by other data to abandon the assumption* (see chapter 2). *Note that if there is any biogeographic congruence among clades, it is necessary to invoke this assumption in order to find the congruence. Conversely, if there is no congruence, use of the principle will not force spurious congruence to result.* The next example demonstrates how this is done.

1. Figure 7.23 depicts the geographic distributions of the members of two clades (1–6 and 12–18) in six areas (A–F). Areas A, B, D, E, and F contain one member of each clade, while area C contains one member of clade 1–6 (species 3) and two members of clade 12–18 (species 14 and 18).

2. Table 7.13 is the binary data matrix, using inclusive ORing, for the areas.

3. Phylogenetic analysis of the data matrix produced one area cladogram

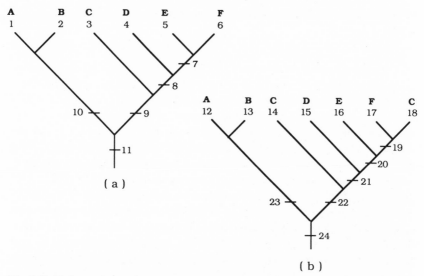

Fig. 7.23. Phylogenetic trees for members of two clades (1–6 and 12–18), with internal branches numbered for cospeciation analysis and with geographic distributions (in areas A–F) listed above species numbers.

Table 7.13 Matrix listing hypothetical areas, the species that inhabit them, and the binary codes representing those species and their phylogenetic relationships.

Area	Taxon	Binary Code		
A	1, 12	1000000001	1100000000	0011
B	2, 13	0100000001	1010000000	0011
C	3, 14, 18	0010000010	1001000111	1101
D	4, 15	0001000110	1000100000	1101
E	5, 16	0000101110	1000010001	1101
F	6, 17	0000011110	1000001011	1101

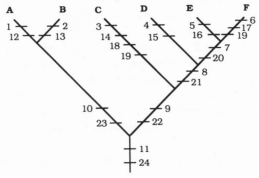

Fig. 7.24. Area cladogram for six hypothetical areas, based on phylogenetic relationships of two clades (1–6 and 12–18). Note that taxon 19 (ancestor of species 17 and 18) occurs twice, once in area F and once in area C.

with a consistency index of 96% (fig. 7.24). Note that taxon 19, the ancestor of species 17 and 18, occurs twice, and that area C has two representatives of clade 12–18 (species 14 and 18).

4. As a consequence of the dual occurrence of clade 12–18 in area C, we recode, listing area C twice, C_1 for species 14 and C_2 for species 18. However, should we put the codes for species 3 with area C_1 or with area C_2? In this case, we place species 3 with species 14 in the same area (C_1) because those two species are phylogenetically congruent with respect to their geographic distributions. Species 18 (more accurately, its ancestor 19) is associated with the incongruent portions of the area cladogram, so we place species 18 in an area by itself (C_2) and list "missing taxa" codes for clade 1–6 in area C_2. In this case, the "missing taxa" are created by our splitting area C into two different areas. Table 7.14 is the binary matrix reflecting this recoding.

5. Phylogenetic analysis of this data matrix produces one area cladogram with a consistency index of 100% (fig. 7.25).

6. Note that we still must provide an a posteriori explanation for the fact

Table 7.14 Matrix listing hypothetical areas, the species that inhabit them, and the binary codes representing those species and their phylogenetic relationships.

Area[a]	Taxon	Binary Code[b]		
A	1, 12	1000000001	1100000000	0011
B	2, 13	0100000001	1010000000	0011
C_1	3, 14	0010000010	1001000000	0101
C_2	18	??????????	?000000111	1101
D	4, 15	0001000110	1000100000	1101
E	5, 16	0000101110	1000010001	1101
F	6, 17	0000011110	1000001011	1101

[a]Area C is split in two.
[b]? = taxa missing from an area.

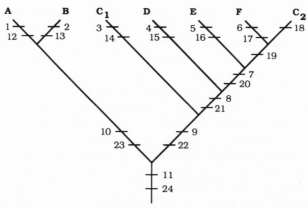

Fig. 7.25. Area cladogram for six hypothetical areas, based on phylogenetic relationships of two clades (1–6 and 12–18), with area C coded as two different areas. Note absence of clade 1–6 from area C_2.

that area C_2 has no representative of clade 1–6. There are two general possibilities; first, that species 18 results from an episode of allopatric speciation mode II in which clade 1–6 did not participate, and second, that there was once a member of clade 1–6 (which would have been the sister species of 6) in area C_2, which has become extinct. If the secondary extinction hypothesis is correct, we may be dealing with allopatric speciation mode I or mode II (refer to chapter 4 for what to do in such cases).

Case Studies

North American freshwater fishes: pre- or post-Pleistocene origins?

Our first example involves the North American freshwater fish fauna. This fauna, containing many localized (endemic) species, is extremely diverse.

Gilbert (1976) noted that North America contains approximately 1,000 described and undescribed species of freshwater fishes, compared with 230 in Australia, 250 in Europe, 1,500 for Asia, 1,800 for Africa, and 2,200 for South America. This diversity has traditionally been attributed to the dispersal of fishes into North America from Europe, Asia, and South America. If this is true, then the North American fish communities are younger than their counterparts elsewhere. The patterns of this diversity also vary across the continent. In general, the drainages east of the Rocky Mountains are species-rich and dominated by the members of only a few genera in a small number of families (Cyprinidae, Percidae, Centrarchidae, and Ictaluridae). By contrast, the western fauna is somewhat depauperate and composed of a different assortment of fishes (distinctive cyprinid and catostomid genera, Salmonidae, Cyprinodontidae, and Cottidae). Researchers have sought explanations for these patterns by identifying the putative center of origin for a particular group, then invoking a variety of dispersals and extinctions (see Mayden 1988 for a detailed discussion). Once again, such explanations were often predicated on the hypothesis that most species had dispersed into this continent from elsewhere. The lack of explicit phylogenetic hypotheses for the evolution of individual clades has made it difficult to formulate rigorous explanations about the origins of this fauna. However, due to the recent efforts of a variety of researchers, building upon the taxonomic legacy bequeathed by ichthyologists such D. S. Jordan and C. L. Hubbs, a solid foundation for such evolutionary investigations is currently under construction (see, e.g., Nelson 1969, 1972; Wiley 1976; Rosen 1978, 1979; Mayden 1985, 1987a,b, in press; Wiley and Mayden 1985).

Mayden (1988) examined the historical biogeography of fishes in seven different clades: the *Notropis leuciodus, Luxilus zonatus-coccogenis, Etheostoma variatum, Nocomis biguttatus,* and *Fundulus catenatus* species groups and the subgenera *Etheostoma (Ozarka)* and *Percina (Imostoma).* These fishes inhabit drainage systems within the Central Highland region (fig. 7.26). The Central Highland consists of three currently disjunct regions. To the west of the Mississippi River lie the Ozark and Ouachita highlands, separated by the floodplain of the Arkansas River. The western highlands, in turn, are separated from their expansive counterpart, the eastern highlands, by the floodplain of the Mississippi River. Prior to the disruptive influences of Pleistocene glaciation, these regions were continuous (see references in Mayden 1988).

Two hypotheses have been proposed to explain the diversity patterns of the freshwater fish fauna in this region. The first hypothesis is based upon observations that the same or related species often occur on the eastern and western sides of the Mississippi. Given this, several researchers proposed that much of the current diversity was produced by the fragmentation and isolation of populations during Pleistocene glaciation. According to this scenario, speciation occurred subsequent to this isolation and has been accompanied by widespread dispersal of the new species, following the retreating glaciers. The second hypothesis postulates that current diversity existed before the Pleistocene glaciation. According to this scenario, the glaciers fragmented the freshwater fauna existing in the expansive and continuous "highland" province. While this does not rule out a role for glaciation in the production of

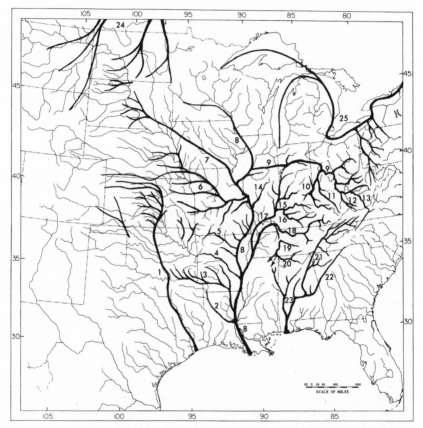

Fig. 7.26. Preglacial river drainages in North America superimposed on existing drainages. *1* = Plains Stream; *2* = Old Red River; *3* = Old Ouachita River; *4* = Old Arkansas River; *5* = White River; *6* = Old Grand-Missouri River; *7* = Ancestral Iowa River; *8* = Old Mississippi River; *9* = Old Teays-Mahomet River; *10* = Old Kentucky River; *11* = Old Licking River; *12* = Old Big Sandy River; *13* = Kanawha River; *14* = Kaskaskia River; *15* = Wabash River; *16* = Green River; *17* = Old Ohio River; *18* = Old Cumberland River; *19* = Old Duck River; *20* = Old Tennessee River; *21* = Appalachian River; *22* = Old Tallapoosa River; *23* = Mobile basin; *24* = Hudson Bay drainage; *25* = St. Lawrence River. (From Mayden 1988.)

the current disjunct distribution patterns of freshwater fishes in the Central Highland region, it minimizes glaciation's role as the stimulus for widespread and recent speciation by these organisms. Phylogenetic analysis has uncovered speciation events in multiple groups of fishes, which are correlated with geological events predating the Pleistocene in both glaciated and unglaciated areas, supporting the second hypothesis. Mayden (1988) focussed his attention on the Central Highland drainages in an attempt to shed further light on the problem.

> For Central Highland fishes one may examine the origin of the fauna by comparing the history of the drainage basins involved and the history of the fishes, inferred from geologic data and phylogenetic relationships, respectively. If congruence is obtained between the phylogenetic relationships and drainage relationships existing prior to the Pleistocene then one may predict that the fish groups existed prior to glaciation and the vicariance hypothesis would be supported. However, if relationships of fishes are congruent with drainage patterns developed after glaciation, then an explanation of dispersal during and after the Pleistocene glaciation may be appropriate.

Mayden performed a phylogenetic analysis using thirty-four river drainages as area "taxa," and the phylogenetic relationships from the seven different clades as "characters" (in all, thirty-seven "characters"). This analysis produced thirty-three equally parsimonious area cladograms. We will base the following discussion upon a simplified version of Mayden's consensus tree for those cladograms (fig. 7.27). This area cladogram highlights four points of agreement among all the thirty-three alternate hypotheses. First, all area cladograms place the rivers of the Interior Highlands in a single clade. That clade, in turn, is the sister group to the drainages of the upper Mississippi River and the old Teays River. Second, prior to the Nebraskan and Kansan glaciations, the Teays River flowed west into the upper Mississippi. Following these glacial advances, the course of the Teays was shifted southwards to flow into the Ohio River, which is the present-day (postglaciation) configuration. All thirty-three area cladograms place the rivers of the old Teays system with the upper Mississippi drainages; the Teays system never clusters with drainages forming part of the lower Ohio River system. Third, the Mobile basin, which was separate from the Mississippi drainage long before the Pleistocene, retains its independent status in this analysis, clustering at the base of the area cladogram. And finally, one clade encompasses all the Mississippi offshoots thought to have formed the drainages for the pre-Pleistocene Central Highland, while yet another clade reconstructs the pre-Pleistocene Teays-Mississippi river system.

In summary, all of these distributions coincide with *pre-Pleistocene*, rather

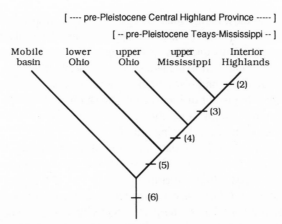

Fig. 7.27. Area cladogram for the Central Highland drainage systems, reconstructed from the phylogenetic relationships of various freshwater fishes. The original area cladogram, based on thirty-four contemporaneous rivers, has been simplified to highlight five groups of rivers that indicate preglacial origins of the fish that inhabit them: Mobile basin (drainages 21–23 in fig. 7.26); lower Ohio (drainages 15–20 in fig. 7.26); upper Ohio (part of the Old Teays River drainage; drainages 10–14 in fig. 7.26); upper Mississippi (drainages 7–9 in fig. 7.26); and Interior Highlands (drainages 2–6 in fig. 7.26). The number of characters supporting each of the five groupings is listed in *parentheses*. (Redrawn and modified from Mayden 1988.)

than *post-Pleistocene* or contemporary, drainage patterns (for a detailed discussion of individual rivers see Mayden 1988). This suggests that there was a diverse and widespread Central Highland ichthyofauna prior to the Pleistocene glaciation, corroborating the studies of speciation patterns reported by Wiley and Mayden (1985; see chapter 4). This does not mean that dispersal and glaciation have been unimportant in this system. For example, in seven cases Mayden was able to demonstrate that specific instances of "geographic homoplasy" coincided with episodes of Pleistocene glacial alterations in river-flow patterns that apparently resulted in some faunal mixing. Not surprisingly, then, current distributions reflect an interaction between the relatively ancient origins and diversification of the fauna and the recent effects of large-scale environmental changes.

Parrots and toucans

Our second example is based on the distribution patterns of some neotropical birds. Since many birds possess the potential for widespread dispersal, it seems counterintuitive to expect speciation patterns to show any degree of congruence with area cladograms. Nonetheless, it is also true that even a champion flier such as the ruby-throated hummingbird is not found in every area, so it is possible that some remnants of historical associations are retained even in the most highly mobile organisms. With respect to many neo-tropical areas, the potential for dispersal appears high, due to the lack of any obvious geographical disjunctions in the huge Amazon basin. In a manner analogous to the traditional explanations for the North American freshwater fish fauna, the diversity and distribution of species in the Amazon basin has been explained as the result of Pleistocene glaciation. In this case, instead of physically intruding into the area, the glaciers were responsible for a general drying of the Amazon basin, leaving relatively small "islands" of forests sur-rounded by "seas" of xeric habitat. Lynch (1988) reexamined this scenario and concluded that the current data base did not provide strong support for the neotropical "forest refugia" theories.

Cracraft and Prum (1988) investigated the evolutionary divergence of four South American avian clades, the parrot genus *Pionopsitta* and the toucan genera *Selenidera* and *Pteroglossus,* living in the Amazon basin. Since the Pleistocene forest refugia theory was based originally on data from avian distributions (Haffer 1969; see also references in Lynch 1988 and in Cracraft and Prum 1988), this seems like an appropriate test case.

1. The phylogenetic trees for these clades are shown in figure 7.28.

2. The data matrix for the areas is presented in table 7.15. Southeastern Brazil is divided into two different areas, C_1 and C_2, because two species in the study group, *Selenidera maculirostris* (species 4), which is a relatively highly derived member of its clade, and *Pionopsitta pileata* (species 13), which is the basal member of its clade, occur there. Notice that we have omitted the intermediate step of producing multiple area cladograms based on inclusive ORing of the data (described in Cracraft 1988).

Table 7.15 Matrix listing eight neotropical areas and the binary codes for members of four clades of birds that inhabit them, based on their phylogenetic relationships.

Area[a]	Binary Code[b]			
A	0100000000	1100100000	0011110001	??????????
B	0010000011	1100010000	1111101011	1110000101
C_1	??????????	??10000000	000001?????	??????????
C_2	0001000011	11????????	??????????	??????????
D	0000100001	1100001001	1111100111	0001000011
E	0000010101	1100000101	1111100111	0000101011
F	0000001101	1100000101	1111100111	0000011011
G	1000000000	0101000000	00011?????	??????????
H	??????????	??00000010	01111?????	??????????

[a]A = Guyana–northeast of Amazon; B = southeast Amazon; C_1 and C_2 = southeast Brazil; D = southwest Amazon; E = northwest Amazon; F = west Amazon; G = Panama–northeast South America; H = north Colombia–Maracaibo.

[b]? = taxa missing from an area.

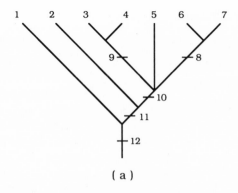

(a)

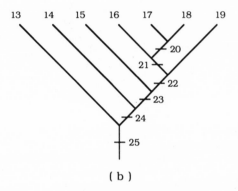

(b)

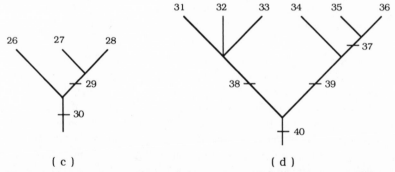

(c)   (d)

Fig. 7.28. Phylogenetic trees for four clades of neotropical birds, with internal branches numbered for cospeciation analysis. (a) Members of the toucan genus *Selenidera: 1 = S. spectabilis;
2 = S. culik; 3 = S. gouldii; 4 = S. maculirostris; 5 = S. nattereri; 6 = S. langsdorffi; 7 =
S. reinwardtii.* (b) Members of the parrot genus *Pionopsitta: 13 = P. pileata; 14 = P. coccinicollaris + P. pulchra + P. hematotis; 15 = P. caica; 16 = P. vulturina; 17 = P. aurantiigena;
18 = P. barrabandi; 19 = P. pyrilia.* (c) Toucans in the *Pteroglossus viridis* group: *26 = P.
viridis; 27 = P. inscriptus; 28 = P. humboldti.* (d) Toucans in the *Pteroglossus bitorquatus*
group: *31 = P. sturmii; 32 = P. reichenowi; 33 = P. bitorquatus; 34 = P. azara; 35 = P.
mariae; 36 = P. flavirostris.* (Redrawn and modified from Cracraft and Prum 1988.)

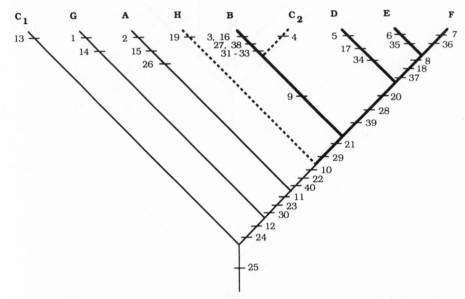

Fig. 7.29. Area cladogram for eight neotropical areas, based on phylogenetic relationships of members of four bird clades. *Light lines* indicate areas outside Amazonia that contain basal members of the bird clades. *Bold lines* indicate four areas of endemism in Amazonia in which members of all four bird clades reside. *Dotted lines* indicate two areas in which relatively derived species of one bird clade occur as a result of two episodes of independent colonization. A = north of the eastern half of the Amazon; B = southern edge of the eastern half of the Amazon; C_1, C_2 = southeastern Brazil; D = southern portion of the Amazon basin; E = northern portion of the Amazon basin; F = western portion of the Amazon basin; G = northeastern South America–Panama; H = northern part of Colombia to the Lake Maracaibo region.

3. The area cladogram produced by this analysis (fig. 7.29) is congruent with the area cladogram presented by Cracraft and Prum (1988) and has a consistency index of 100%. This surprising result suggests that the phylogenetic patterns for the birds are explained by a single historical sequence of associations among the areas. Of course, each of the bird clades exhibits some degree of evolutionary independence, as you can tell from the data matrix. There are three important components to the evolution of this avian fauna depicted by the area cladogram. We will refer to these as (1) the origin of the clades, or "from where did the Amazonian groups come?" (2) evidence of cospeciation, and (3) departures from cospeciation.

The area cladogram suggests that the neotropical avifauna was assembled in a variety of ways. First, there is an interesting pattern involving the sequential addition of basal clade members which, once they begin co-occurring with other avian groups, exhibit patterns of vicariant speciation common to the other groups. Southeastern Brazil (area C_1) is inhabited by only one study species, *Pionopsitta pileata,* the basal parrot species in its clade. Central America plus the Choco (area G) occurs next on the area cladogram because it contains *Selenidera spectabilis,* the basal member of that toucan clade, and *Pionopsitta coccinicollaris + P. pulchra + P. hematotis,* the second basal parrot group. Finally, the Guyanan area (A) is inhabited by the basal member of the *Pteroglossus viridis* species group (*P. viridis*), the second member of *Selenidera* (*S. culik*), and the third member of *Pionopsitta* (*P. caica*). It is not until we come to the first Amazonian area (area B) that we find all four clades represented.

A cornerstone of the Pleistocene forest refugia theory is that the Amazonian diversity results from multiple independent episodes of parapatric and peripheral isolates allopatric speciation. However, the Amazonian portion of the area cladogram provides the strongest support for vicariant speciation (allopatric speciation mode I) in these four avian clades. The members of all four clades exhibit complete congruence with the sequence of relationships among areas B, D, E, and F on the area cladogram (i.e., evidence of cospeciation). This departs markedly from the expectations of the refugia theory.

The distribution patterns of birds in the Amazonian regions demonstrate two instances of independence from this background of vicariant speciation. First, the occurrence of two sets of sister species, *Selenidera langsdorffi* (6) plus *S. reinwardtii* (7), and *Pteroglossus mariae* (35) plus *P. flavirostris* (36), in the sister areas E and F, conforms to a hypothesis of allopatric speciation. However, *Pionopsitta barrabandi* (18) and *Pteroglossus humboldti* (28) occur in both areas E and F. This suggests that the vicariant event affecting the *Selenidera* and *Pteroglossus bitorquatus* species-group representatives in those areas did not affect the members of *Pionopsitta* and the *Pteroglossus viridis* species group. Second, *Pteroglossus humboldti* (28) occurs in areas D, E, and F, suggesting that the speciation events giving rise to *S. nattereri* (5), *Pionopsitta aurantiigena* (17), and *Pteroglossus azara* (34) did not affect the *Pteroglossus viridis* species group. Thus, the members of *Selenidera* and the *Pteroglossus bitorquatus* species group exhibit three instances of vicariant speciation in the Amazon, members of *Pionopsitta* exhibit two instances, and members of the *Pteroglossus viridis* species group show only one.

Differential speciation rates among members of clades living in a single area represent another form of evolutionary independence in the assembling of this neotropical avifauna. The area along the southern edge of the eastern Amazon (B) is inhabited by birds from all four study clades. Three of these clades are represented by a single, widespread species, *S. gouldii* (3), *Pionopsitta vulturina* (16), and *Pteroglossus inscriptus* (27). The fourth clade, the

Pteroglossus bitorquatus species group, is represented by three species, *P. sturmii* (31), *P. reichenowi* (32), and *P. bitorquatus* (33), displaying the most restricted distribution of all the birds in this area. Cracraft and Prum (1988) reported that the species status of these toucans is uncertain at the moment. However, if they are differentiated species, the *Pteroglossus bitorquatus* species group exhibits an interesting coupling of decreased distribution and increased diversity, compared to its co-inhabitants along the banks of the Amazon.

Areas H (northern Colombia and Maracaibo) and C_2 (southeastern Brazil) also represent departures from the general pattern of area relationships depicted by figure 7.29. Both of these areas are placed among the Amazonian regions on the cladogram, due to the relationship of a single species to the other members of its genus: the parrot *Pionopsitta pyrilia* (area H) and the toucan *Selenidera maculirostris* (area C_2). These distributions appear to be the result of two independent speciation episodes involving some form of dispersal (allopatric speciation mode II). If this is the case, then we have identified two examples of the speciation mode associated with Pleistocene forest refugia theories. However, the theories were designed to explain species diversity and distribution patterns within the Amazon basin, and these birds live along the periphery of that area. In Lynch's (1982) study of ceratophryine frogs (see chapter 4), the forest-dwelling species of *Ceratophrys* represented two separate cases of dispersal into the Amazon basin from peripheral habitats. These findings are also contrary to the expected movements of species out of forest refugia into surrounding habitats.

This study thus indicates that this neotropical avifauna has been assembled by a combination of (1) the sequential addition of basal members of the clades into areas surrounding the Amazon, (2) vicariant speciation among the toucans and parrots within the Amazonian areas, (3) two cases of peripheral isolates allopatric speciation on the periphery of the Amazon basin, and (4) factors unique to each bird clade that affected the form and nature of the clade's response to the vicariant effects. Mayden (1988) concluded that Pleistocene refugia did not explain the diversification and distribution of North American freshwater fishes. Cracraft and Prum (1988) came to the same basic conclusions, as did Lynch (1988), about neotropical Pleistocene forest refugia theories.

> These generalized historical patterns . . . [arose] via fragmentation (vicariance) of a widespread ancestral biota. . . . These vicariance events could have originated. . . . at various times during the Cenozoic. The inference that diversification of the Neotropical biota is primarily the result of . . . isolation within Quaternary forest refugia, is unwarranted, given present data.

Evolution of minnows, characoids, and catfish

The majority of species of freshwater fishes belong to a group with the impressive name Ostariophysii. This group, in turn, comprises (1) the cypriniforms, a large collection of fishes including the cyprinoids, the ubiquitous northern temperate minnows and their African and European relatives, the gymnotoids (knife-fishes, including the famous South American "electric eel"), and the characoids (e.g., the tetras and piranhas), and (2) the widespread siluriforms, or catfish, which exhibit tremendous diversity primarily in Africa and South America. Novacek and Marshall (1976) examined a number of explanations that had been proposed for the origin, diversification, and current distribution of ostariophysans within a phylogenetic framework for the five major groups of these fishes. They optimized the distributions of the fishes (South America, Africa, Europe, North America, and Asia) on the phylogenetic tree and concluded that (1) the Ostariophysii probably did not originate in Gondwana, because there are no Australian representatives; (2) the Ostariophysii probably originated in South America after the breakup of Gondwana but prior to the breakup of South America and Africa; (3) the ancestral cypriniforms and siluriforms diverged in South America and dispersed into Africa; and (4) the South American and African ostariophysans evolved independently following the rifting of South America from Africa, with representatives from the African groups dispersing into Europe, Asia, and North America. We have reanalyzed this example using the methods outlined in the preceding sections.

1. The phylogenetic tree for the five major groups of ostariophysans is shown in figure 7.30.

2. Table 7.16 lists the data matrix for the areas inhabited by ostariophysans

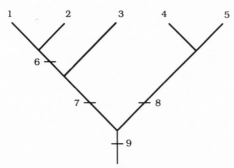

Fig. 7.30. Phylogenetic tree for the major groups of ostariophysan fishes, with internal branches numbered for cospeciation analysis. *1* = Characoidea; *2* = Gymnotoidea; *3* = Cyprinoidei; *4* = Diplomystidae; *5* = Siluroidei.

Table 7.16 Matrix listing areas, the ostariophysan groups that inhabit them, and the binary codes representing those groups and their phylogenetic relationships.

Area	Taxon	Binary Code
Africa	1, 3, 5	101011111
South America	1, 2, 4, 5	110111111
Europe	3, 5	001010111
North America	3, 5	001010111
Asia	3, 5	001010111

and the binary codes for the taxa themselves and their phylogenetic relationships.

3. Phylogenetic analysis of this data matrix produces two equally parsimonious area cladograms (fig. 7.31), each with a consistency index of 81.8%. Both of the area cladograms support Novacek and Marshall's general conclusion that much of the early diversification of these fishes occurred before the separation of South America and Africa. The first area cladogram (fig. 7.31a) proposes that the divergence of the common ancestor of the cypriniforms (7) from the common ancestor of the siluriforms (8), as well as the evolution of the earliest siluroids (5), occurred prior to the South American–African split. This explains the independent divergence of African and South American catfish. However, this area cladogram also postulates that the evolution of the characoids (6) involved dispersal, either from South America to Africa or from Africa to South America, producing the two independent characoid lines in these areas. The second area cladogram (fig. 7.31b) provides stronger support for Novacek and Marshall's hypothesis, suggesting that the cyprinoid (3)/characoid (6) divergence and the South American–African characoid (1) divergence took place prior to the rifting of South America and Africa. However, there is a catch; according to this interpretation, early representatives of the characoid line occurred in North America, Europe, and Asia, and became extinct in those areas (reversals for 1 and 6) subsequent to the rifting of South America from Africa.

The two area cladograms also provide different explanations for the absence of characoids from the Northern Hemisphere. Of the five major ostariophysan taxa, two (the gymnotoids and the diplomystids) are endemic to South America. The other three taxa occur in Africa and were thus theoretically capable of dispersing into the Northern Hemisphere; however, only the siluroids and cyprinoids occur there. The first area cladogram suggests that either the characoids originated in Africa and dispersed from there into South America rather than into the Northern Hemisphere, or the characoids originated in South America, dispersed from there into Africa while the African cyprinoids and siluroids were dispersing from Africa into the Northern Hemi-

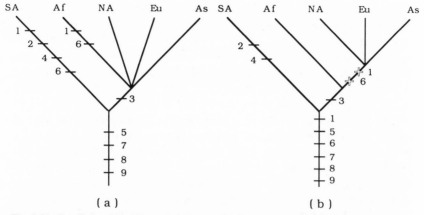

Fig. 7.31. Equally parsimonious area cladograms for five areas, based on phylogenetic relationships of the major ostariophysan fishes. *1* = Characoidea; *2* = Gymnotoidea; *3* = Cyprinoidei; *4* = Diplomystidae; *5* = Siluroidei; *6–9* = ancestors; *cross* = extinction.

sphere, and simply got no farther north. The second area cladogram suggests that the dispersal of the characoid, cyprinoid, and siluroid lineages from Africa into the Northern Hemisphere was followed by the extinction of the characoids in that area. The discovery of characoid fossils in the Northern Hemisphere would resolve this question (it would allow us to change the coding in the matrix for the area cladogram).

The freshwater stingrays of South America

Sharks, skates, and rays are an old and diverse group of cartilaginous fishes that inhabit areas ranging across the full spectrum of environmental salinity, from marine to euryhaline to freshwater. Because of this ubiquitous distribution, physiologists have a long-standing fascination with elasmobranch osmoregulation. Marine elasmobranchs retain urea and other organic substances, in their blood and tissue fluids, creating an internal environment comparable to the surrounding seawater. This minimizes the osmotic loss of that most precious quantity, water. The rectal gland supplements kidney function, secreting salt (NaCl) in a fluid that is twice the NaCl concentration of body fluids. Euryhaline elasmobranchs, coastal and inshore species, are osmoregulatory chameleons. They function like marine species under conditions of high salinity; however, in less saline waters their urea concentration drops to 20–50% of marine levels, and rectal gland function is either reduced or stopped. Some of these species, such as the bull sharks and sawfish of Lake Nicaragua, may even spend extended periods of time in fully freshwater habitats.

Cruising throughout the major river systems of eastern South America are members of the stingray family Potamotrygonidae, the only elasmobranchs that are permanently adapted to freshwater habitats. They lack the ability to concentrate urea, although they produce some of the necessary enzymes. Their rectal glands are very small and apparently nonfunctional. No amount of physiological conditioning can induce these rays to concentrate urea or excrete NaCl (see references in review by Thorson, Brooks, and Mayes 1983). These species are faced with the problems of operating an internal system that is hyperosmotic to the environment. Like freshwater teleosts, these rays are capable of producing massive amounts of dilute urine to compensate for this problem. The highly evolved nature of stingrays relative to sharks and skates, and the absence of totally freshwater species in any other elasmobranch group, have led most physiologists to assume that potamotrygonids are derived from marine ancestors; that is, they are secondarily adapted to a freshwater existence. Additionally, the presence of a rectal gland, albeit vestigial, and of some of the enzymes for producing urea has been accepted as evidence that potamotrygonids are not a particularly old group. Because potamotrygonids are restricted to rivers that empty into the Atlantic Ocean, ichthyologists have tended to assume that the ancestor of the potamotrygonids was an Atlantic marine or euryhaline stingray that dispersed into freshwater, adapted to the new surroundings, and then dispersed throughout eastern South America, speciating along the way.

Brooks, Thorson, and Mayes (1981) uncovered a new perspective on the origins of this enigmatic group of stingrays in their phylogenetic investigations of the ecological association between potamotrygonids and their helminth parasites. They began with a historical biogeographic analysis to discover how long the potamotrygonids had been in fresh water. If they arrived relatively recently and speciated as a result of independent dispersal, potamotrygonids and their parasites *should not show correlated patterns of speciation with organisms that evolved in the freshwater habitats.* The geographic distribution patterns for the parasites of potamotrygonids are complex: some of the parasite species appear to be restricted to single river systems, while others are more widespread (table 7.17).

1. A total of twenty-three species of parasitic worms have been found inhabiting potamotrygonids thus far. Thirteen of these species are members of three clades (fig. 7.32), while the remaining ten species each represent a different clade.

2. Table 7.18 is the data matrix for the six localities based on inclusive ORing of the phylogenetic relationships and distributions of the stingray helminths.

3. Phylogenetic analysis of the data matrix produces a single area cladogram (fig. 7.33) with a consistency index of 80%. This area cladogram sup-

Table 7.17 Geographic distribution of twenty-three species of parasitic worms inhabiting South American freshwater stingrays.

Parasite[a]	Locality[b]					
	1	2	3	4	5	6
1. *Acanthobothrium quinonesi*	0	0	0	0	+	+
2. *Acanthobothrium regoi*	0	0	0	+	0	0
3. *Acanthobothrium amazonensis*	0	0	+	0	0	0
4. *Acanthobothrium terezae*	+	0	0	0	0	0
8. *Potamotrygonocestus magdalenensis*	0	0	0	0	0	+
9. *Potamotrygonocestus orinocoensis*	0	0	0	+	0	0
10. *Potamotrygonocestus amazonensis*	0	0	+	+	+	0
13. *Rhinebothroides moralarai*	0	0	0	0	0	+
14. *Rhinebothroides venezuelensis*	0	0	0	+	+	0
15. *Rhinebothroides circularisi*	0	0	+	0	0	0
16. *Rhinebothroides scorzai*	+	0	0	+	0	0
17. *Rhinebothroides freitasi*	0	+	0	0	0	0
18. *Rhinebothroides glandularis*	0	0	0	+	0	0
24. *Eutetrarhynchus araya*	+	+	0	+	0	0
25. *Rhinebothrium paratrygoni*	+	+	0	+	0	0
26. *Paraheteronchocotyle tsalickisi*	0	0	+	0	0	0
27. *Potamotrygonocotyle amazonensis*	0	0	+	0	0	0
28. *Echinocephalus daileyi*	0	0	+	+	0	0
29. *Paravitellotrema overstreeti*	0	0	0	0	0	+
30. *Terranova edcaballeroi*	0	0	0	+	0	0
31. *Megapriapus ungriai*	0	0	0	+	0	0
32. *Leiperia gracile*	+	0	0	0	0	0
33. *Brevimulticaecum* sp.	+	0	0	0	0	0

[a]Species are numbered for phylogenetic and biogeographic analysis.

[b]1 = upper Paraná River, including the lower Mato Grosso; 2 = mid-Amazon, near Manaus; 3 = upper Amazon, near Leticia; 4 = Orinoco Delta; 5 = Lake Maracaibo tributaries; 6 = mid- to lower-Magdalena River.

ports a qualitative assessment that approximately 80% of the species composition in these communities is a reflection of common phylogenetic history, which is more consistent with a history of vicariant speciation (allopatric speciation mode I) than with a history of speciation via dispersal (allopatric speciation mode II). Interestingly, the areas of endemism for the potamotrygonids are also areas of endemism for species of ostariophysan fishes, members of groups that historical biogeographic analysis suggests are ancient rather than recent residents of neotropical freshwater systems (see preceding example). The area cladogram is congruent with the hypothesized geological history of the region, which links the origins of the major South American river systems with the uplifting of the Andes beginning early in the Cretaceous (see Brooks, Thorson, and Mayes 1981 for references to the geological

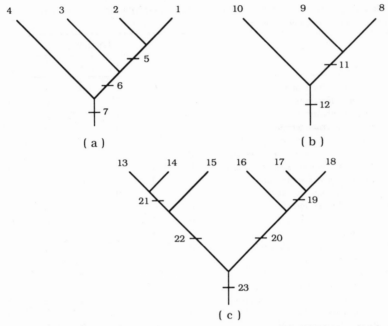

Fig. 7.32. Phylogenetic trees for members of three tapeworm genera inhabiting neotropical fresh-water stingrays, with internal branches numbered for cospeciation analysis. (*a*) *1* = *Acantho-bothrium quinonesi; 2* = *A. regoi; 3* = *A. amazonensis; 4* = *A. terezae.* (*b*) *8* = *Potamotry-gonocestus magdalenensis; 9* = *P. orinocoensis; 10* = *P. amazonensis.* (*c*) *13* = *Rhinebothroides moralarai; 14* = *R. venezuelensis; 15* = *R. circularisi; 16* = *R. scorzai; 17* = *R. freitasi; 18* = *R. glandularis.*

Table 7.18 Matrix listing six river systems in eastern South America and the binary codes for members of the helminth parasite groups inhabiting freshwater stingrays residing in those areas.

Area	Binary Code			
Upper Paraná River	0001001000	0000010001	0011100000	011
Manaus, mid-Amazon	0000000000	0000001011	0011100000	000
Leticia, upper Amazon	0010011001	0100100000	0110011100	000
Orinoco Delta	0100111011	1101010111	1111100101	100
Lake Maracaibo	1000111001	0101000000	1110000000	000
Magdalena River	1000111100	1110000000	1110000010	000

evidence supporting this interpretation). It appears, then, that potamotrygo-nids are not relatively recent invaders of the neotropics.

A large portion of this phylogenetic component in parasite community composition may be obscured when attention is focussed upon individual communities (refer to fig. 7.33). For example, the Paraná, mid-Amazon, up-per Amazon, Orinoco, and Magdalena systems all contain species whose phylogenetic relationships correspond to the geological history of the areas in which they occur. In addition to species endemic to the area, the Orinoco community contains species that have colonized from three other systems, the upper Amazon (species 10 and 28), the Paraná (species 16, 24, and 25) and the mid-Amazon (species 19, the ancestor of species 18). The Orinoco community thus has the highest diversity, although it is not the oldest. The Maracaibo community also has representatives from three different source areas, the Magdalena (species 1), Orinoco (species 14), and the upper-Amazon (species 10). Clearly, these communities have been assembled in different ways historically, and in ways that could not be deduced from observations of contemporaneous species distributions alone. We will discuss this in more detail in chapter 8. Such complex distribution patterns also support the conclusion that potamotrygonids have been in the neotropics for a long time.

On the basis of this initial analysis, the Orinoco and Maracaibo localities appear to be composite areas. So, as suggested in the first part of this chapter, let us recode the data matrix, separating the Orinoco into three areas (D_1, D_2, and D_3) and the Maracaibo area into three areas (E_1, E_2, and E_3; table 7.19).

Phylogenetic analysis of this matrix produces a new area cladogram (fig. 7.34) with a consistency index of 97%. This new area cladogram highlights four evolutionary components that have contributed to the helminth commu-

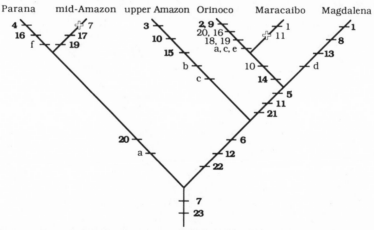

Fig. 7.33. Area cladogram for six river systems in eastern South America, based on phylogenetic relationships of the helminth parasites that inhabit stingrays living in those rivers. For the names of the species corresponding to the numbers, see table 7.17. *Numbers in bold type* = species whose phylogenetic relationships agree with the geological history of the areas in which they occur; *nonbold numbers* = instances of colonization (dispersal); *letters* = species that are the sole representatives of a single clade: a = 24, 25; b = 26, 27; c = 28; d = 29; e = 30, 31; f = 32, 33.

Table 7.19 Matrix listing six river systems in eastern South America and the binary codes for members of the helminth parasite groups inhabiting freshwater stingrays residing in those areas, based on recoding the Orinoco Delta (area D) and Lake Maracaibo (area E) three times.

Area	Binary Code[a]			
Upper Paraná River	0001001???	??00010001	0011100000	011
Manaus, mid-Amazon	??????????	??00001011	0011100000	000
Leticia, upper Amazon	0010111001	0100100000	0110011100	000
Orinoco Delta (D_1)	0100111011	1101000011	1110000001	100
Orinoco Delta (D_2)	??????????	??00010111	0011100000	000
Orinoco Delta (D_3)	???????001	012???????	???0000100	000
Lake Maracaibo (E_1)	???????001	012???????	???0000000	000
Lake Maracaibo (E_2)	??????????	??01000000	1110000000	000
Lake Maracaibo (E_3)	1000111???	??????????	???0000000	000
Magdalena River	1000111100	1110000000	1110000010	000

[a] ? = taxa missing from an area.

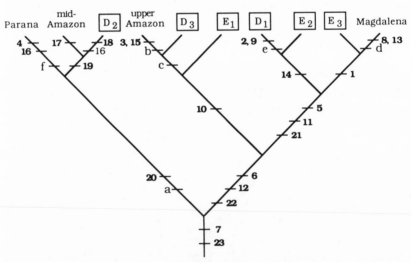

Fig. 7.34. Area cladogram for six river systems in eastern South America, based on phylogenetic relationships of the helminth parasites that inhabit stingrays living in those rivers. The Orinoco (*D*) and the Lake Maracaibo region (*E*) are listed three times, indicating the number of different historical faunal elements that have influenced the contemporaneous diversity. For a description of the species corresponding to the numbers, see table 7.17. *Numbers in bold type* = species whose phylogenetic relationships agree with the geological history of the areas in which they occur; *remaining numbers* = instances of colonization (dispersal).

nity composition in the neotropical freshwater stingrays (fig. 7.35). The first component is the historical geological, or vicariant, backbone linking the Paraná, upper Amazon, Orinoco (D_1), and Magdalena areas. These areas all contain species whose phylogenetic relationships correspond to the geological history of the regions. The remaining components of the parasite distribution patterns involve three sequences of dispersal from these areas along the following routes: from the Paraná to the mid-Amazon to the Orinoco (D_2); from the upper Amazon into the Orinoco (D_3); and from the upper Amazon, the Orinoco, and the Magdalena, forming the Maracaibo fauna (upper Amazon to E_1; Orinoco to E_2; Magdalena to E_3).

Fig. 7.35. Map of South America, highlighting six areas in which freshwater stingrays and the communities of helminths inhabiting them reside. *Dashed lines* indicate vicariant relationships among the Magdalena River, Orinoco River, upper Amazon River, and upper Paraná River. The Lake Maracaibo and the middle Amazon River communities appear to have been derived mostly by colonization from one or more of the four vicariant areas.

As we stated above, assumptions about the evolutionary sequence of osmoregulatory modifications and observations of current geographic distribution led many biologists to think that potamotrygonids were the recently derived descendants of marine stingrays. The most common evolutionary
scenario postulated that the potamotrygonid ancestor moved from the Atlantic
Ocean into the Amazon basin during the Pliocene marine ingression. Subsequent to this invasion, a population was isolated from the ancestor, progressively adapting to freshwater and spreading throughout South America by
stream capture during the past 3–5 million years. The congruence between
the area cladogram based on the phylogenetic relationships of the parasitic
worms inhabiting potamotrygonids, areas of endemism for ostariophysan
fishes inhabiting the neotropics, and the geological history of the river systems in which they occur, pushes the origin of the parasite communities back
to at least the mid-Miocene. Given that the parasite fauna is an old one, there
are two alternative explanations for the origins of the parasite-stingray associations. First, the parasitic fauna that now inhabits the potamotrygonids occurred in South America prior to the appearance of these stingrays. In this
case the parasites or their closest relatives should be found in freshwater organisms, such as ostariophysan fishes, that were living in South America
during the mid-Miocene or earlier. This explanation supports the proposal
that the potamotrygonids are "recently" derived. Alternately, the freshwater
stingrays may be older than biologists once believed, and their parasites reflect that ancient ancestry. If this is the case, the parasites inhabiting potamotrygonids, or their closest relatives, should inhabit marine stingrays whose
geographic distribution is consistent with a hypothesis that the group originated as a result of marine invasion of South America no later than the mid-
Miocene.

In order to examine these alternate explanations, we must expand the scope
of the study, to include the closest relatives (and their hosts) of the parasites
found in the freshwater stingrays. Four of the twenty-three parasite species
listed in table 7.17 inhabit either teleosts (*Paravitellotrema overstreeti, Ter-
ranova edcaballeroi*) or crocodilians (*Leiperia gracile, Brevimulticaecum* sp.)
in particular areas, where local potamotrygonids have picked them up. The
remaining nineteen species of parasites are restricted to stingray hosts. The
closest relatives of these species inhabit marine stingrays (with the exception
of *Megapriapus ungriai*, the only acanthocephalan known to inhabit elasmobranchs of any kind, and whose relationships to other acanthocephalans is
uncertain). Hence, it seems likely that most of the parasite groups inhabiting
potamotrygonids were brought into neotropical freshwater habitats with the
ancestor of the stingrays themselves. Once again, we have support for the
second explanation: the potamotrygonids are older than previously thought.

If we entertain the possibility that the ancestor of the potamotrygonids (and their parasites) arrived in neotropical freshwater habitats no later than the mid-Miocene, we must reevaluate our ideas about the source of those marine ancestors. The geography of South America prior to the mid-Miocene differed in three significant ways from what we see today: Africa and South America were joined (i.e., there was no Atlantic Ocean at the mouth of the Amazon), the Andes began sweeping upwards from the south in the early Cretaceous and moving northward, and the Amazon River flowed into the Pacific Ocean until the mid-Miocene, when it was blocked by Andean orogeny, becoming an inland sea and eventually opening to the Atlantic Ocean. *This leads us to the startling conclusion that if potamotrygonids are a relatively old component of neotropical freshwater diversity east of the Andes, they must have come from the Pacific Ocean, which is west of the Andes!*

Now if we enlarge the spatial scale of this study to include the geographic distribution of the marine relatives of the parasites inhabiting potamotrygonids, we find additional support for the hypothesis that these stingrays and their parasites originated from marine ancestors that were isolated in South America from the Pacific Ocean by the Andean orogeny. The closest relatives of the parasites inhabiting potamotrygonids occur in Pacific marine stingrays (fig. 7.36). A similar origin has been suggested for Amazonian freshwater anchovies (Nelson 1984) and possibly for neotropical freshwater needlefish (Collette 1982). In addition, each of the parasite species inhabiting potamotrygonids requires a mollusc or arthropod intermediate host, so it seems likely that mollusc and arthropod species derived from marine ancestors also moved into neotropical freshwater habitats along with the ancestor of the potamotrygonids. As a consequence, we now recognize the possibility that *a sizeable component of current neotropical freshwater diversity might be derived from Pacific marine ancestors.*

Overall then, the current data base indicates that potamotrygonids and their parasites (1) are older (no later than mid-Miocene rather than Pliocene), (2) came from a different source (moved into the Amazon River from the Pacific rather than the Atlantic Ocean), and (3) have been affected more strongly by phylogenetic influences (i.e., allopatric speciation mode I) on their diversification and distribution, than previously thought. These findings could not have been achieved without phylogenetic analysis and historical biogeography.

Comments on Historical Biogeographic Studies

Choice of spatial scale greatly influences the type of questions asked and the analytical methods used in ecological biogeography (Brown and Gibson 1984). Advocates of macroecology (Brown and Maurer 1989) have suggested

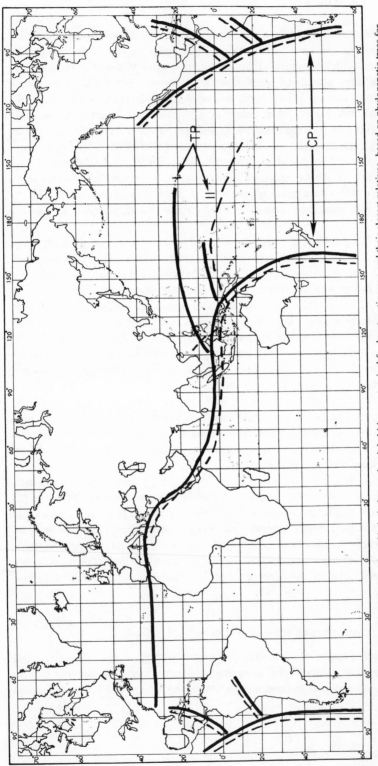

Fig. 7.36. Historical biogeographic relationships of helminth parasites inhabiting neotropical freshwater stingrays and their closest relatives, based on phylogenetic trees for species groups in the tapeworm genera *Acanthobothrium*, *Eutetrarhynchus*, and *Rhinebothrium* (*solid lines*) and in the roundworm genus *Echinocephalus* (*dotted lines*). Note both circum-Pacific and trans-Pacific distribution patterns, with species most closely related to those occurring in freshwater stingrays being part of the circum-Pacific pattern. (Redrawn and modified from Brooks 1988b and from Brooks and Deardorff 1988.)

that researchers studying ecological associations should search for regular patterns of distribution and abundance by expanding the spatial scale of their studies. Brooks (1988b) recently suggested that the degree and form of phylogenetic influence in historical biogeography may also be influenced by the spatial scale chosen. Specifically, the larger the spatial scale chosen for study, (1) the more likely we are to find evidence of replicated allopatric speciation events (allopatric speciation mode I), (2) the greater the phylogenetic effects on the diversity examined, (3) the older the origins of the biotas studied, and (4) the more complicated the historical explanations for the biotic composition. The stingray example illustrates the importance that spatial scale may have on historical biogeographic explanations.

Recent discussions of the literature and methods employed in historical biogeography (Wiley 1988a,b; Cracraft 1988; Noonan 1988) have warned against approaches that eliminate or minimize the effects of any evolutionary process a priori. After all, although "corroboration" and "refutation" are part of the scientific process, the attraction of that process lies beyond hypothesis testing in the realm of discovery. Based on the preceding examples, it is evident that there are historical components in the current distributions of many groups of species. It is also evident that evolutionary independence in terms of dispersal and speciation can be manifested within a single historical sequence of area relationships. Not surprisingly then, the geographical distribution patterns of species and clades have apparently been molded by the evolutionary interactions among a variety of historical and nonhistorical processes. We hope we have demonstrated that the methodology presented herein is sensitive to these diverse influences.

For those of you with a taste for more-complicated examples there are numerous studies available: for example, North and Central American coleopteran insects (Whitehead 1972, 1976; Noonan 1988; Liebherr 1988); neotropical leptodactylid frogs (Lynch 1975); a variety of fish groups to examine Caribbean biogeography (Rosen 1975); fossil and recent gars (Wiley 1976); African caddisflies (Morse 1977); Central American poeciliid fishes (Rosen 1979; see also Zandee and Roos 1987; Funk and Brooks 1990); neotropical microteiid lizards (Presch 1980); the southern beeches (*Nothofagus*: Humphries 1981); cyprinodontiform fishes worldwide (Parenti 1981); fishes, frogs, turtles, birds, insects, plants, and marsupials to examine the relationships of biotas on North America, South America, Europe, Australia, and New Zealand (Patterson 1981); Australian birds (Cracraft 1982a, 1983a, 1986); elements of the Central American herpetofauna (Savage 1982); neotropical gymnopthalmid lizards (Hillis 1985); xantusiid lizards (Crother, Miyamoto, and Presch 1986); Indo-Pacific cicadoid insects (Duffels 1986); the high Andean herpetofauna (Lynch 1986); members of *Eucalyptus* (Ladiges and Humphries 1986; Ladiges, Humphries, and Brooker 1987); Indo-Pacific

mirid (Heteroptera) insects (Schuh and Stonedahl 1986); harpacticoid copepods associated with hermit crabs (Ho 1988); some cyprinid fishes from West
Africa (Howes and Teugels 1989); lunulate sand dollars (*Mellita* spp.: Harold
and Telford 1990). There are also other examples, many cited in Wiley
(1988b) and some presented in other parts of this book. This list is by no
means exhaustive, but it is a good starting point!

Cospeciation among Ecological Associates: How Did These Particular Species Come to Be Associated with One Another?

In the preceding sections we discussed a variety of methods used to
differentiate between historical and nonhistorical components in *the distribution patterns of extant organisms* (cospeciation with respect to geography).
In this section we will explore the historical (cospeciation) and nonhistorical
(departures from cospeciation) components in *the patterns of ecological associations of organisms*. We will begin with studies based on a variety of
different groups, each of which exhibits a different type of ecological association. These examples will highlight two important points: the methods of
historical analysis for ecological associations parallel the methods of historical biogeographic analysis (only here we use hosts rather than areas as
"taxa"), and these methods are not constrained by particular types of associations. We will then turn our attention to four groups of flatworms parasitizing
vertebrates. These examples, drawn from members of the monophyletic subclass called the Cercomeromorphae, are presented to demonstrate that it is
possible to assess macroevolutionary patterns of cospeciation among members of large clades.

Sharks and rays with copepods up their noses

Deets (1987) presented a phylogenetic systematic study of a monophyletic
group of seven species of copepods that attach themselves with their large,
prehensile second antennae to the nasal lamellae of a variety of sharks and
stingrays. His study produced a single phylogenetic tree for the genus *Kroeyerina* plus *Prokroyeria meridionalis* (the sister species to *Kroeyerina* + *Kroyeria*), based on ninety-one characters with a consistency index of 92.86%
(fig. 7.37).

1. Table 7.20 is the data matrix produced when the binary codes for the
parasite species are listed with their associated hosts.

2. Phylogenetic analysis of the hosts based on the phylogenetic tree for the
copepod parasites (using the binary codes listed in table 7.20) produces a
single host cladogram with a consistency index of 100% (fig. 7.38).

So far everything should look familiar. Instead of using the phylogenetic
relationships of organisms to reconstruct the historical relationships of geo-

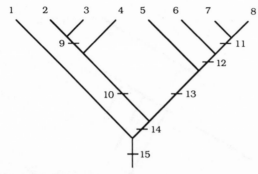

Fig. 7.37. Phylogenetic tree for parasitic copepods *Prokroyeria meridionalis* and seven species of *Kroeyerina*, coded for cospeciation analysis. *1* = *P. meridionalis; 2* = *K. mobulae; 3* = *K. nasuta; 4* = *K. deborahae; 5* = *K. elongata; 6* = *K. scottorum; 7* = *K. cortezensis; 8* = *K. benzorum.*

Table 7.20 Matrix listing chondrichthyan hosts and binary codes for the phylogenetic relationships of the copepods *Prokroyeria meridionalis* and species of *Kroeyerina*.

Host	Parasite	Binary Code
Callorhinchus callorhynchus	1	100000000000001
Mobula japonica	2	010000001100011
Mobula lucasana	2	010000001100011
Dasyatis centroura	3	001000001100011
Rhinobatus productus	4	000100000100011
Prionace glauca	5	000010000000111
Galeocerdo cuvier	5	000010000000111
Sphyrna lewini	6	000001000001111
Sphyrna zygaena	6	000001000001111
Carcharhinus falciformis	7	000000100011111
Isurus oxyrhynchus	8	000000010011111
Alopias vulpinus	8	000000010011111

graphical areas (new area cladogram), we are using the phylogenetic relationships of one group of organisms to reconstruct the historical relationships of another group of organisms (new "associate" or "host" cladogram). In this way, we get a picture of the histories of particular ecological associations. However, we are missing a critical piece of information in this example, a phylogenetic tree for the hosts, the equivalent of an area cladogram based on geological evidence. This is an important component of an ecological association study. Although the consistency index for the host cladogram is 100%, the relationships of the hosts indicated by the parasite data do not necessarily reflect the "actual" host phylogeny. The position of the ratfish *Callorhinchus callorhynchus* in figure 7.38 agrees with the hypothesis that chimaeroids are the sister group of the elasmobranchs (sharks and rays). Likewise, *Mobula,*

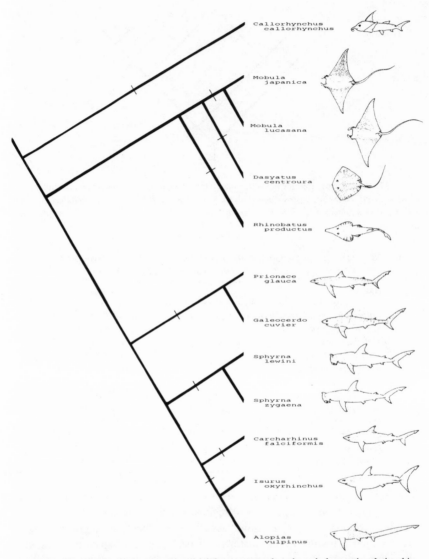

Fig. 7.38. Host cladogram for some chondrichthyan groups, based on phylogenetic relationships of parasitic copepods *Prokroyeria meridionalis* and species of *Kroeyerina*. (From Deets 1987.)

Dasyatis and *Rhinobatus* are "rays," and their relative relationships in figure 7.38 also agree with current estimates of elasmobranch phylogeny. The other hosts are all "lamnoid" (*Prionace, Galeocerdo,* and *Sphyrna*) and "carchar-hinoid" (*Carcharhinus, Isurus,* and *Alopias*) sharks. Each of those groups is considered monophyletic on the basis of current taxonomy, but the parasite data support a paraphyletic status for the "lamnoids." In the absence of a phylogenetic tree for the hosts, or phylogenetic trees for other parasites inhabiting the same elasmobranchs, we are left with some inconclusive portions of this study. Remember, the ultimate goal in cospeciation studies is to delineate the historical and nonhistorical components of biological association and distribution patterns. This can only be accomplished by comparing the new host (area) cladogram with the "actual" historical relationships of the hosts (areas).

Monkeys and mites

O'Connor (1988) presented a phylogenetic systematic analysis of seven species of psoroptid mites (subfamily Cebalginae), whose members inhabit the hair follicles and fur of New World monkeys. His phylogenetic tree for six genera was based on seventeen characters and had a consistency index of 100% (fig. 7.39).

1. Table 7.21 is the data matrix produced when the binary codes for the parasite species are listed with their associated hosts.

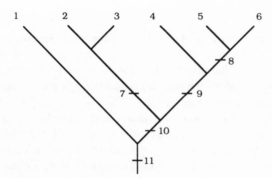

Fig. 7.39. Phylogenetic tree for six genera of cebalgine mites inhabiting New World monkeys, coded for cospeciation analysis. *1 = Procebalges; 2 = Schizopodalges; 3 = Alouattalges; 4 = Cebalgoides; 5 = Cebalges; 6 = Fonsecalges.*

Table 7.21 Matrix listing New World monkeys and binary codes for the phylogenetic relationships of the cebalgine mites.

Host	Mite	Binary Code
Pithecia	1	1000000001
Lagothrix	2	0100010011
Alouatta	3	0010010011
Sanguinus	4	0001000111
Cebus	4, 5	0001101111
Saimiri	6	0000101111
Callithrix	6	0000101111

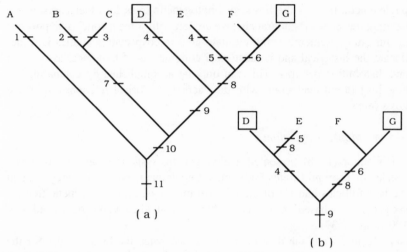

Fig. 7.40. Host cladograms for seven genera of New World monkeys, based on phylogenetic relationships of genera of cebalgine mites inhabiting them. *A* = *Pithecia; B* = *Lagothrix; C* = *Alouatta; D* = *Sanguinus; E* = *Cebus; F* = *Saimiri; G* = *Callithrix.*

2. Phylogenetic analysis of the data matrix produced two equally parsimonious host cladograms, both with consistency indices of 91.67% (fig. 7.40). In both cases, the relative phylogenetic relationships among the Cebidae genera postulated by the parasite relationships are congruent with most current taxonomic hypotheses. That is, *Pithecia* (A) is the sister group of *Lagothrix* (woolly monkeys), *Alouatta* (howler monkeys), *Cebus* (capuchins), and *Saimiri* (squirrel monkeys); *Lagothrix* (B) and *Alouatta* (C) are each other's closest relatives; and *Cebus* (E) and *Saimiri* (F) are each other's closest relatives. However, *Sanguinus* (D) and *Callithrix* (G) are members of the Callithricidae (marmosets and tamarins), a monophyletic group distinct from the Cebidae; therefore, their placement on the host cladogram represents two episodes of host-switching by species within this group of mites. O'Connor noted that there was some ambiguity about the exact sequence of host switching. One host cladogram (fig. 7.40a) supports an interpretation that *Sanguinus* may have picked up its mite from a species ancestral to *Cebus* and *Saimiri;* the other (fig. 7.40b) supports the notion that *Sanguinus* picked up its mite from *Cebus,* and *Callithrix* picked up its mite from *Saimiri.* Problems in interpreting the exact sequence of host switching parallel problems in determining the sequence of geographic dispersal in historical biogeographic studies.

Cospeciation patterns in multispecies ecological associations can be analyzed in the same way as historical biogeographical studies using multiple groups. Consider the following example based on the association between great apes and some of the nematodes that infect them.

Nematode parasites of great apes

Nematodes are small parasitic worms whose members swell the ranks of one of the most species-rich phyla on earth. Within this group, hookworms are a notorious bane of humans in all corners of the world, creating medical and economic havoc wherever they appear. Pinworms reside at the other end of the medical spectrum, making up for their innocuous nature by their ubiquitous distribution; over 500 million people are estimated to be infected with one species alone, *Enterobius vermicularis* (Schmidt and Roberts 1988). Needless to say, although much research has focussed on the medical relationship between *Homo sapiens* and nematodes, these worms also inhabit other organisms. Interestingly for students of historical ecology both nematode groups are associated with a wide variety of other great ape species.

Phylogenetic trees for two groups of nematodes inhabiting primates have been presented for (1) the pinworm genus *Enterobius* (thirty-two characters, with a consistency index of 89%: Brooks and Glen 1982) and (2) the hookworm genus *Oesophagostomum* (twenty-five characters, with a consistency index of 80%: Glen and Brooks 1985). Glen and Brooks (1986) evaluated the degree of cospeciation for this multispecies assemblage. In each group of worms, there is a clade of species almost exclusively restricted to great apes,

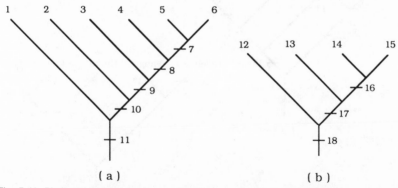

(a)   (b)

Fig. 7.41. Phylogenetic trees for two groups of nematode parasites inhabiting great apes. (*a*) Members of the hookworm subgenus *Oesophagostomum (Conoweberia): 1 = O. (C.) blanchardi; 2 = O. (C.) raillieti; 3 = O. (C.) ovatum; 4 = O. (C.) pachycephalum; 5 = O. (C.) stephanostomum; 6 = O. (C.) ventri.* Species 6 is known only from jaguars in South America, and species 4 is found only in cercopithecid primates; therefore, they represent instances of host switching. (*b*) Members of the pinworm genus *Enterobius: 12 = E. vermicularis; 13 = E. buckleyi; 14 = E. anthropopitheci; 15 = E. lerouxi.*

including humans. Brooks (1988a) presented a simplified version of the study, examining the portions of the two trees composing the species inhabiting the great apes (fig. 7.41).

1. Table 7.22 is the data matrix produced when the binary codes for the parasite species are listed with their associated great ape hosts.

Table 7.22 Matrix listing genera of great apes and the binary codes for the phylogenetic relationships of two groups of nematode parasites.

Host	Parasite	Binary Code
Hylobates	1, 2, 3, 12	111000001111000001
Pongo	1, 13	100000001110100011
Homo	5, 12	000010111111000001
Pan	5, 14	000010111110010111
Gorilla	5, 15	000010111110001111

2. In this case, phylogenetic analysis of the data matrix produces two equally parsimonious host cladograms, one of which (fig. 7.42a) is congruent with the most widely accepted hypothesis of great ape phylogeny. The other (fig. 7.42b) groups *Hylobates* (gibbons) and *Pongo* (orangutans) together, something no primate biologist has ever suggested. Therefore, we will accept

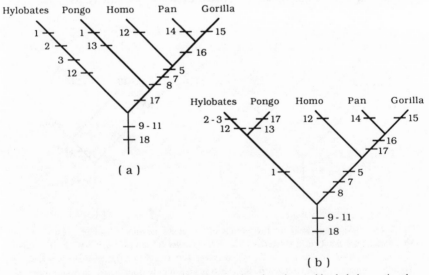

Fig. 7.42. Host cladograms for genera of great apes, based on the combined phylogenetic relationships of two groups of nematodes that parasitize them.

figure 7.42a as the host cladogram for this cospeciation study. The consistency index is less than 100% (88.9%), due to two postulated cases of host switching, one between *Hylobates* and *Pongo* (hookworm taxon 1: *Oesophagostomum (C.) blanchardi*) and the other between *Hylobates* and *Homo* (pinworm taxon 12, the ubiquitous *Enterobius vermicularis*). It appears, then, that these nematodes ape the phylogeny of their hosts quite closely.

Historical Congruence: "Real" or Fortuitous?

Incongruence between a "host cladogram" (reconstructed from the associate's phylogenetic relationships) and the hosts' phylogenetic tree (reconstructed from host characters) is attributed to colonization events by the associate species. Now, is the reverse situation, congruence between the "host cladogram" and host phylogenetic tree, always an indication of cospeciation? The preliminary answer to this question is no, because, theoretically at least, it is possible that the members of an associate group evolved as a result of sequential host switching that coincidentally mirrored the phylogenetic relationships of the hosts (see chapter 8). In such a case we would find congruence between host and associate phylogenies that was not indicative of a historical association between the groups. This is an important consideration because it clouds the distinction between historical and nonhistorical influences on the evolution of ecological associations. In recent years, there has been some concern on the part of parasite ecologists (e.g., Holmes and Price 1980) that patterns of congruence between host and parasite phylogenies might not always imply cospeciation. Specifically, if we find a situation in which the phylogenetic relationships of a group of parasites are congruent with the relative phylogenetic relationships of their hosts, but the hosts that are actually inhabited represent only a small portion of the members of the host clade(s), how do we know that it is not simply a fortuitous outcome of host switching?

Brooks and Bandoni (1988) proposed a research protocol for distinguishing ecological associations that represent relictual episodes of cospeciation from those that had been assembled by sequences of host switching that fortuitously mirrored host phylogenetic relationships. They proposed that the first step out of the maze is the recognition that the associations are less diverse than expected. This is accomplished by asking two questions: Are the associations depauperate with respect to associations exhibited by sister groups? Is the depauperate group old enough to have achieved a relictual status? Notice that this is simply another application of Mayden's (1985) criteria for studying species diversity within clades (see chapter 4). The methods of historical biogeography are well suited to investigating the second question.

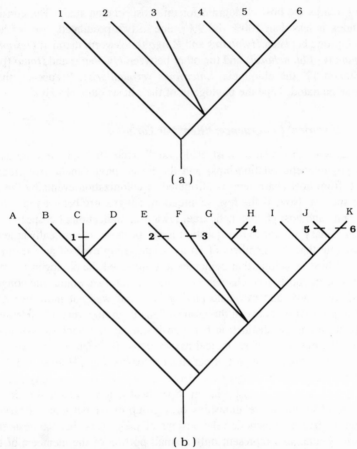

Fig. 7.43. Phylogenetic trees for the digenean family Liolopidae and its vertebrate hosts. (*a*) Liolopid genera: *1* = *Liolope copulans; 2* = *Moreauia; 3* = *L. dollfusi; 4* = *Dracovermis; 5* = *Harmotrema; 6* = *Helicotrema.* (*b*) Eleven major groups of tetrapod vertebrates hosting liolopids: *A* = Sarcopterygii (lungfishes, coelacanths); *B* = Anura (frogs); *C* = Caudata (salamanders); *D* = Gymnophiona (caecilians); *E* = Mammalia (mammals); *F* = Chelonia (turtles); *G* = Aves (birds); *H* = Crocodilia (crocodilians); *I* = Rhynchocephalia (tuatara); *J* = Ophidia (snakes); *K* = Sauria (lizards). *B–D* = Amphibia; *E–K* = Amniota; *F–K* = Reptilomorpha; *G* + *H* = Archosauria; *J* + *K* = Squamata.

Numerical relicts

Let us consider the Liolopidae, a trematode family, comprising fewer than fifteen species allocated to five genera, inhabiting the intestines of a variety of tetrapod vertebrates (fig. 7.43). Mapping the phylogeny of the trematodes onto the phylogeny of the hosts reveals complete congruence between these two markers of evolutionary history, but closer examination of the distribution patterns casts some doubts on our original interpretation. Five of the eleven major groups of tetrapods do not host liolopids, and the vast majority of species within the inhabited tetrapod groups are also not associated with liolopids. It is therefore tempting to explain the phylogenetic "fit" as a coincidence and invoke sequential host switching. However, it is possible that the congruence reflects a long-standing association in which one or more members are relictual groups. Is there a way out of this maze?

The liolopids are much less diverse than their sister group, the strigeoid digeneans (comprising the families Cyathocotylidae + Proterodiplostomatidae + Strigeidae), and exhibit biogeographic patterns coinciding with the breakup of Pangaea (fig. 7.44). Hence, it would appear that the liolopids are a very old and very depauperate group. Biological data support the ancient picture painted by the geographical distribution patterns. Liolopids are generally associated with a wide range of archaic vertebrate hosts, including cryptobranchid salamanders, sideneck turtles, the duck-billed platypus, crocodilians, and iguanid lizards. It appears that all the evidence collected to date implies that they are relicts of some sort. According to Simpson's (1944) definitions (see chapter 4), the liolopids are either **numerical relicts,** the few remaining survivors of a group that was once more diverse, or **phylogenetic relicts,** "living fossil" species that originated a long time ago and have never become very diverse. In chapter 5 we discovered that ecological and behavioral diversification tend to be phylogenetically conservative. Given this, it may be possible to distinguish numerical relictual associations from phylogenetic relictual associations, based on the degree of ecological diversification among the associations. Since liolopids inhabit freshwater, estuarine, and terrestrial hosts, indicating a fair amount of diversification in life cycle characters, we suggest that current associations represent the survivors of a group of associations that were once more diverse; that is, the liolopids are numerical relicts.

Phylogenetic relicts

There are two types of phylogenetic relicts in ecological associations. The first type involves cases in which neither of the associated groups ever became very diverse. The gyrocotylid flatworms are prime candidates for this category. They are ecologically conservative, being restricted as adults entirely to the spiral intestines of chimaeroid fishes, which are themselves phylogenetic relicts. Like their hosts, the gyrocotylids are less species-rich than their sister group (in this case the amphilinideans + the tapeworms). In the second case one of the associates becomes highly diverse while the other does not. For example, the amphilinidean flatworms are much less diverse than their sister group, the true tapeworms, and also exhibit a high degree of ecological uni-

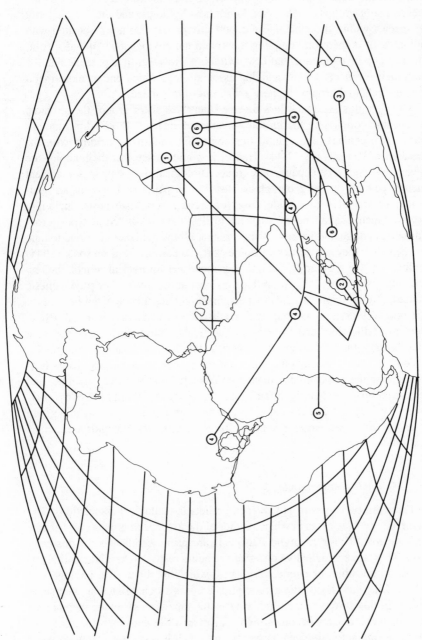

Fig. 7.44. Map depicting continental configurations prior to the breakup of Pangaea beginning in the early Cretaceous, showing distribution of liolopid digeneans connected by their phylogenetic relationships. *1 = Liolope copulans; 2 = Moreauia; 3 = L. dollfusi; 4 = Dracovermis; 5 = Harmotrema; 6 = Helicotrema.* (From Brooks and Bandoni 1988.)

formity. They all occur as adults in the body cavity of their hosts, and six of the eight known species inhabit freshwater ray-finned fishes. We have shown biogeographic evidence of their antiquity (see the beginning of this chapter), and we will show a high degree of congruence between amphilinidean phylogeny and the relative phylogenetic relationships of their hosts (see the next section). Amphilinids differ from gyrocotylids because they do not inhabit hosts that are themselves phylogenetic relicts; therefore, they are an example of a group that failed to become as diverse as the other member of its association.

Spurious congruence due to host switching

Entepherus laminipes is a parasitic copepod species inhabiting the branchial filters of mantid stingrays, including the manta ray (*Manta birostris*), the spinetail mobula (*Mobula japonica*), the vacatilla (*Mobula tarapacana*), the smoothtail mobula (*Mobula thurstoni*), and the devilfish (*Mobula hypostoma*) from the Sea of Cortez, as well as *Mobula rochebrunei* from Madagascar. It is the sister species of four other genera: *Leutkenia,* occurring on louvars, epipelagic teleostean fishes of the genus *Luvarus;* and *Philorthagoriscus, Orthagoriscola,* and *Cecrops,* all of which inhabit the ocean sunfish (*Mola mola*).

Phylogenetic systematic analysis (Benz and Deets 1988) of the family Cecropidae, based on forty characters, produced a single phylogenetic tree with a consistency index of 90.9% (fig. 7.45).

1. Table 7.23 is the data matrix produced when the binary codes for the parasite species are listed with their associated hosts.

2. Phylogenetic analysis of the data matrix produced a host cladogram with a consistency index of 100%. The host cladogram (fig. 7.46) depicts the

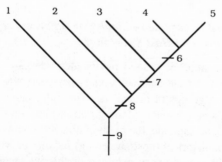

Fig. 7.45. Phylogenetic tree for five genera of parasitic copepods, representing the family Cecropidae, inhabiting mesopelagic fish, with internal branches numbered for cospeciation analysis. *1 = Entepherus; 2 = Luetkenia; 3 = Philorthagoriscus; 4 = Orthagoriscola; 5 = Cecrops.*

Table 7.23 Matrix listing hosts for members of the copepod family Cecropidae and the binary codes indicating the phylogenetic relationships of the parasites.

Hosts	Parasite	Binary Code
Mobula	1	100000001
Luvarus	2	010000011
Mola mola	3, 4, 5	001111111

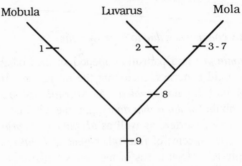

Fig. 7.46. Host cladogram based on phylogenetic relationships of parasitic copepods of the family Cecropidae.

elasmobranch hosts (*Mobula* spp.) as the sister groups of the louvar and the ocean sunfish. This is congruent with the relative phylogenetic relationships of the hosts. However, among the vast numbers of elasmobranchs and teleosts, only these three taxa are known to host these copepods. Additionally, the hosts that are inhabited are epipelagic organisms. This is an independently derived trait within each of the host groups. We believe, therefore, that the current host-parasite associations result from a series of host-switching events.

Do Related Groups Show Similar or Different Proportions of Cospeciation and Host Switching?

Once we have established a solid phylogenetic data base for single-clade associations, we can expand our evolutionary perspective to cospeciation patterns among related groups of organisms. This information will allow us to ask whether members of monophyletic groups within a larger clade have all been influenced to the same degree by the interaction between historical (cospeciation) and nonhistorical (host-switching) factors, or whether each group represents a unique evolutionary outcome of this interaction. Such investigations are already underway for the Cercomeromorphae, the group of parasitic platyhelminths containing the Eucestoda, or true tapeworms; their sister

group the Amphilinidea; the Gyrocotylidea, which is the sister group of the
Eucestoda plus Amphilinidea; and the Monogenea, which is the sister group
of the other three taxa (Brooks, O'Grady, and Glen 1985b; Brooks 1989a,b;
fig. 7.47).

Monogeneans and catfish

The monogeneans are among the smallest, most host-specific, and most
diverse groups of parasitic flatworms. They enjoy the dubious distinction of
being an extremely well studied parasitic group, not because of their inherent
beauty as living organisms, but because of their negative impact on commer-
cial fisheries projects. Monogeneans exhibit direct development and have
generation times much shorter than those of their vertebrate hosts; hence, it
is possible for a single individual to establish a viable deme, and produce
colonizing offspring, while residing on one host. Additionally, these flatworms are easily transferred between hosts; all that is required is that the
hosts come into contact with each other from time to time. Based on these
life cycle characteristics, then, the evolutionary diversification of this group
has traditionally been consigned to the realm of sympatric speciation via
widespread host switching.

Members of the genus *Ligictaluridus* (five species) inhabit the gills of a
variety of ictalurid catfish hosts in North America. Klassen and Beverly-
Burton (1987) presented a phylogenetic systematic analysis of *Ligictaluridus,*
based on ten characters, that produced a single tree with a consistency index
of 100% (fig. 7.48).

1. Table 7.24 is the data matrix produced when the binary codes for the
parasite species are listed with their associated hosts.

2. Phylogenetic analysis of the data matrix produced one host cladogram
with a consistency index of 100% (fig. 7.49). The perfect fit of the data to
the cladogram is due to the marked host specificity of the parasites. However,

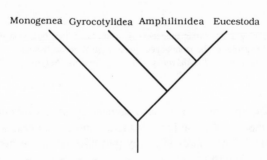

Fig. 7.47. Phylogenetic tree for the four major groups of parasitic flatworms composing the
Cercomeromorphae.

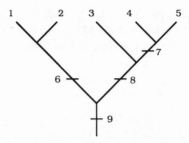

Fig. 7.48. Phylogenetic tree for five species of *Ligictaluridus*, with internal branches numbered for cospeciation analysis. *1* = *L. pricei; 2* = *L. monticellii; 3* = *L. posthon; 4* = *L. floridanus; 5* = *L. mirabilis.*

Table 7.24 Matrix listing ictalurid catfish hosts (by subgenus) and the binary codes for the phylogenetic relationships of species of the monogenean genus *Ligictaluridus.*

Host	Parasite	Binary Code
Ictalurus (Ameiurus)	1, 2	110001001
Ictalurus (Ictalurus)	4, 5	000110111
Noturus (Noturus)	3	001000011
Noturus (Schilbeodes)	1	100001001

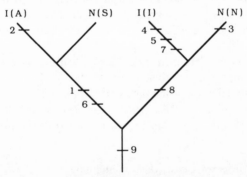

Fig. 7.49. Host cladogram for four subgenera of ictalurid catfish, based on the phylogenetic relationships of five species of the monogenean genus *Ligictaluridus* that parasitize them. *I(I)* = *Ictalurus (Ictalurus); I(A)* = *I. (Ameiurus); N(N)* = *Noturus (Noturus); N(S)* = *N. (Schilbeodes).*

pronounced host specificity does not guarantee phylogenetic congruence, as evidenced by the fact that the host cladogram mixes *Ictalurus* and *Ameiurus* with *Noturus* and *Schilbeodes* in a way that does not correspond to current hypotheses of relationships.

3. When the parasite phylogeny data are mapped onto either of two phylogenetic hypotheses for the ictalurid taxa (fig. 7.50) the "fit" between the

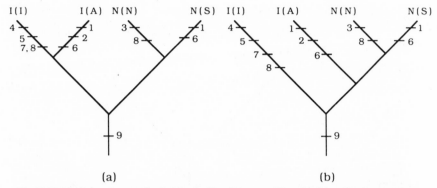

Fig. 7.50. Phylogenetic trees for four ictalurid catfish taxa. *I(I)* = *Ictalurus (Ictalurus); I(A)* = *I. (Ameiurus); N(N)* = *Noturus (Noturus); N(S)* = *N. (Schilbeodes).* (*a*) Phylogenetic tree redrawn from Taylor 1969. (*b*) Phylogenetic tree redrawn from Lundberg 1970.

parasite data and the host phylogeny is reduced to 75%. A number of host-switching scenarios may be postulated to explain the given host-parasite relationships.

Now, reexamine table 7.24. Notice that the bullheads, *Ameiurus*, and the channel catfish, *Ictalurus*, are associated with more than one species of parasite. This situation is analogous to the problem arising in biogeography when more than one member of the same clade occurs in one area (see table 7.6).

1. So, paralleling the biogeographical resolution to the problem, let us recode *Ictalurus* and *Ameiurus* as separate taxa for each parasite species (table 7.25).

2. Phylogenetic analysis of this matrix produces one tree with a consistency index of 100% (fig. 7.51).

This new host cladogram supports the following explanation. The common ancestor of *Ligictaluridus* (species 9) parasitized the common ancestor of *Ictalurus*. The separation of the two major clades within *Ligictaluridus* is

Table 7.25 Matrix listing ictalurid catfish taxa and the binary codes for the members of the monogenean genus *Ligictaluridus* that inhabit them, with each host group listed separately for each occurrence of a member of the parasite group.

Host	Parasite	Binary Code
Ictalurus (Ameiurus)	1	100001001
Ictalurus (Ameiurus)	2	010001001
Ictalurus (Ictalurus)	4	000100111
Ictalurus (Ictalurus)	5	000010111
Noturus (Noturus)	3	001000011
Noturus (Schilbeodes)	1	100001001

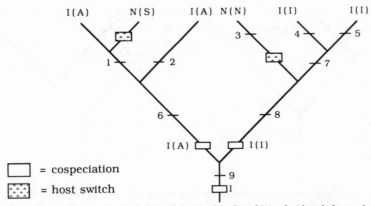

Fig. 7.51. Host cladogram for ictalurid catfish taxa, based on *Ligictaluridus* phylogenetic relationships and listing each host group separately for each occurrence of a member of the parasite group. *I(I)* = *Ictalurus (Ictalurus); I(A)* = *I. (Ameiurus); N(N)* = *Noturus (Noturus); N(S)* = *N. (Schilbeodes).*

congruent with the divergence of the channel cats and the bullheads from their common ancestor. These associations were maintained through the speciation events producing *L. pricei and L. monticellii* (associated with bullheads) and *L. floridanus and L. mirabilis* (associated with channel cats). The occurrence of *L. pricei* on the madtom *Noturus (Schilbeodes)* and of *L. posthon* on the stone catfish *Noturus (Noturus)* represent two incidents of host switching. Thus, the majority of the evolutionary divergence of *Ligictaluridus* occurred in association with the ancestral host group. Interestingly, ecological evidence can shed some light on one of the proposed host switches. Both bullheads and madtoms prefer shallow-water, muddy-bottomed, heavily vegetated habitats (Klassen and Beverly-Burton 1987). If these fishes also occur together, then the transfer of *L. pricei* between these unrelated hosts is relatively easy to envision. The transfer of some members of ancestor 8 from a channel cat to a stone cat and subsequent speciation of *L. posthon* in association with a stone cat is more problematical. The fishes do not overlap in their ecological preferences today, stone cats living on the bottom of rapidly flowing deep streams and rivers, while channel cats are members of the pelagic community in rivers and lakes. Further data are required to explain this anomalous example.

Gyrocotylids and ratfish

The gyrocotylideans are a small group of parasitic flatworms, often resembling pieces of sea lettuce, that live as adults only in the spiral intestines of chimaeroid fishes (ratfish). Very little is known of their biology or natural history, beyond the intriguing phenomenon that nearly 90% of all *individual* ratfish examined are infected with gyrocotylids, and almost 90% of those are infected with exactly two specimens of gyrocotylids. In addition, four of the six species of ratfish hosts are inhabited by more than one species of gyrocotylid (three of them host two species and one hosts three). Bandoni and Brooks (1987b) presented a phylogenetic analysis of the ten gyrocotylidean species, based on twenty-four characters that produced the phylogenetic tree, with a consistency index of 87.5%, shown in figure 7.52.

1. Table 7.26 is the data matrix produced when the binary codes for the parasite species are listed with their associated hosts.

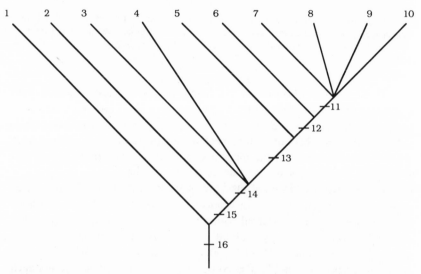

Fig. 7.52. Phylogenetic tree for ten species of gyrocotylideans, with internal branches numbered for cospeciation analysis. *1 = Gyrocotyle rugosa; 2 = G. confusa; 3 = G. parvispinosa; 4 = G. nybelini; 5 = G. abyssicola; 6 = G. maxima; 7 = G. major; 8 = G. nigrosetosa; 9 = G. urna; 10 = G. fimbriata.*

Table 7.26 Matrix listing chimaeroid fish hosts for gyrocotylid flatworms and the binary codes for the phylogenetic relationships of the parasites.

Host	Parasite	Binary Code
Callorhinchus callorhynchus	1, 6	1000010000011111
Chimaera monstrosa	2, 4, 9	0101000010111111
Chimaera phantasma	10	0000000001111111
Hydrolagus colliei	3, 10	0010000001111111
Hydrolagus affinis	5, 7	0000101000111111
Hydrolagus ogilbyi	8	0000000100111111

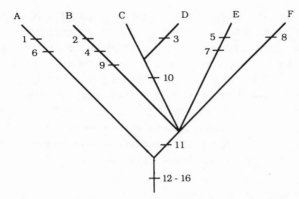

Fig. 7.53. Host cladogram for six species of chimaeroid fish, based on the phylogenetic relationships of the gyrocotylid flatworms that inhabit them. *A* = *Callorhinchus callorhynchus; B* = *Chimaera monstrosa; C* = *Chimaera phantasma; D* = *Hydrolagus colliei; E* = *H. affinis; F* = *H. ogilbyi.*

2. Phylogenetic analysis of the data matrix produces a host cladogram with a consistency index of 100% (fig. 7.53). However, this cladogram links *Chimaera phantasma* and *Hydrolagus colliei* together, rather than placing them with the other members of their respective genera. In addition, it recognizes only a large polytomy for the members of *Chimaera* and *Hydrolagus*. And finally, note that four of the six ancestral taxa for the parasite group (i.e., characters 12–15) are placed at the base of the host cladogram, rather than close to their descendant contemporaneous species. This indicates that massive host switching has occurred within the ratfish. For example, *Callorhinchus callorhynchus* hosts gyrocotylid species 1 and 6, *Chimaera monstrosa* hosts species 2, 4, and 9, *Hydrolagus colliei* hosts species 3 and 10, and *H. affinis* hosts species 5 and 7. The co-occurring parasite species are not each other's closest relatives and represent one relatively primitive and one (or two) relatively derived species. It is the secondary presence of the derived species in the relatively primitive hosts that accounts for the codes for ancestors 12–15 being at the base of the tree.

Amphilinids, fish, and turtles

The sister group of the true tapeworms, the Amphilinidea, is an esoteric group of flatworms living in the body cavities of freshwater and estuarine ray-finned fishes and in one species of freshwater turtle. We have already discussed the biogeography of this group (see, e.g., fig. 7.6). Figure 7.54 depicts the phylogenetic tree for the eight species of amphilinids (again).

1. Table 7.27 is the data matrix produced when the binary codes for the parasite species are listed with their associated hosts.

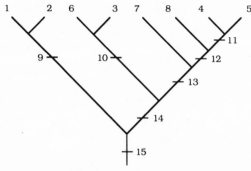

Fig. 7.54. Phylogenetic tree for eight species of amphilinid flatworms, with internal branches numbered for cospeciation analysis. *1 = Amphilina foliacea; 2 = A. japonica; 3 = Gigantolina elongata; 4 = Schizochoerus liguloideus; 5 = S. africanus; 6 = G. magna; 7 = S. paragonopora; 8 = S. janickii.*

Table 7.27 Matrix listing hosts for amphilinid flatworms and the binary codes for the phylogenetic relationship of the parasites.

Host	Parasite	Binary Code
Acipenseriformes	1	100000001000001
Huso huso		
Acipenser spp.[a]		
Acipenseriformes	2	010000001000001
Huso dauricus		
Acipenser medirostris		
Acipenser schrenki		
Acipenser transmontanus		
Siluriformes	7	000000100000111
Aorichthys seenghala		
Aorichthys aor		
Bagarius bagarius		
Osteoglossiformes	8	000000010001111
Arapaima gigas		
Osteoglossiformes	4	000100000011111
Arapaima gigas		
Osteoglossiformes	5	000010000011111
Gymnarchus niloticus		
Perciformes	3	001000000100011
Plectorhynchus niger		
Plectorhynchus pictus		
Chelonia	6	000001000100011
Chelodina longicollis		

[a]*Acipenser* spp. = *A. baeri, A. guldenstadtii, A. naccari, A. nudiventris, A. ruthensis, A. stellatus,* and *A. sturio.*

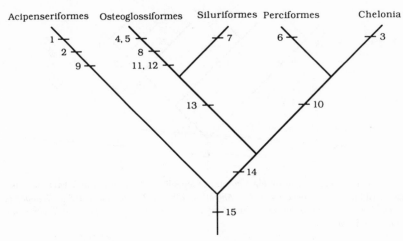

Fig. 7.55. Host cladogram based on phylogenetic relationships of amphilinid flatworm parasites.

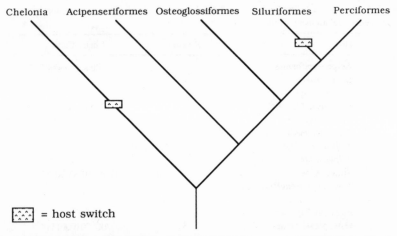

= host switch

Fig. 7.56. Optimization of all eight species of amphilinid flatworms and their phylogenetic relationships onto the phylogenetic tree of their hosts.

   2. Phylogenetic analysis of this data matrix produces one host cladogram with a consistency index of 100% (fig. 7.55). The species of parasites inhabiting acipenseriforms are a monophyletic group, as are those inhabiting osteoglossiforms. However, unless there is something amiss with current theories of evolution, the turtle *Chelodina longicollis* is not the sister group of perciform teleostean fishes, so we must interpret the presence of *Gigantolina elongata* in this turtle as being the result of a host switch. In addition, contrary to the current phylogenetic analysis of the actinopterygians, this clado-

gram places siluriform fishes with the osteoglossiforms rather than with the perciforms; therefore, the presence of *Schizochoerus paragonopora* in a siluriform host must be the result of a host switch as well.

3. Mapping the amphilinidean phylogeny onto the relative phylogenetic relationships for the fish and turtle hosts produces a fit of the parasite and host phylogeny of 75% (estimated by the consistency index; fig. 7.56). This reemphasizes the interaction between cospeciation and host switching (two events) in the evolution of the amphilinids.

Tapeworms and seabirds

The monophyletic group Eucestoda includes the fascinating, diverse, and persistently maligned true tapeworms. Adult tapeworms are widespread throughout vertebrate intestines around the world. They range in size from the minute *Echinococcus multilocularis* (1.2 to 3.7 mm long) in coyotes, dogs, and wolves to *Polygonoporus,* the thirty-meter-long associate of sperm whales (Schmidt and Roberts 1988). Although researchers have tended to concentrate on the medically relevant species, data concerning tapeworm relationships with nonhuman hosts have been accumulating for more than a century. We will investigate this association from the perspective of the colorful and interesting Alcidae (Charadriiformes), the family of seabirds including puffins, murres, and guillemots.

As you might expect from the name, many alcids are routinely parasitized by tapeworms of the genus *Alcataenia.* This genus is currently composed of nine species, seven of which inhabit alcids and two of which inhabit sea gulls

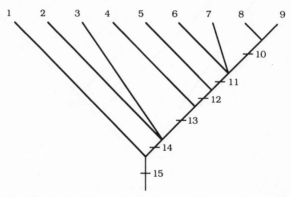

Fig. 7.57. Phylogenetic tree for nine species of the tapeworm genus *Alcataenia,* coded for cospeciation analysis. *1* = *A. larina pacifica; 2* = *A. l. larina; 3* = *A. fraterculae; 4* = *A. cerorhincae; 5* = *A. pygmaeus; 6* = *A. armillaris; 7* = *A. longicervica; 8* = *A. meinertzhageni; 9* = *A. campylacantha.*

(Laridae, also Charadriiformes). Phylogenetic systematic analysis of *Alcataenia* (Hoberg 1986), based on twenty characters, produced a phylogenetic tree with a consistency index of 77% (fig. 7.57).

1. As always, table 7.28 is the data matrix produced when the binary codes for the parasite species are listed with their associated hosts.

Table 7.28 Matrix listing alcid bird hosts and binary codes for the phylogenetic relationships of species of the tapeworm genus *Alcataenia*.

Host	Parasite	Binary Code
Laridae	1, 2	110000000000011
Fratercula	3	001000000000011
Cerorhinca	4	000100000000111
Aethia	5	000010000001111
Uria aalge	6, 7, 8	000001110111111
Uria lomvia	6, 7, 8	000001110111111
Cepphus carbo	9	000000001111111
Cepphus columba	9	000000001111111
Cepphus grylle	9	000000001111111

2. Phylogenetic analysis of the data matrix produces one host cladogram with a consistency index of 100% (fig. 7.58). Once again, this does not necessarily indicate complete cospeciation. For example, the current classification of charadriiform birds suggests that larids are not the sister group of alcids. Because the oldest tapeworm species is found in larids, it appears that

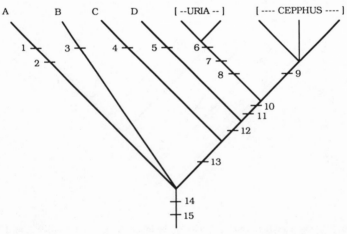

Fig. 7.58. Host cladogram for six larid and alcid bird groups, based on phylogenetic relationships of tapeworms of the genus *Alcataenia* that inhabit them. *A* = Laridae; *B* = *Fratercula; C* = *Cerorhinca; D* = *Aethia*.

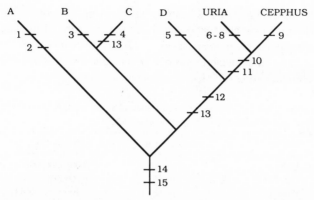

Fig. 7.59. Optimization of phylogenetic information about *Alcataenia* species onto phylogenetic tree of larid and alcid bird hosts. Birds: *A* = Laridae; *B* = *Fratercula*; *C* = *Cerorhinca; D* = *Aethia*. Tapeworms: *1* = *Alcataenia larina pacifica; 2* = *A. l. larina; 3* = *A. fraterculae; 4* = *A. cerorhincae; 5* = *A. pygmaeus; 6* = *A. armillaris; 7* = *A. longicervica; 8* = *A. meinertzhageni; 9* = *A. campylacantha.*

the occurrence of *Alcataenia* in alcids must have involved at least one ancestral host-switching event from sea gulls to alcids (possibly two, because the *Alcataenia* species in *Fratercula* could have been derived from a host-switching event different from the one that resulted in the association between *Alcataenia* ancestor 13 and the ancestor of the remaining alcids). In addition, the current classification of the alcids places *Cerorhinca* and *Fratercula* as sister groups in the same clade, while the parasite-based host cladogram indicates that the two genera are paraphyletic.

3. Optimizing the parasite phylogeny onto the host phylogeny proposed by Strauch (1985) produces a consistency index of 93.75%, reflecting the differences in the positioning of *Cerorhinca* (fig. 7.59).

We are not certain that the differences between the trees is due to host switching, because the phylogeny proposed by Strauch was not produced using phylogenetic systematics; hence, the two trees are not directly comparable. Nevertheless, the host relationships indicated by the phylogenetic relationships of their parasites are very close to the phylogeny proposed by Strauch. Normally this would be cause for celebration. However, there is something amiss here because, combining the codes of the paraphyletic group of species including *Alcataenia armillaris, A. longicervica,* and *A. meinertzhageni* makes it appear that they are a monophyletic group supporting the monophyly of *Uria*. This, in turn, occurs because *Uria* is the sister group of *Cepphus,* which hosts the species of *Alcataenia* missing from the paraphyletic group. Ironically, then, we managed to obtain the "right" answer for the "wrong" reasons.

This example demonstrates another shortcoming of combining species codes when more than one species inhabits the same hosts. This should sound familiar. It is equivalent to the problems encountered when two or more members of the same clade inhabit the same area (see figs. 7.9–7.12).

1. We recode, listing each species separately (table 7.29).

Table 7.29 Matrix listing larid and alcid bird hosts and binary codes for the phylogenetic relationships of species of the tapeworm genus *Alcataenia*.

Host[a]	Parasite	Binary Code
Laridae	1	100000000000001
Laridae	2	010000000000011
Fratercula	3	001000000000011
Cerorhinca	4	000100000000111
Aethia	5	000010000001111
Uria aalge (1)	6	000001000011111
Uria lomvia (1)	6	000001000011111
Uria aalge (2)	7	000000100011111
Uria lomvia (2)	7	000000100011111
Uria aalge (3)	8	000000010111111
Uria lomvia (3)	8	000000010111111
Cepphus carbo	9	000000001111111
Cepphus columba	9	000000001111111
Cepphus grylle	9	000000001111111

[a]Each host is listed separately for each parasite species that inhabits it.

2. Phylogenetic analysis of this new data matrix produces one cladogram, also with a consistency index of 100% (fig. 7.60), which gives us a better picture of the possible history of cospeciation between *Alcataenia* and alcids. Note that despite marked host specificity, there may have been as many as five host-switching events during the evolutionary elaboration of the tapeworm-alcid association. As previously discussed, one or both of the occurrences of the older tapeworm species in the older alcids may be the result of a host switch from larids to alcids (species 3 and 13). If, as Strauch suggested, *Cerorhinca* is the sister group of *Fratercula,* then the presence of *Alcataenia fraterculae* may represent a host switch and subsequent speciation by a population of ancestor 13 in *Cerorhinca.* And finally, the paraphyletic status of *A. armillaris, A. longicervica,* and *A. meinertzhageni* is highlighted by the separation of *Uria* into three groups; however, only two of the three associations between *Alcataenia* and *Uria* need be explained by host switching.

Comments on Cospeciation in an Ecological Context

It is evident from the studies presented in this section that historical components can be found in a variety of specialized ecological associations. It is also evident that considerable evolutionary independence, in terms of host switching, can be manifested within a single historical sequence of host relationships. When ecological associations are examined at the level of the Cercomeromorphae clade, cospeciation and host switching each account for about half of the observed patterns. However, the degree of cospeciation varies considerably among closely related groups within members of this clade, as does the importance of host switching. For example, the influence of history is (1) strong in the association between amphilinids and fish and turtles (two host switches in fifteen speciation events, or 86.7% cospeciation); (2) strong in the association between monogeneans and catfish (two host switches in nine speciation events, 77.8% cospeciation); (3) moderate in the association between *Alcataenia* tapeworms and alcid birds (five host switches in fifteen speciation events, 66% cospeciation); and (4) weak in the association between gyrocotylids and ratfish (about 50% cospeciation). The picture is even more complicated when we examine the taeniids, highly host-specific tapeworms that inhabit a variety of carnivores, including mustelids, canids, and felids. A phylogenetic systematic analysis of fifteen taeniid species reported virtually no congruence between host and parasite phylogenies (Moore and Brooks 1987). This study was based on nineteen characters and produced

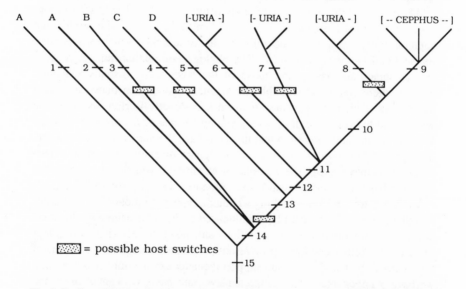

Fig. 7.60. Host cladogram for larid and alcid birds, based on phylogenetic relationships of members of the tapeworm genus *Alcataenia*, with each host group listed separately for each species of *Alcataenia* that inhabits it. Birds: *A* = Laridae; *B* = *Fratercula*; *C* = *Cerorhinca*; *D* = *Aethia*. Tapeworms: *1* = *Alcataenia larina pacifica*; *2* = *A. l. larina*; *3* = *A. fraterculae*; *4* = *A. cerorhincae*; *5* = *A. pygmaeus*; *6* = *A. armillaris*; *7* = *A. longicervica*; *8* = *A. meinertzhageni*; *9* = *A. campylacantha*.

four different trees, each with a consistency index of 38%. However, because of the paucity of characters and the ambiguity of the results (multiple trees with a low consistency index), we are not particularly confident that the results represent a robust phylogenetic hypothesis for the group. In addition, the fifteen species analyzed represent only a fraction of all the taeniids. Therefore, we are not certain how much of the disagreement between host and parasite phylogenies is due to rampant host switching and how much is due to a poorly resolved parasite phylogeny.

Interestingly, the relative contributions of cospeciation and host switching also show considerable variation within groups. As discussed above, the associations between monogeneans in the genus *Ligictaluridus* and their catfish hosts are tightly constrained by history. Boeger and Kritsky (1989) discovered a different pattern in their investigations of the twelve monogenean genera making up the family Hexabothriidae whose members inhabit ratfish, sharks, and rays. They used ninety-two characters to produce a single phylogenetic tree with a consistency index of 81.2%. They then compared the fit of the parasite phylogeny with three phylogenetic hypotheses of elasmobranch relationships discussed by Compagno (1977). The fit of the hexabothriid genera to the various host phylogenies ranged from 32.7 to 45.7%, suggesting widespread host switching among ancestral groups. Klassen and Beverly-Burton (1988) examined the phylogenetic relationships of yet another group of monogeneans, with the imposing description "ancyrocephalids with articulating haptoral bars," inhabiting the gills of the centrarchid fishes *Micropterus* (basses) and *Lepomis* (sunfishes). The phylogenetic relationships of parasite species inhabiting basses and those of the hosts were virtually identical, whereas there was no discernible phylogenetic association between sunfish species and their monogeneans. Klassen and Beverly-Burton discussed the widespread hybridization that occurs among species of *Lepomis*, in contrast to *Micropterus*, and suggested that this facilitated the numerous host transfers that apparently occurred during the evolutionary diversification of these parasites. This may be a prime example of extensive diversification in a group resulting from repeated sympatric speciation by means of host switching.

We began this section by asking whether members of monophyletic groups within a larger clade have all been influenced to the same degree by the interaction between historical (cospeciation) and nonhistorical (host-switching) factors, or whether each group represents a unique evolutionary outcome of this interaction. The data at hand support the latter explanation; however, until we have a larger and more comprehensive data base, we cannot draw any generalizations about the relative importance of historical and nonhistorical influences on the evolution of close ecological associations. At the moment, there are very few detailed studies of large groups once we move outside the parasitic flatworms and arthropods, and even there the coverage is pretty thin.

Brooks (1988a) recently reviewed the literature and methods employed in past studies comparing host and parasite phylogenies. For those who wish to pursue this research program further, other studies are listed in table 7.30.

Table 7.30 Studies of cospeciation in an ecological context, using phylogenetic systematics listed by associate group, with host groups and references following.

Protists
Coccidians in cricetid rodents (Reduker, Duszynski, and Yates 1987)

Helminths
Platyhelminths: Digeneans in vertebrates (Brooks 1979a; Brooks and Macdonald 1986); tetrapods (Brooks and Overstreet 1978); anurans (Brooks 1977); crocodilians (Brooks 1980b, 1981); North American freshwater turtles (Platt 1988; Macdonald and Brooks 1989). **Aspidobothreans** in vertebrates (Brooks, Bandoni, Macdonald, and O'Grady 1989). **Monogenea** on elasmobranchs (Boeger and Kritsky 1989); North American catfish (Klassen and Beverly-Burton 1987); North American centrarchid fishes (Klassen and Beverly-Burton 1988). **Gyrocotylidea** in chimaeroid fishes (Bandoni and Brooks 1987a). **Amphilinidea** in vertebrates (Bandoni and Brooks 1987b). **Eucestoda** in tetrapods (Brooks 1978); neotropical catfish (Brooks and Rasmussen 1984); carnivore mammals (Moore and Brooks 1987); alcid birds (Hoberg 1986).

Nematoda: oxyurids in Old World primates (Brooks and Glen 1982); **strongyloids** in Old World primates (Glen and Brooks 1985); **trichostrongyloids** in North American ruminants (Lichtenfels and Pilitt 1983); **metastrongyloids** in North American cervids (Platt 1984).

Digeneans + nematodes in crocodilians (Brooks and O'Grady 1989). **Digeneans + eucestodes + nematodes** in hominoid primates (the great apes; Glen and Brooks 1986). **Digeneans + eucestodes + monogeneans + nematodes** in neotropical freshwater stingrays (Brooks, Thorson, and Mayes 1981).

Arthropods
Chelicerata: mites on primates (O'Connor 1984); cormorant birds (O'Connor 1985); New World primates (O'Connor 1988).

Mandibulata: Crustacea: pinnotherid crabs on echinoderms (Griffith 1987); **copepods** on marine teleosts (Ho and Do 1985); pelagic marine fishes (Benz and Deets 1988); scomberomorph marine fishes (Cressey, Collette, and Russo 1983; Collette and Russo 1985); elasmobranchs (Deets 1987; Deets and Ho 1988; Dojiri and Deets 1988). **Insecta: dipterans** on various plants (Roskam 1985); **agaonid wasps** on figs (Ramirez 1974); **beetles** on termites (Jacobson, Kistner, and Pasteels 1986); **lice** on carnivore mammals (Kim 1985); **fleas** on neotropical mammals (Linardi 1984); **variety of groups** on *Nothofagus* (Humphries, Cox, and Neilson 1986).

Summary

Cospeciation studies are important because they allow us to estimate the ages of biotas and to reconstruct the historical sequence by which they have been assembled. This, in turn, sets the stage for historical ecological studies of coevolution and of community evolution, which we will discuss in chapter 8. One basic theme underlying this and subsequent chapters is that spatial and resource allocation are important components of community evolution. Spatial allocation patterns are revealed by studies of cospeciation in a geographical context (historical biogeography), whereas resource allocation patterns are revealed by studies of cospeciation in an ecological context. As a consequence, historical ecologists investigating both aspects of cospeciation will uncover the extent to which phylogenetic influences have shaped these components of community and biotic structure.

For this reason, it is important to emphasize two generalities that emerge from this chapter. First, we have presented a single methodological approach for documenting patterns of both spatial and resource allocation (the latter in terms of co-occurring species). Wiley (1988a) has termed this approach BPA, for *Brooks parsimony analysis,* because it was first outlined for use in studies of host-parasite associations (resource allocation) by Brooks (1981) and extended to studies of biogeography (spatial allocation) by Brooks (1985). As we have noted above, BPA has required substantial modification, most recently by Wiley (1988a,b) and Brooks (1990), from the original formulation. As a result of this modification, BPA is now robust enough to be used as a general analytical tool for documenting macroevolutionary patterns of spatial and resource allocation. However, beware of the assumptions that it is either a "perfect" method (something not yet produced by scientists) or the best possible formulation. Both Page (1987, 1988) and Simberloff (1987, 1988) have called for statistical tests of cospeciation hypotheses (see the discussion in chapter 6). Since these tests are designed to examine a different set of questions (degrees of congruence among phylogenetic trees) than those addressed by BPA (pinpointing particular instances of incongruence), the development of and interaction between both methodologies will add depth to our evolutionary explanations.

The second generalization that emerges from this chapter is that this modified version of BPA is sensitive to a variety of evolutionary influences (see, for example, the study of Amazonian birds by Cracraft and Prum 1988). This is an encouraging result, for it frees us of concerns that BPA might be a reductionist approach that attempts to force data to conform to an "all cospeciation" model. In fact, the results of the numerous studies presented in this book imply that entire clades do not generally evolve as a result of a single speciation mode. We therefore do not expect all members of an association to conform to a single cospeciation scenario, but rather to represent the unique interaction of historical (vicariance/cospeciation) and nonhistorical (dispersal/resource-switching) events.

8   Coadaptation

Studies of cospeciation attempt to uncover the patterns of geographical or ecological association between and among clades. When we asked questions about how species came to be geographically and ecologically associated, we were investigating the most obvious characteristic of ecological association patterns, species composition (cospeciation; chapter 7). We will now build upon that base to show you that it is possible to uncover the influences of adaptive interactions between species in shaping these macroevolutionary patterns (coadaptation). In essence, we will be embarking upon a search for causal explanations of cospeciation and resource/host switching.

Coadaptation can be investigated from two different, but not mutually exclusive, perspectives: coevolution and community evolution. We will use the term "coevolution" to encompass both phylogenetic co-occurrence and mutual adaptive interactions between species. Thus construed, coevolution is not a "process" in the classical sense; it is a descriptive term applied to a variety of evolutionary forces produced from the interplay between macroevolutionary (cospeciation and host switching) and microevolutionary (mutual adaptive responses by members of associated species) processes. Cospeciation and host switching (discussed in chapter 7) produce the phylogenetic context within which coadaptation occurs; and coadaptation, in turn, provides information about the processes involved in the evolutionary diversification of biological associations within this historical structure. *Coadaptation is manifested in the degree to which the coevolving species affect, or have affected, each other's genetic makeup, or the way in which they influence each other's ecology.*

The other perspective on coadaptation comes from research aimed at discovering the ways in which multispecies ecological associations evolve and are maintained. Unlike coevolutionary research, studies in community evolution are not based on the assumption that strong and often highly specialized ecological interactions are occurring between species (although this may be the case). Because they are not individual entities tied together by the bonds of reproduction and development, communities do not evolve in the same way species evolve; they are "assembled" through time. If they are more

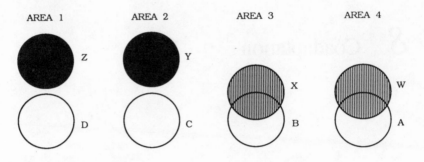

(a) DISTRIBUTION OF FISH SPECIES IN WATER COLUMN

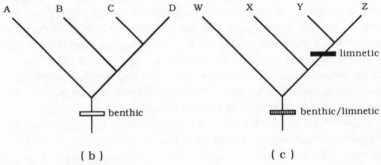

(b)   (c)

Fig. 8.1. The influence of history at the community level. (*a*) Distribution of foraging preferences. *White circle* = species foraging in the benthos; *striped circle* = species foraging in both
the benthos and the limnos; *black circle* = species foraging in the limnos. (*b*) Phylogenetic
relationships for clade A–D, based on nonecological data. (*c*) Phylogenetic relationships for clade
W–Z, based on nonecological data.

than arbitrary units, randomly dispersed through time and space, *we should
be able to document the effects of both macro- and microevolutionary processes on their evolutionary assemblage.*

Consider the following hypothetical example of the interaction between
past and present at the community level (elaborated from Mayden 1987a).
Imagine discovering that two fish species in a large lake (area 1; fig. 8.1a) do
not overlap ecologically; say, for example, one is a benthic forager (species
D) and one is limnetic (species Z). A possible explanation for such habitat
separation is that it represents the effects of competition between these two
species in the past. Is this a reasonable hypothesis? The short answer to this
question is, we don't know. The longer answer is, without a phylogeny for
the fishes and a record of their relatives' interactions with each other, it is

impossible to ascertain whether the association in our research lake is a result of interactions between Z and D or a historical legacy of interactions between their ancestors. So, after extensive fieldwork, we uncover species C (benthic) and Y (limnetic) in area 2, species X (demonstrates both foraging modes) and B (benthic) in area 3, and species W (demonstrates both foraging modes) and A (benthic) in area 4 (fig. 8.1a). As luck would have it, phylogenies exist for the two clades, based on morphological data. When the foraging modes are optimized on the trees, we discover that foraging on the benthos is plesimorphic for all members of the A + B + C + D clade. These species have not changed their foraging habits, interactions with members of the other clade notwithstanding. Conversely, foraging on both benthic and limnetic prey was primitive for the W + X + Y + Z clade, but something happened during the interaction between the ancestor of Y + Z and the ancestor of C + D, and the former moved out of the benthic into the limnetic realm. So, while this does rule out a role for interspecific competition between past populations of species D and Z in shaping the current foraging modes in these fishes, it does not rule out the possibilities that competition was either involved in the habitat shift in the appropriate ancestors or is maintaining the divergent foraging habits today.

At the moment, little of the research in either coevolution or community ecology has used phylogenetic information; therefore, this chapter will serve more to indicate future research possibilities than to present a data base from which generalizations can be derived.

Coevolutionary Dynamics: How Are the Members of an Association Interacting with One Another?

Coevolutionary associations have been studied with increasing intensity ever since Ehrlich and Raven (1964) published their pioneering work on butterfly–host plant interactions. The debate concerning the evolutionary processes underlying extant association patterns has been particularly vigorous, but, until recently (Futuyma and Kim 1987), the discussions generally have not incorporated phylogenetic components into the testing protocol for the various coevolutionary models (see, e.g., Futuyma and Slatkin 1983). Many examples have been assigned coevolutionary status simply because of the complexity of the ecological interactions (e.g., Sussman and Raven 1978; Moran 1989). Most of the models of coevolution are based on microevolutionary (population genetical and population ecological) processes. This state of affairs is hardly surprising given the wealth of experimental and field data available at that level, compared to which the number of macroevolutionary studies places a distant, but nevertheless optimistic, second. Three major classes of coevolutionary models have emerged from this vast data base. If these models are realistic representations of the processes that have affected coevolutionary interactions, we should be able to find some phylogenetic patterns characteristic of each model.

Allopatric Cospeciation

Allopatric cospeciation (Brooks 1979b), or the "California model," is based on the assumption that hosts and associates are simply sharing space and energy. As the null model for historical ecological studies of coadaptation, it predicts *congruence between host and associate phylogenies* based solely upon simultaneous allopatric speciation in associate and host lineages, that is, vicariance events (fig. 8.2). Like any null model, *support for the hypothesis of cospeciation offers relatively weak explanatory power.* For example, discovering that a particular set of associations has resulted from allopatric cospeciation eliminates coevolutionary models based on host switching, but does not allow us to distinguish the effects of a historical correlation from the effects of some mutual interaction that maintains or promotes the association and its diversification. And even if we assume the latter, the delineation of cospeciation patterns by themselves, no matter how detailed, does not allow us to differentiate among a variety of ways in which the associated species might be causally, rather than casually, intertwined.

One possible way to untangle correlation from causation with respect to cospeciation patterns is to examine the basis of the specificity in the associa-

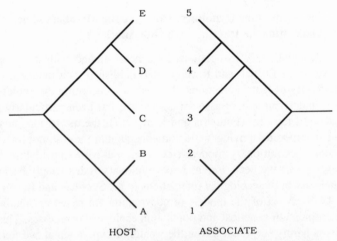

HOST   ASSOCIATE

Fig. 8.2. The allopatric cospeciation, or California model, of coevolution. Complete congruence between the phylogeny for the hosts (taxa represented by *letters*) and the phylogeny for their associates (taxa represented by *numbers*) is due solely to simultaneous cospeciation.

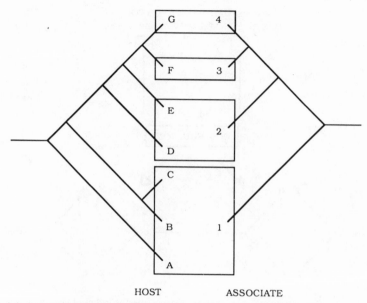

HOST   ASSOCIATE

Fig. 8.3. Interaction between cospeciation and specificity displayed by the associate species. In this scenario, the degree of cospeciation increases through time due to some inherent drive towards increasing host specificity in the associate lineage.

tions. For example, Smiley (1978) presented a version of the cospeciation model based on the assumption that there is a general evolutionary tendency towards specialization, in this case specialization in host choice, by associated species. This model postulates that as an evolving associate lineage becomes progressively more specialized, it will be associated with fewer and fewer hosts, and cospeciation events will increase accordingly (fig. 8.3). In this case the degree of cospeciation is the result of an evolutionary trend inherent in the associated species and not of mutual adaptive modifications between host and associate.

Alternatively, the association may be maintained by the presence of certain host resources that are of adaptive value to an associated species. If an associate species requires a specific resource and/or a resource with a restricted distribution among host groups, then it is possible that the phylogenetic association between the host group and the associated group is due to the distribution of the resource. Consider a hypothetical ancestral beetle species X characterized by trait q_0 that serves as a cue for members of a parasitic wasp species A (other examples of q_0 might be a chemical cue that triggers feeding in a phytophagous insect, a chemical cue that triggers settling behavior in larvae, or a visual cue that triggers orientation in a vertebrate or insect polli-

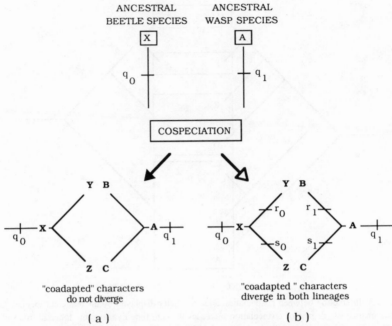

Fig. 8.4. Examining the potential for coadaptation within the framework of cospeciation. In both of the scenarios, preliminary phylogenetic analysis has uncovered complete congruence between host and associate phylogenies. (*a*) Host (X to Y + Z) and associate (A to B + C) speciate simultaneously; descendant species retain the plesiomorphic coadapted trait complex (q_0 in the host; q_1 in the associate). (*b*) Host and associate speciate simultaneously, and diversification in the coadapted trait complex (q_0 to $r_0 + s_0$ in the host; q_1 to $r_1 + s_1$ in the associate) is congruent with the speciation events.

nator). The wasp's reliance on q_0 is denoted by its possession of the coevolved trait q_1. Now let us follow two evolutionary scenarios involving speciation in the beetle lineage producing descendant species Y and Z, as well as speciation in the wasp lineage producing descendant species B and C (fig. 8.4).

The evolutionary association between species Y and B, on the one hand, and between Z and C, on the other, represents an instance of cospeciation. Although both scenarios depicted in figure 8.4 produce congruent host and associate phylogenies, the processes underlying this congruence are somewhat different in each case. In the first scenario (fig. 8.4a), q_0 and q_1 are phylogenetically conservative so the associations are maintained, at least in part, by the common coadapted trait complex q_0q_1. We have evidence here for "adaptively constrained" coevolution because speciation in both lineages has occurred without changes in the causal basis for the primitive host-associate interaction. In the second scenario (fig. 8.4b), the cospeciation of the beetle and wasp clades is matched by the diversification of the adaptively significant trait complex. Species Y and B now interact through the coadapted characters r_0 and r_1, while the association between Z and C is maintained by traits s_0 and s_1. This scenario, representing "adaptively driven" coevolution, provides ecological evidence that the observed cospeciation patterns are more than casual historical associations.

In figure 8.4a we have evidence that the critical host resource is a plesimorphic trait. If this primitive resource is phylogenetically widespread, then the potential for the associated species to switch among hosts from different clades is increased dramatically. Such host switching, based on the "tracking" of a resource common to hosts that do not form a clade, is the province of the "resource-tracking" models of coevolution.

Resource Tracking

This class of models, which can also be referred to as **colonization** models, is based on the concept that hosts represent patches of necessary resources which associates have "tracked" through evolutionary time (Kethley and Johnston 1975). Three pieces of information are required in order to thoroughly examine the explanatory power of these models: a phylogeny for the hosts, a phylogeny for the associates, and an explicit description of the resource. Depending on the phylogenetic distribution of the resource, one of three general macroevolutionary patterns will be produced.

The **sequential colonization model** (Jermy 1976, 1984), originally designed to explain insect-plant coevolution, proposes that the diversification of phytophagous insects took place *after* the radiation of their host plants. The insects are hypothesized to have colonized new host plants many times during their evolution. In each case the colonization was the result of the evolution of insects responding to a particular biotic resource that already existed in at least one plant species. That resource, in turn, is postulated to have been either **plesiomorphically** (fig. 8.5a) or **convergently** (fig. 8.5b) widespread, so the predicted macroevolutionary pattern is that *host and associate phylogenies will show no congruence* (fig. 8.5c).

From the perspective of resource-tracking models, the explanation for *congruence between portions of host and associate phylogenies* is that the associates are specialized on a host resource that is restricted to the host clade and has evolved in a manner congruent with the phylogenetic diversification of the host clade. This model differs from sequential colonization because it proposes that the resources are distributed in an apomorphic, rather than a plesiomorphic or convergent, fashion. It differs from allopatric cospeciation by assuming that hosts and associates do not speciate simultaneously (association by descent); rather, the hosts evolve first and then associates colonize

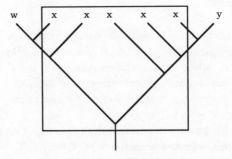

(a) RESOURCE IS PLESIOMORPHIC

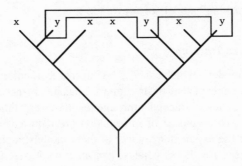

(b) RESOURCE IS CONVERGENT

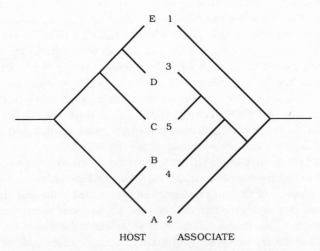

HOST   ASSOCIATE

(c) LACK OF CONGRUENCE BETWEEN PHYLOGENIES

Fig. 8.5. Sequential colonization model. (*a*) Target area for colonization sequences (*box*) if the resource being used is plesiomorphic and widespread. (*b*) Target area for colonization sequences if the resource being used is convergent and widespread. (*c*) Predicted lack of congruence between the host and associate phylogenies.

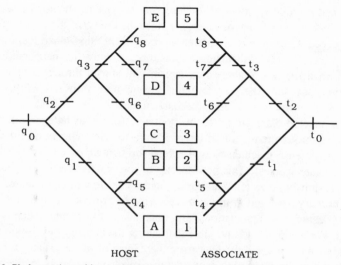

HOST   ASSOCIATE

Fig. 8.6. Phylogenetic tracking model. Every speciation event in the host lineage is accompanied by a modification of the resource (changes in q). Members from the ancestral associate colonize the new hosts, adapt to the new resource (changes in "tracking ability," t), and eventually produce a new associate species responsive to that resource.

them. The evolutionary sequence of events is as follows (fig. 8.6): A new host species evolves, characterized, in part, by an evolutionarily modified form of the required resource. This new species of host is then colonized by individuals from the species associated with the ancestral host. Some of these individuals adapt to the new form of the resource, eventually producing, in their turn, a new species of associate. And so the cycle continues.

In order for this evolutionary scenario to occur, the ancestral host species *must* persist after the speciation event that produces descendants bearing the modified resource; otherwise, the ancestral associate species would have no resource base to support the population through the colonization phase. As we discussed in chapter 4, ancestral species can persist following sympatric, parapatric, and peripheral isolates allopatric speciation. So, in order to distinguish "phylogenetic tracking" from allopatric cospeciation (fig. 8.4b) we require (1) an understanding of the manner in which the host group speciated (see chapter 4), (2) a method for distinguishing persistent ancestors on a phylogenetic tree (see chapters 2 and 4), and (3) a detailed mapping of the host-resource and associate tracking characters (if present) on the appropriate phylogenetic trees.

Evolutionary Arms Race

This is the classical coevolution model (Mode 1958; Ehrlich and Raven 1964; Feeny 1976; Berenbaum 1983), sometimes termed the **exclusion** model, and originally proposed for insect-plant systems. It may be summarized as follows: phytophagous insects reduce the fitness of their hosts. Plants that, by chance, acquire traits (defense mechanisms) that make them unpalatable to these insects will increase their fitness relative to their undefended brethren, and the new defense mechanism will spread throughout the plant population (new host species). Eventually some mutant insects will, in their turn, overcome the new defense mechanism and be able to feed on the previously protected plant group. If this confers a fitness advantage on the mutants (e.g., through reduction of inter- or intraspecific competition for food), the counterdefense mechanism will spread throughout the insect population (new associate species). This new species of insect will be able to specialize on the previously protected plant group, and the cycle will begin anew.

The primary assumption in "arms race" models of coevolution is that coevolving ecological associations are maintained by mutual adaptive responses. For example, it is possible that during the course of evolution novel traits arise that "protect" the host from the effects of the associate. It is also possible that traits countering such "defense mechanisms" may evolve in the associate lineage. The macroevolutionary patterns that result depend upon the time scale on which the adaptive responses occur. In systems for which the "defense" and "counterdefense" traits arise on a microevolutionary scale, we would expect *fully congruent host and associate phylogenies, with appro-*

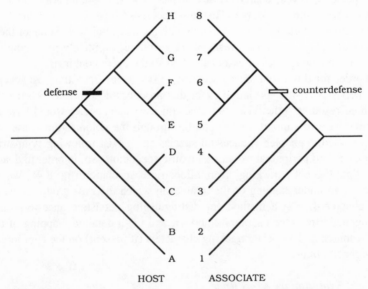

Fig. 8.7. Evolutionary arms race, type I. Host and associate phylogenies are congruent. Defense and counterdefense traits appear at the same point in the common phylogenies.

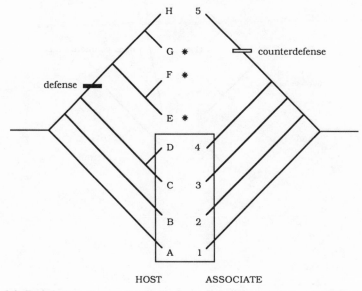

Fig. 8.8. Evolutionary arms race, type II. The host and associate phylogenies are congruent (*box*) up to the point at which the defense trait appears. If the origination of a counterdefense lags behind this, and if the hosts continue to speciate, associate species will be prohibited from interacting with any of the new host species (*asterisk*). Once a counterdefense appears, the host and associate phylogenies rejoin, and the cycle continues.

priate "defense" and "counterdefense" traits appearing at the same point in the common phylogeny (fig. 8.7). In such cases, the coevolutionary arms race would not affect the patterns of macroevolutionary associations between host and associate clades, leading us to expect phylogenetic congruence. This type of evolutionary arms race can be differentiated from allopatric cospeciation by the presence in both lineages of co-originating, mutually adaptive traits.

Evolutionary arms-race models generally assume that, in many cases, the time scale on which the "defense" and "counterdefense" traits originate in response to reciprocal selection pressure is longer than the time between speciation events. Given this, we might expect to find macroevolutionary patterns similar to those shown in figure 8.8, in which *the associate group is missing from most members of the host clade characterized by possession of the "defense" trait.*

A third possible macroevolutionary pattern results when one or more relatively plesiomorphic members of a host clade are colonized by more recently derived members of the associate group bearing the "counterdefense" trait. In this case, *host and associate phylogenies will demonstrate some degree of incongruence, and we would expect to find evidence that some associates*

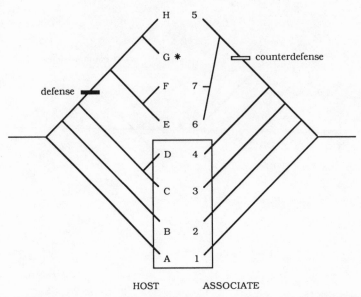

HOST   ASSOCIATE

Fig. 8.9. Evolutionary arms race, type III. Once the counterdefense trait has appeared in the associate lineage, continued speciation will produce new species capable of colonizing host species bearing only the old defense. The situation depicted here is an intermediate version of the large number of possible patterns that may be produced, ranging from all associates that bear the counterdefense (species 5, 6, and 7) back-colonizing on all members of the host lineage, to no back-colonization.

have "back-colonized" hosts in the clade diagnosed by the presence of the "defense" trait (fig. 8.9).

And finally, Ehrlich and Raven (1964) postulated that coevolutionary patterns similar to ones expected for the sequential colonization resource-tracking model depicted in figure 8.5b would result when host shifts by insects with a counterdefense trait occurred between plants that had **convergently** evolved similar secondary metabolites in response to insect attack. In both cases, there is a departure from phylogenetic congruence between hosts and associates; however, the resource-tracking model requires only that the associates be opportunistic. The host resource can be widespread due to either plesiomorphic occurrence or convergence, but its evolutionary patterns are not affected by the presence or absence of the associates. In contrast, the mechanism by which the host resource evolves in the Ehrlich and Raven case requires a high degree of convergent mutual modification on the part of the host and associate groups. Differentiation between the two models requires information about the evolutionary elaboration of putative defense and counterdefense traits.

Case Studies

Birches and the midges that gall them

Dipteran insects of the genus *Semudobia* (family Cecidomyiidae, subfamily Cecidomyiinae) are commonly called gall midges. The association between gall midges and their hosts is more intimate than a simple dinner/diner relationship. Females lay their eggs in bracts or fruits of various species of birches (genus *Betula*), and the larvae develop in situ, drawing both sustenance and shelter from their host, inducing a thickening in the plant tissue (gall formation) in return. Research on members of other gall midge tribes indicates that characteristics of the host plant have considerable influence on larval development (Åhman 1981, 1985; Skuhravy, Skuhrava, and Brewer 1983). This, in turn, may provide a barrier to speciation via host switches in these insects (Roskam 1985). Although as many as three species of *Semudobia* may co-occur in the same host, none of the five species composing the genus is found on plants other than birches. This specificity led Roskam (1985) to investigate the coevolutionary aspects of the association between birches and their gall midges.

The birch genus *Betula* comprises four sections, *Costatae*, *Humiles*, *Acuminatae*, and *Exelsae* (fig. 8.11b). Members of the sections *Costatae* and *Humiles* bear erect pendulous flowers, called catkins, and retain their fruits over the winter, while their relatives, the *Acuminatae* and *Exelsae*, display pendulous catkins and drop their fruits in the autumn. *Semudobia skuhravae*, the sister species of the rest of the gall midges (fig. 8.10a), induces galls in the bracts of *Costatae*, *Humiles*, and *Exelsae* birches. The remaining gall midge species all lay their eggs in the fruits of various members within the *Exelsae* section. Since these birches drop their fruits in the fall, the fruit-galling *Semudobia* develop and overwinter in the soil. Gall midges are never found on birches in the section *Acuminatae*.

There are pronounced differences in host preferences among the gall midges. *S. skuhravae* occurs commonly in association with members of sections *Costatae* and *Humiles*, and less so with members of *Exelsae*. Within the Palearctic *Exelsae*, *S. betulae* is regularly found on species in the series *Verrucosae* and only occasionally associated with the series *Pubescentes*. The reverse situation occurs for *S. tarda*, a gall midge displaying a marked preference for *Pubescentes* and a secondary preference for *Verrucosae*. Relationships between the insects and their host plants produce a similar pattern of preferences in the Nearctic *Exelsae*, where *S. steenisi* and *S. brevipalpis* occur commonly in association with birches within the *Verrucosae* and less commonly with *Pubescentes*. Mapping the host preferences of the five contemporaneous species of *Semudobia* onto the host phylogeny (fig. 8.11a) re-

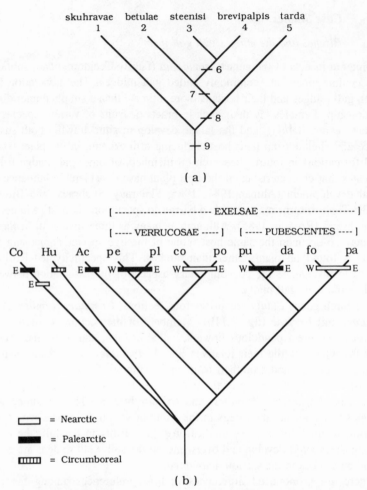

Fig. 8.10. Phylogenetic trees for the birches and their gall midges. (*a*) The gall midges, genus *Semudobia*, coded for coevolutionary analysis. (*b*) The host birches, genus *Betula*. Sections: *Co = Costatae; Hu = Humiles; Ac = Acuminatae*. Species: *pe = Verrucosae pendula; pl = V. platyphylla; co = V. coerulea; po = V. populifolia; pu = Pubescentes pubescens; da = P. davurica; fo = P. fontinalis; pa = P. papyrifera*. The distributions of the birches are mapped onto their phylogenetic tree: *W* = west; *E* = east.

veals that no two species of *Semudobia* show their greatest preference for the same birches. Based on these distributions, and the assumption that the associations of greatest preference are those of longest evolutionary duration, Roskam postulated that the majority of the birch–gall midge associations could be explained by allopatric cospeciation (fig. 8.10b).

In brief, Roskam proposed that the initial speciation event producing *S.*

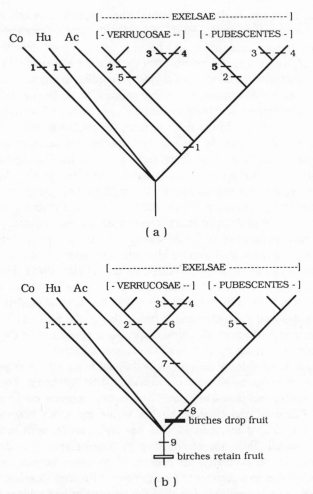

Fig. 8.11. Of birches and gall midges. *Co = Costatae; Hu = Humiles; Ac = Acuminatae. 1 = Semudobia skuhravae; 2 = S. betulae; 3 = S. steenisi; 4 = S. brevipalpis; 5 = S. tarda.* (*a*) Host preferences of the five gall midge species mapped onto the host phylogeny. *Bold numbers* = preferred birch host; *nonbold numbers* = secondarily preferred hosts. (*b*) Phylogeny for the gall midges, mapped onto the phylogeny for the host birches.

skuhravae and the ancestor of the remaining gall midges (ancestor 8; fig. 8.11b) represented a host shift from the fruit-retaining birches, *Costatae* and *Humiles,* to the fruit-dropping birches, *Exelsae.* This change in host preference was accompanied by a shift from the old resource (retained bracts) to the new resource (dropped fruit), and was reinforced by a selective advantage to individuals overwintering in the soil, compared to midges that remained within the birch tree itself (Möhn 1961). The congruence of this portion of the host and associate phylogenies, coupled with the associate's response to an apomorphic resource, appears to be an example of phylogenetic resource tracking. However, it is also possible that the ancestor of *Semudobia* (ancestor 9) cospeciated with the ancestor of *Betula*. According to this second interpretation, one of the descendant gall midge species, *S. skuhravae*, retained the ancestral (plesiomorphic) "bract present/lay eggs in bracts" association with the new birch species. In the sister lineage, the evolutionary appearance of the character "drop fruit" in the new host species was accompanied by the evolutionary appearance of the character "lay eggs in fruit" in the ancestor of the remaining gall midges (ancestor 8). Resolution of this problematical event requires more details about the manner in which the host group speciated.

The remaining speciation events occurred within a common ecological context (eggs laid in dropped fruit). The second speciation event, producing *S. tarda* and the ancestor of the remaining three midge species (ancestor 7; fig. 8.11b), was associated with the allopatric speciation of the *Exelsae* series *Verrucosae* and *Pubescentes* (Roskam and van Uffelen 1981). Finally, two vicariance events occurred. The first isolated *S. betulae* in the Palearctic and the ancestor of *S. steenisi* and *S. brevipalpis* in the Nearctic (ancestor 6), and the second isolated *S. steenisi* and *S. brevipalpis* in the west and east Nearctic, respectively. Although allopatric cospeciation appears to be the predominant mode underlying the pattern of insect-host associations in this system, gall midges were able to incorporate additional hosts into their preference repertoire under circumstances of secondary host sympatry. The biogeographic distributions shown in figure 8.10b support apparent convergent sympatry of Palearctic and Nearctic birches within the series *Verrucosae* and *Pubescentes,* and of members of *Costatae* and *Humiles* with members of *Exelsae.* Overall, then, the combination of biogeographical evidence supporting convergent sympatry, the patterns of phylogenetic association, and the differences in host preference tend to support Roskam's explanation. Additional research is required to explain the absence of gall midges from any member of *Betula* section *Acuminatae.* This appears to be an excellent test case for an evolutionary arms race, or exclusion, explanation, that is, one in which the plants appear to have "won" the arms race, at least for the moment. If this is true, members of *Acuminatae* may produce an unrecognizable cue, a substance that repels *Semudobia,* or a substance that affects development of gall midge larvae adversely.

Gyrocotylids and ratfish revisited

In chapter 7 we discovered that the evolutionary association between the leaflike gyrocotylid flatworms and their ratfish hosts has involved a large

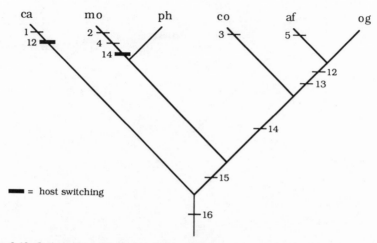

Fig. 8.12. Optimization of five relatively most plesiomorphic species of gyrocotylid flatworms and their phylogenetic relationships onto the phylogenetic tree of their hosts. Flatworms: *1* = *Gyrocotyle rugosa; 2* = *G. confusa; 3* = *G. parvispinosa; 4* = *G. nybelini; 5* = *G. abyssicola.* Hosts: *ca* = *Callorhinchus callorhynchus; mo* = *Chimaera monstrosa; ph* = *Chimaera phantasma; co* = *Hydrolagus colliei; af* = *H. affinis; og* = *H. ogilbyi.*

amount of host switching (see fig. 7.53). Which of the coevolutionary models of host switching best explains the observed interactions between these organisms? We can approach this question by optimizing the phylogenetic relationships for gyrocotylid species 1–5 onto a host phylogenetic tree separately from those for gyrocotylid species 6–10 (fig. 8.12). This portion of the gyrocotylid phylogeny has a consistency index of 83.3% on the host phylogeny. Two putative host-switching events are highlighted by this analysis. The first of these is the apparent colonization of *Chimaera monstrosa* by ancestor 14, producing gyrocotylid species 4. The second is the colonization of *Callorhinchus callorhynchus* by ancestor 12.

Now let us consider the distribution of taxa 6–10 and their phylogenetic relationships (fig. 8.13). This mapping has a consistency index of 53.8% on the host cladogram, due to apparent host switches involving taxa 10 (one switch), 11 (four switches), and 12 (one switch). This highlights a fascinating sequence that was overlooked in our original analysis. Apparently populations of ancestor 11 have become established on new hosts quite frequently throughout the history of the gyrocotylids. Once on a new host, these populations have speciated at least four times, producing extant species 7, 8, 9, and 10. It is the widespread occurrence of descendants of ancestor 11 that accounts for most of the ambiguity in the original host cladogram based on the parasite data.

Combining the last two figures produces a mapping onto the host clado-

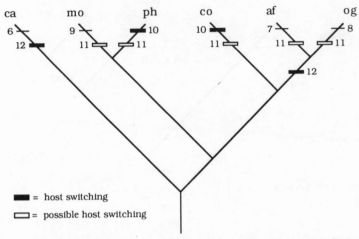

Fig. 8.13. Optimization of five relatively most apomorphic species of gyrocotylid flatworms and their phylogenetic relationships onto the phylogenetic tree of their hosts. Since it is impossible to distinguish cases of cospeciation from cases of host switching, all occurrences have been depicted as host switching. For example, the occurrence of species 10 with two species of ratfish represents one case of cospeciation from ancestor 11 and one case of host switching, but we do not know which is which. Flatworms: 6 = *Gyrocotyle maxima;* 7 = *G. major;* 8 = *G. nigrosetosa;* 9 = *G. urna;* 10 = *G. fimbriata.* Hosts: *ca* = *Callorhinchus callorhynchus; mo* = *Chimaera monstrosa; ph* = *Chimaera phantasma; co* = *Hydrolagus colliei; af* = *H. affinis; og* = *H. ogilbyi.*

gram with a consistency index of 69.5% (fig. 8.14). The evolution of the particular host-parasite associations in this system thus seems to have involved two phases, an early phase that was mostly cospeciation followed by a later wave of host switching. The relatively high consistency index is due to the fact that nine of the ten contemporaneous species of gyrocotylids inhabit only a single host. Since the consistency index for the contemporaneous species (taxa 1–10) is 91%, whereas the consistency index for the ancestral taxa is 50%, a better picture of the widespread influence of host switching in the evolution of gyrocotylids is obtained by considering the ancestral taxa.

The resource being tracked by these colonizing gyrocotylids could be a biochemical or physiological property of ratfish, or it could be an ecological property (such as common feeding habits leading to the ingestion of prey items containing infective gyrocotylid larvae). We do not know anything about the physiological or biochemical requirements of gyrocotylids in their ratfish hosts, nor do we know how gyrocotylids are transmitted one generation to the next from host to host. Because the later wave of host switching always involved transfers from one species of ratfish to another, it is tempting to suggest that the colonizing gyrocotylids were tracking the same resource required by the resident species, a resource that is unique to ratfish but also plesiomorphic and widespread among the group (sequential colonization). Under such conditions we might expect host transfers whenever two or more species of ratfish became sympatric. If this were the case, however, it would be difficult to imagine how host switching would lead to speciation (it certainly did not in the *Semudobia* example, where colonization simply led to an increase in the number of hosts used by each species). Furthermore, in each case of host switching, the newly colonized host is already inhabited by a resident gyrocotylid species, which might suggest speciation as a result of ecological divergence mediated by interspecific competition. This assumes that the resource required by gyrocotylids is the same for resident and for colonizing species, and that it is limited enough to lead to competitive interactions between the resident and the colonizing parasite species. Of course, it is also possible that the colonizers were tracking a (host) resource not used by the resident species. And yet, there is the maddeningly unexplained observation that the vast majority of ratfish are infected with only two individual gyrocotylids each, leaving vast amounts of the intestine unoccupied (anathema to parasitologists!). For now we have reached the limits of description and explanation that can be attained solely by phylogenetic analysis, and must await ecological, life cycle, physiological, and biochemical studies to resolve the enigma of these worms.

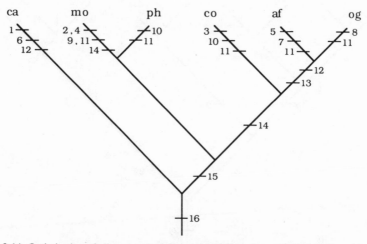

Fig. 8.14. Optimization of all ten species of gyrocotylid flatworms and their phylogenetic relationships onto the phylogenetic tree of their hosts. Flatworms: *1 = Gyrocotyle rugosa; 2 = G. confusa; 3 = G. parvispinosa; 4 = G. nybelini; 5 = G. abyssicola; 6 = G. maxima; 7 = G. major; 8 = G. nigrosetosa; 9 = G. urna; 10 = G. fimbriata.* Hosts: *ca = Callorhinchus callorhynchus; mo = Chimaera monstrosa; ph = Chimaera phantasma; co = Hydrolagus colliei; af = H. affinis; og = H. ogilbyi.*

Crabs and Echinoderms

Small crabs within the genus *Dissodactylus* can be found wandering through littoral and sublittoral habitats of the western Atlantic Ocean from Massachusetts to Brazil and the eastern Pacific Ocean from Baja California to Peru. Within these habitats, the crabs commonly associate with and browse upon a variety of echinoid echinoderms, including mellitid sand dollars, sea urchins, and sea biscuits (Telford 1982). Griffith (1987) presented a phylogenetic systematic analysis of the thirteen species within the genus based on twenty-eight characters. His study produced a single phylogenetic tree with a consistency index of 93.3% (fig. 8.15).

Cospeciation analysis of the data matrix produces one host cladogram (fig. 8.16) with a consistency index of 85.1%. However, this cladogram severs the relationship between the sea biscuits *Clypeaster rosaceus* and *C. subdepressus*, and separates the keyhole sand dollar *Mellita longifissa* from the other mellitids (*Mellita* spp. and *Encope* spp.). In addition, contrary to current taxonomic hypotheses, the heart urchins (*Meoma* and *Plagiobrissus*) are shown as the most highly derived members of a clade including sea biscuits (*C. subdepressus*) and sand dollars (*M. longifissa*). Given the marked rear-

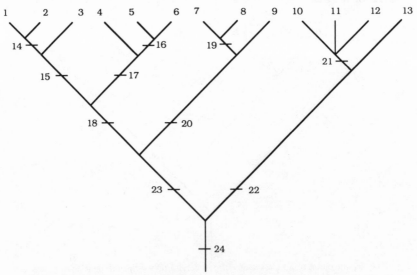

Fig. 8.15. Phylogenetic tree for thirteen species of pinnotherid crabs composing the genus *Dissodactylus*, with internal branches numbered for cospeciation analysis. *1 = D. primitivus; 2 = D. schmitti; 3 = D. latus; 4 = D. glasselli; 5 = D. mellitae; 6 = D. crinitichelis; 7 = D. lockingtoni; 8 = D. nitidus; 9 = D. xantusi; 10 = D. rugatus; 11 = D. juvenilis; 12 = D. usufructus; 13 = D. stebbingi.*

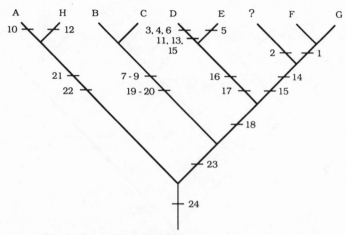

Fig. 8.16. Host cladogram for various echinoderms, based on phylogenetic relationships of species of associated pinnotherid crabs, representing the genus *Dissodactylus*. Crabs: *1 = D. primitivus; 2 = D. schmitti; 3 = D. latus; 4 = D. glasselli; 5 = D. mellitae; 6 = D. crinitichelis; 7 = D. lockingtoni; 8 = D. nitidus; 9 = D. rugatus; 10 = D. xantusi; 11 = D. juvenilis; 12 = D. usufructus; 13 = D. stebbingi.* Hosts: *A = Clypeaster rosaceus; B = Encope* spp.; *C = Mellita* spp.; *D = C. subdepressus; E = Mellita longifissa; F = Meoma* spp.; *G = Plagiobrissus* spp.; *H = C. speciosus; ? =* unknown.

rangement of echinoderm relationships postulated by the crabs' phylogeny, the association between the two groups does not appear to have included much cospeciation. If we optimize host types on the crab tree, we obtain a different perspective (fig. 8.17).

According to this depiction, the association between these crabs and the various echinoderms began with an interaction between sea biscuits (*C. subdepressus,* D in fig. 8.17) and the ancestral *Dissodactylus.* More than half (fourteen out of twenty-four branches) of the subsequent *Dissodactylus* speciation occurred within this ancestral association (thicker branches in fig. 8.17). Superimposed upon this historical framework are at least four host switches from the sea biscuit to (1) heart urchins (by *Dissodactylus primitivus*), (2) a mellitid sand dollar (by *D. mellitae*), (3) a variety of mellitid sand dollars (by the clade comprising *D. lockingtoni, D. nitidus, and D. xantusi*), and (4) another sea biscuit (by *D. rugatus*). The high consistency index in the crab-based host cladogram is due to the speciation of *Dissodactylus* that occurred in association with *C. subdepressus.* However, there is little evidence that the original association between echinoderms and *Dissodactylus* or any of the changes in host association subsequent to that event involved cospeciation. These findings suggest that at least some, if not all, of the echinoid hosts used by these crabs had evolved prior to the time the crabs

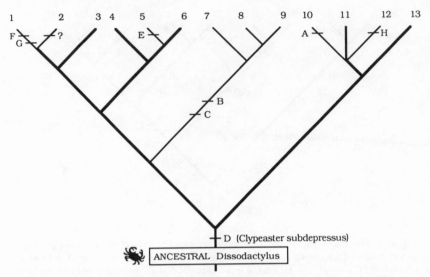

Fig. 8.17. Phylogenetic tree for species of pinnotherid crabs composing the genus *Dissodactylus*, with hosts optimized onto it. Crabs: *1 = D. primitivus; 2 = D. schmitti; 3 = D. latus; 4 = D. glasselli; 5 = D. mellitae; 6 = D. crinitichelis; 7 = D. lockingtoni; 8 = D. nitidus; 9 = D. xantusi; 10 = D. rugatus; 11 = D. juvenilis; 12 = D. usufructus; 13 = D. stebbingi*. Hosts: *A = Clypeaster rosaceus; b = Encope* spp.; *C = Mellita* spp.; *D = C. subdepressus; E = Mellita longifissa; F = Meoma* spp.; *G = Plagiobrissus* spp.; *H = C. speciosus; ? =* unknown.

decided they were tasty morsels. If so, this is an example of sequential colonization analogous to some models of the colonization of plants by insects (Jermy 1976, 1984; Mitter, Farrel, and Wiegemann 1988).

Of butterflies, magnolids, and rosids

The Papilionidae is a relatively small (for insects) group of exotic, widespread, swallowtailed butterflies which feed upon a variety of plants within the subclasses Magnoliidae and Rosidae. Miller (1987) examined the phylogenetic relationships of the papilionids on three different levels in his search for macroevolutionary patterns of butterfly-plant associations. Figure 8.18 depicts his phylogenetic tree, based on forty-four morphological characters with a consistency index of 70%, for the six tribes of swallowtails. Unfortunately, there is a paucity of phylogenetic information about the relationships among the major plant groups upon which these butterflies feed (table 8.1). Hickey and Wolfe (1975) suggested that the Magnoliidae is a paraphyletic group; postulating that the Magnoliales are the sister group of the Aristolochiales + Laurales + Rosidae, and the Aristolochiales are the sister group of the Laurales + Rosidae. Dahlgren and Bremer (1985) reexamined the

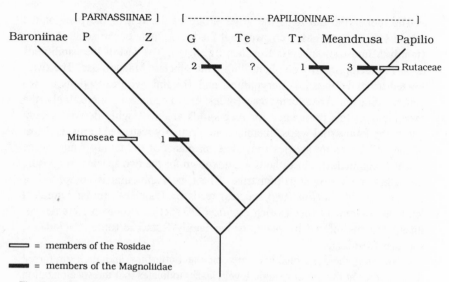

Fig. 8.18. Phylogenetic relationships within the butterfly family Papilionidae. The Baroniinae. Parnassiinae, and Papilioninae are subfamilies. Tribes: *P* = Parnassiini; *Z* = Zerynthiini; *G* = Graphiini; *Te* = Teinopalpini; *Tr* = Troidini. *Meandrusa* and *Papilio* are genera within the tribe Papilionini. The host-plant preferences, at the family level, have been mapped onto this cladogram. *1* = Aristolochiales; *2* = Magnoliales; *3* = Laurales; *?* = dietary data missing.

relationships phylogenetically and found a number of equally parsimonious trees. They agreed, however, with the previous authors' conclusions that the Magnoliidae are paraphyletic, adding that three orders within this "subclass," the Annonales, Magnoliales, and Laurales, are also probably paraphyletic. Without an adequate phylogeny for the host plants, studies of coevolutionary relationships between swallowtails and their preferred foodstuffs are at best preliminary; but nonetheless, they are interesting.

Table 8.1 Classification of the two major plant subclasses of interest to swallowtail butterflies.

Subclass Magnoliidae		Subclass Rosidae	
Magnoliales	**Laurales**	**Rosales**	**Fabales**
Winteraceae	Monimiaceae	Crassulaceae	Mimoseae
Magnoliaceae	Lauraceae	Saxifragaceae	
Annonaceae	Hernandiaceae	Rosaceae	
Piperales	**Aristolochiales**	**Saphindales**	**Apiales**
Piperaceae	Aristolochiaceae	Rutaceae	Araliaceae
		Zygophyllaceae	Apiaceae

Source: Classification based on Cronquist 1981; redrawn and modified from Miller 1987.
Note: Bold type denotes orders, regular type, families.

The monotypic *Baronia,* the sister group of the rest of the papilionids, inhabits west-central Mexico, where it feeds on plants of the genus *Acacia,* (family Mimoseae, subclass Rosidae; fig. 8.18). The rest of the papilionids appear to feed primarily on plants within the subclass Magnoliidae. The swallowtail tribes Parnassiini, Zerynthiini, and Troidini, with exceptions in two genera, dine on Aristolochiaceae (order Aristolochiales). Interestingly, the exceptions to this preference for magnoliids are two highly derived genera within the Parnassiini whose members are primarily restricted to rosids. Most of the 147 Graphiini species specialize on plants of the family Annonaceae (order Magnoliales). Host plants are unknown for the two species composing the rather modest Teinopalpini tribe. The tribe Papilionini is comprised of two genera. *Meandrusa* (two species) feeds on Lauraceae (order Laurales) while the extremely species-rich *Papilio* (220 species strong) prefers the culinary delights offered by plants in the family Rutaceae (order Sapindales, subclass Rosidae).

Examining the relationships between the butterflies and their preferred food sources at this phylogenetic level clearly indicates that the specialization of *Papilio* species on rosids represents a colonization event. In addition, it appears that some degree of host switching among three of the magnoliid "orders" (Magnoliales, Laurales, and Aristolochiales) accounts for much of the diversification within the subfamilies Parnassiinae and Papilioninae. However, this host switching has not been random, and the patterns depicted in figure 8.18 suggest a certain amount of host specificity at the tribal level in swallowtail butterflies. Although the Parnassiinae and the Troidini are not sister groups, their species feed primarily on members of the Aristolochiaceae. Ehrlich and Raven (1964) argued that this similarity was the result of maintenance of the plesiomorphic host preference in these butterflies, while Miller proposed that preference for Aristolochiaceae evolved twice (convergence). In light of the questionable monophyletic status of the Magnoliales and Laurales (host switches 2 and 3 in fig. 8.18), resolution of this problem awaits a detailed phylogeny for the host plants, as well as information about food preferences in the Teinopalpini.

Miller moved next to the second level of his analysis: examination of the phylogenetic relationships among the five genera within the tribe Graphiini. The species within this tribe are predominantly distributed throughout tropical and neotropical African/Indo-Australian regions. *Iphiclides* provides the exception; both species inhabit the Palearctic. *Graphium* is the largest genus (eighty-nine species), with *Eurytides* (fifty-three species) a close second and the remaining genera, *Iphiclides* and *Lamproptera* (two species) and *Protographium* (one species) out of the running. Miller's phylogenetic tree for this group (fig. 8.19) is based on fifty-six morphological characters and has a consistency index of 89%. Mapping the distribution of food preferences on

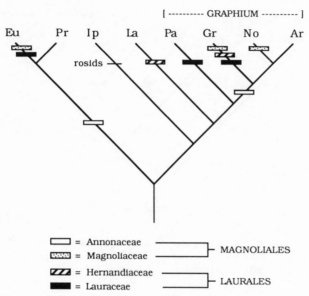

Fig. 8.19. Dietary preferences mapped onto the phylogenetic tree for the swallowtail butterfly tribe Graphiini. Genera: *Eu* = *Eurytides; Pr* = *Protographium; Ip* = *Iphiclides; La* = *Lamproptera*. Subgenera within the genus *Graphium: Pa* = *Pazala; Gr* = *Graphium; No* = *Nomius; Ar* = *Arisbe*. Bars = host preferences.

this tree suggests that a fair amount of host switching has occurred at this level. With the exception of *Iphiclides,* who feed on rosids (family Rosaceae, order Rosales), all members of this tribe prefer some type of magnoliid plant. Three of the genera feed primarily on Annonaceae (order Magnoliales); however, although these are the only swallowtail butterflies to use this resource, it is impossible to determine whether this is a primitive or derived preference within the group because the sister group to the Graphiinae, the Parnassiinae, do not eat any of the plants eaten by the Graphiini. Past this point a number of explanations for the patterns within the remaining genera are possible, and as many as ten sequential colonizations can be postulated by this analysis. Of these, only the *Iphiclides* preference for rosids is a concrete example of host switching. The remaining putative cases of sequential colonization cannot be further evaluated without resolution of the phylogenetic relationships among the host plants, since the suspect "groups" Annonaceae, Magnoliales, and Laurales are included in this analysis.

A different pattern is found when host preferences are optimized on the phylogenetic tree (fig. 8.20). This analysis suggests that these butterfly genera appear to be primitively associated with plants of the family Annonaceae. Two speciation events on the phylogenetic tree are associated with host

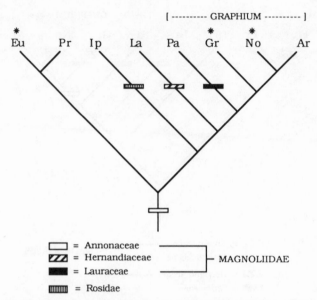

Fig. 8.20. Dietary preferences optimized onto the phylogenetic tree for the swallowtail butterfly tribe Graphiini. Genera: *Eu* = *Eurytides;* *Pr* = *Protographium;* *Ip* = *Iphiclides;* *La* = *Lamproptera*. Subgenera within the genus *Graphium: Pa* = *Pazala; Gr* = *Graphium; No* = *Nomius; Ar* = *Arisbe. Bars* = host preferences; * = generalist on many hosts.

switches from Annonaceae to one other Magnoliidae order (the Laurales), and one speciation event is associated with a switch from Magnoliidae to Rosidae. The widespread host preferences displayed by three of the subgenera (denoted by an asterisk in fig. 8.20: *Graphium, Nomius,* and *Eurytides*) are postulated to represent colonization events subsequent to the original point of cladogenesis. This scenario thus proposes that the initial diversification within the Graphiini resulted from a combination of three instances of host switching and four speciation events associated with retention of the plesiomorphic food type.

Evidence bearing on the second scenario can be sought by investigating the food preferences within a phylogenetic context for each Graphiini genus. If this scenario is correct, primitive members of each genus should display a preference for Annonaceae plants. Miller undertook just such an analysis of the genus *Graphium*. The relative phylogenetic relationships for the species within this genus whose host plants have been identified is depicted in figure 8.21 (see Saigusa et al. 1982, cited in Miller 1987). Once again, mapping the distribution of plant preferences on this tree suggests that a fair amount of host switching has occurred within this genus. However, as was the situation at the generic level, the butterfly-plant association pattern portrayed in the preceding figure is only one of several potential representations of these data. Optimization of host types on the phylogenetic tree suggests that the species in this genus are primitively associated with plants of the family Annonaceae (fig. 8.22).

Two speciation events on the tree are associated with host switches from Annonaceae to other Magnoliidae orders (the Laurales and the Piperales), and one speciation event is associated with a switch within the order Laurales from Lauraceae to Hernandiaceae. The remaining speciation patterns are congruent with these initial three host switches across magnoliid orders. The widespread host preferences displayed by five of the butterfly taxa (denoted

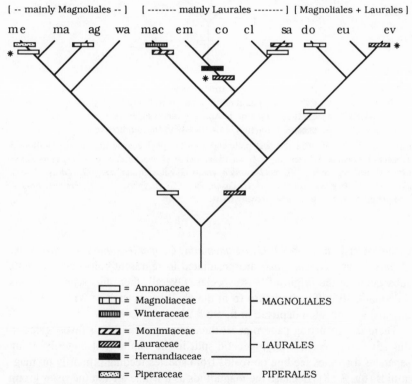

Fig. 8.21. Dietary preferences mapped onto the phylogenetic tree for the swallowtail butterfly subgenus *Graphium*. All the host plants depicted on this cladogram are members of the Magnoliidae; however, *G. macleayanum* and *G. sarpedon* also feed on rosids. Species: *me = G. mendana; ma = G. macfarlanei; ag = G. agamemnon; wa = G. wallacei; mac = G. macleayanum; em = G. empedovana; co = G. codrus; cl = G. cloanthus; sa = G. sarpedon; do = G. doson; eu = G. euryplus; ev = G. evemon.* Bars = host preferences; * = generalist on many hosts.

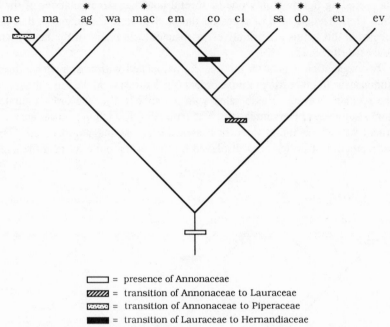

Fig. 8.22. Optimization of the host preferences on the phylogenetic tree for the swallowtail subgenus *Graphium*. Species: *me* = *G. mendana; ma* = *G. macfarlanei; ag* = *G. agamemnon; wa* = *G. wallacei; mac* = *G. macleayanum; em* = *G. empedovana; co* = *G. codrus; cl* = *G. cloanthus; sa* = *g. sarpedon; do* = *G. doson; eu* = *G. euryplus; ev* = *G. evemon. Bars* = host preferences; * = generalist on many hosts.

by an asterisk in fig. 8.22: *G. agamemnon, G. macleayanum, G. sarpedon, G. doson,* and *G. euryplus*) are postulated to represent colonization events subsequent to the origin of these species. Overall, then, this scenario paints a dramatically different picture from the ebullient portrait of widespread sequential colonization depicted in figure 8.21.

There is a recurring pattern at all three phylogenetic levels investigated in this study. The earliest phylogenetic split in the swallowtail butterfly group separates a species feeding on rosids from species feeding primarily on magnoliids (fig. 8.18). Because the magnoliids as a whole are not the sister group of the rosids, this would appear to be evidence of association between the plants and insects subsequent to the evolution of the plant groups. However, if we consider the Baroniinae and Parnassiinae to be numerical relict groups and examine only the Papilioninae, the relationships of the butterfly tribes and the plant families are potentially congruent, depending upon the details of the relationships within the Magnoliidae: the Graphiini feed mostly on members of the Magnoliales, the Troidini mostly on members of the Aristolochiales, *Meandrusa* feeds on members of the Laurales, and *Papilio* is specialized on rosids.

Diversification at the generic level within the Graphiini occurs primarily in association with members of the magnoliid family Annonaceae, with colonization occurring through switching to either other magnoliid species or to rosids. In this case, there is evidence of a correlation between cladogenesis and host switching to rosids in *Iphiclides*, the only Nearctic member of the group, followed by at least two colonization events from Annonaceae to Laurales. This pattern is repeated within the genus *Graphium*, where most of the phylogenetic diversification has also occurred in association with members of the Annonaceae, and three speciation events are correlated with colonization of different magnoliid orders.

Analyses at the generic and specific levels provide evidence of substantial phylogenetic diversification in some of the insect groups feeding on the members of a single family of plants. We do not have detailed phylogenetic hypotheses for the plant groups, so it is not possible to tell if there are any congruent portions of the host and butterfly phylogenies, or if these represent cases of "phylogenetic tracking." For example, the association between the *Graphium* species and members of the Annonaceae may be similar to the case of dissodactylid crabs and their echinoderm hosts discussed previously. If so, the butterflies speciated while the plants did not, which could be construed as evidence that the plants evolved before the association with the butterflies began. In addition, the relatively restricted range of hosts inhabited suggests some degree of resource tracking, while the high degree of host switching within that context suggests that the resource is widespread (and hence presumably plesiomorphic rather than convergent) among members of the magnoliids. This latter point is corroborated by two observations: (1) major instances of host switching occurring outside the magnoliids involve members of the Rosidae, which contains the hosts of the sister group of the Papilionidae, and (2) the Magnoliidae is a paraphyletic group, and paraphyletic taxa are grouped by shared plesiomorphic characters. Miller (1987) suggested that these data supported a resource-tracking model of sequential colonization more than a coevolutionary arms-race model. We agree that this is the best explanation for those episodes of phylogenetic diversification involving host switching. However, this study must be considered preliminary because of its lack of species-level information concerning host plant and butterfly phylogenetic relationships, and because the episodes of phylogenetic diversification that apparently did not involve host switching have not been identified as episodes of either phylogenetic tracking or of allopatric cospeciation.

Summary

The current data base for phylogenetic analyses of potentially coevolved or coevolving systems is sparse and weighted towards the interests of a few authors. Since the majority of empirical phylogenetic studies involve meta-zoan endoparasites of vertebrates, while most of the dynamical models of coevolution were developed for insect-plant systems, it is inappropriate at this time to form generalizations about the relative merits of particular models. However, it appears that there is considerable overlap in the macroevolution-ary patterns associated with each coevolutionary model. For example, one outcome of a resource-tracking dynamic produces patterns of host and asso-ciate phylogenetic congruence that look like allopatric cospeciation patterns, while another outcome produces patterns that resemble the results of a co-evolutionary arms race. Outcomes of the arms-race model range from strict cospeciation patterns to widespread host-switching patterns. Thus, the ma-croevolutionary effects of coevolutionary processes tend to blur the distinc-tions among the models (table 8.2).

Table 8.2 Comparative summary of models of coevolution.

| | | Model Class | |
| | | Host Switching | |
Phylogenetic Pattern	Allopatric Cospeciation[a]	Resource Tracking[b]	Coevolutionary Arms Race[c]
Congruence	**null model**	apomorphic resource **phylogenetic tracking**	defense/counterdefense traits co-orginate
Incongruence	*	plesiomorphic resource OR convergent resource **sequential colonization**	hosts with no associates OR associates back-colonize hosts defense/counterdefense traits do not co-originate

[a] Adaptive response need not be present in either host or associate.

* = broad host range and selective extinction (this is dangerous to invoke because all incon-gruences could be "explained" this way, reducing all coevolutionary explanations to cospecia-tion).

[b] Adaptive response may be present or absent in host, is present in associate.

[c] Adaptive response is present in both host and associate.

Virtually every association analyzed to date includes some departures from cospeciation, each of which, by definition, must involve some form of host switching. The majority of examples discussed in chapter 7 appear to be combinations of allopatric cospeciation and sequential colonization, the latter identified as such because the wide range of associate specificity indicates that these species are tracking a plesiomorphic resource (e.g., gyrocotylids and almost any ratfish, *Taenia* and almost any carnivore, *Ligictaluridus* and ictalurid catfish from two different subgenera). It is thus likely that apparently discrete "models" of coevolution at the microevolutionary level are influencing coevolving systems in a variety of ways at the macroevolutionary level. If this is true, we should not expect a priori that the coevolutionary history of a given clade will conform completely, or even predominantly, to a single model. Rather, like speciation, each system may represent an aggregate of the differential effects of allopatric cospeciation, resource tracking, and the evolutionary arms race. Unfortunately, at the moment the data base is not large enough to investigate the relative frequencies of these modes in a manner analogous to Lynch's (1989) study of speciation modes (see chapter 4).

Coevolution and Evolutionary Specialization

Having discussed the models and patterns of coevolution in a general sense, we will now turn our attention to some of the characters involved in such interactions. It is possible that differences in the biological attributes of diverse associations may bias evolutionary outcomes. For example, phytophagous insects, by their very behavior, decrease the fitness of their host plants. From the plant's perspective the relationship is straightforward; characters that contribute to foiling insect diners are selectively advantageous. On the other hand, the relationship between insect pollinators and their host plants is founded on a delicate balance between opposing selection pressures. Again from the plant's perspective, the decrease in fitness caused by the pollinator's role as herbivore is balanced against the increased fitness that results from the dissemination of gametes. Some associations track broadly through a single trophic level (e.g., polyphagous phytophagous insects), whereas others track narrowly through several levels (e.g., digenetic trematodes, which use molluscs as first intermediate hosts; plants, invertebrates, or vertebrates as second intermediate hosts; and vertebrates as final hosts). Of all the biological attributes, the degree of specificity, or specialization, exhibited by the members of any given association has consistently been assigned a prominent role in explanations of coevolutionary systems.

Resource Specificity

Concepts of evolutionary specialization and resource ("host") specificity play major roles in all of the coevolutionary models because the extent to which an associate can be expected to colonize new hosts is dependant upon its degree of specialization on the original host resource, and upon the evolutionary diversification of the hosts (see Futuyma and Moreno 1988, and Humphery-Smith 1989 for recent reviews of theories about the evolution of specialization and host specificity). Examination of evolutionary specialization, host specificity, and host switching within a macroevolutionary context produces a vexing paradox (cf. the discussion in chapter 4 concerning models of sympatric speciation; also Futuyma and Mayer 1980). On one hand, it seems reasonable to propose that the possibilities for successful sequential colonization are enhanced in inverse proportion to the degree of host specificity. That is, species that respond to a more general, widespread type of cue or a large number of different cues are afforded a greater opportunity to colonize a variety of hosts than their narrowly focussed counterparts. But on the other hand, if colonization of a new host leads to speciation and the establishment of a unique association, then the host must have acted as a strong directional selection force. The chances of this occurring should be higher for species with pronounced host specificity, because they are theoretically more sensitive to changes in the host component of their environments than their more tolerant, generalist relatives.

The mechanisms underlying patterns of resource specialization depend upon whether the specificity originates from the biology of the associate or the biology of the host. If the association is maintained by the host group's possession of a particular resource that is necessary for the survival of the associate, then the specificity may reflect the resource's distribution among sympatric or parapatric host species and the opportunistic behavior of the associate. Specificity due to some attribute of the associate might result from a general macroevolutionary trend toward ecological specialization in the group, which could be manifested as increasing host specificity (Smiley 1978). Alternatively, it could be due to some characteristic of the associate's deme structure that increases the likelihood that host switching will lead to speciation. For example, groups that are capable of producing a viable deme from a single colonization event would be (1) more likely to speciate as a result of colonization than those requiring a larger founding population and (2) more likely to be members of a clade comprising many host-specific species than of a clade comprising a few generalists. The monogenean parasitic flatworms are excellent examples of both these points. Monogeneans exhibit direct development and have generation times much shorter than those of their vertebrate hosts; hence, it is possible for a single monogenean to establish a viable deme, and produce colonizing offspring, while residing on one host. As predicted, the evolutionary diversification of this species-rich group has been driven by a great deal of host switching (chapter 7), even though individual monogenean species are highly host-specific.

Of course, as in any biological system we expect attributes of both hosts and associates to play roles in determining specificity patterns. For example, in the relationship between ancyrocephalid monogeneans and centrarchid fishes, the hosts exhibit high rates of hybridization. The characteristics of monogenean biology that make them such good colonizers are thus coupled with an increased opportunity for colonization, since, as different species of centrarchid fishes assemble on the fields of courtship, more than just gametes are exchanged. Once again, the result has been the evolutionary diversification of many highly host-specific species, most of which have evolved as a result of host switching.

In the following section, we will consider two types of questions about host specificity and macroevolution: Is there a relationship between host specificity and the degree of cospeciation for a given clade? Are there any macroevolutionary trends in host specificity within clades? A preliminary answer to the first question can be sought among the examples presented in this book. In chapter 7 we discussed numerous studies in which pronounced host specificity was coupled with substantial phylogenetic congruence between hosts and associates. We also presented cases in which pronounced host specificity was coupled with substantial incongruence between host and associate phylogenies (monogenean flatworms and centrarchid fishes, gyrocotylid flatworms and ratfish, tapeworms and alcid birds, and tapeworms and carnivorous mammals). The relationship between *Homo sapiens* and some of their roundworm parasites is an interesting example of this second category. Of the two nematode groups inhabiting the great apes, one (*Enterobius*, the pinworms) is more host-specific than the other (*Oesophagostomum*, hookworms). And yet, host relationships implied by the pinworms place *Homo* between *Hylobates* (gibbons) and *Pongo* (orangutans), whereas host relationships implied by the less-specialized hookworms unite *Homo* with *Pan* (chimpanzees) and *Gorilla*, the consensus view of hominoid relationships. Finally, if the example of the gall midges (Roskam 1985) represents a case of allopatric cospeciation followed by an expansion of the host repertoire through secondary colonization, then differences in host preference provide an essential indicator of the coevolutionary dynamics in this system. Overall, then, it appears that pronounced host specificity may be a necessary, but not sufficient, component of cospeciation. If this is true, then low degrees of host specificity should be correlated with limited phylogenetic congruence, while substantial host specificity will be associated with a wide range of phylogenetic patterns. This may be the reason for all the exceptions to the various "parasitological rules" that have been formulated over the past century (see Brooks 1979b, 1985).

The second question requires that we search for regularities in patterns of host specificity emerging over the evolutionary diversification of a group. There are three possibilities here. First, host specificity can increase during phylogenesis as historical effects progressively constrain host preferences (fig. 8.23a). This is consistent with hypotheses of progressive specialization in coevolving lineages (see, e.g., Smiley 1978). The second possibility involves the reverse process; host specificity decreases during phylogenesis (fig. 8.23b). This is consistent with hypotheses about the evolution of ex-

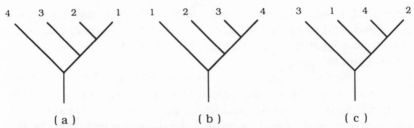

Fig. 8.23. Three potential macroevolutionary trends in host specificity within a clade. *Numbers* = the number of host species inhabited by each species of the associate clade. (*a*) Specificity increases. (*b*) Specificity decreases. (*c*) No patterns in specificity emerge.

treme opportunists, in which traits that facilitate opportunistic behavior are optimized during evolution. Finally, it is possible that no macroevolutionary regularities will emerge during the course of phylogenetic diversification (fig. 8.23c).

According to the current data base, the predominant macroevolutionary trend in changes in host specificity within a clade corresponds to the pattern shown in figure 8.23c. While some members of any given clade of associates are highly host-specific, others are not, and the distribution of these two types of associates within the clade appears to be random. The only example that we have discovered that demonstrates increasing host specificity within an entire clade (fig. 8.23a) is provided by the relationship between telorchiid trematodes and North American turtles (Macdonald and Brooks 1989). To date, there are no phylogenetically based studies that show the pattern depicted in figure 8.23b.

Genetic Diversification

A newly emerging approach to studying coevolution involves attempts to determine the influence that members of an association might have on each other's rate of genetic divergence. A complete investigation of coevolutionary changes in genetic characters must include both phylogenetic and genetic information. Examination of the sister-group relationships within each study group and the degree of phylogenetic congruence (cospeciation) among the study groups allows us to determine whether each association is a historical one. *Once the historical framework has been established, studies of genetic differentiation among members of each clade involved in the association will allow us to determine the degree of correlation between the putative genetic changes.* There are many methods for assessing the degree of genetic differentiation within and among species, the most common of which is the calculation of "genetic distances" based on molecular data. There is currently a great controversy raging around the use of genetic distance data for building phylogenetic trees (if you are interested in this particular issue, see, e.g.,

Farris 1981, 1985; Avise and Aquadro 1982; Felsenstein 1982; Avise 1983; Buth 1984). For the purposes of this chapter, however, all you need to remember is that if you construct phylogenetic trees using genetic distance data, you cannot then use genetic distance data to study the relationship between host and associate genetic divergence without introducing circularity. Although there are numerous studies investigating either the phylogenetic or the genetic relationships between species, very few authors have attempted to combine the two to produce a more robust investigation of coevolution. Therefore, our discussion of the genetic correlates of coevolutionary associations must, at the moment, be based primarily on hypothetical examples and preliminary studies.

Before proceeding with this discussion, we would like to offer two caveats for those who are interested in expanding this promising area of research. First, always remember that estimates of the relationship between host and associate genetic divergence are highly susceptible to sampling errors. For example, it is often difficult to draw a statistically adequate sample of the entire genome for members of either host or associate clade. Second, recall that population biologists have compiled a substantial data base demonstrating that degrees of genetic variability often differ between groups of organisms (see, e.g., Ayala 1982b). Because of this, if we only have access to genetic distance data, we can never be sure if correlations (or their absence) between species are a reflection of a coadaptive interaction (or lack of it) or an artifact of the species genetic structure. *In order to have stronger grounds for postulating coadaptation we need to compare the genetic distances among members of both groups involved in the association with the genetic distances among members of their nonassociated sister groups.* A complete coevolutionary analysis thus incorporates information about (1) the differences in genetic distance between associated species and (2) the macroevolutionary relationship between changes in genetic distances among members of a clade and the appearance of an ecological association between those species. Four potential patterns are produced by the interaction between phylogenetic and genetic processes in coevolving systems (table 8.3).

Table 8.3 Interactions between phylogenetic and genetic factors in coevolving systems.

	Genetic Divergence	
	Equal Rates	Different Rates
Cospeciation	reciprocal coadaptation (fig. 8.24)	no coadaptation (fig. 8.27)
	I	II
Host switch	directional coadaptation (fig. 8.26)	no coadaptation (fig. 8.25)
	III	IV

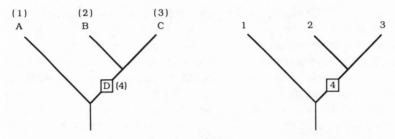

PHYLOGENETIC CONGRUENCE BETWEEN THE TWO GROUPS

MEASURE:
distance (B, C) and distance (2, 3)
distance (A, D) and distance (1, 4)

DISCOVER:
distance (B, C) = distance (2, 3)
distance (A, D) = distance (1, 4)

Fig. 8.24. Interaction type I. Total congruence between the host and associate phylogenies and similarity between the genetic distance measurements for the appropriate host-to-host/associate-to-associate comparisons. *Letters* = host taxa; *numbers* = associate taxa. The distributions of the associates in each host are listed in parentheses above the host tree.

In the first pattern we find phylogenetic association between hosts and associates (cospeciation) coupled with a similar amount of genetic divergence in both clades (fig. 8.24). There are two possible explanations for these observed patterns: interactions between the organisms produces **reciprocal coadaptation** which, in turn, reinforces the association, or, what appears to be reciprocal coadaptation could simply be a manifestation of equivalent evolutionary rates in the host and associate lineage, independent of any interactions between the associated species. In order to distinguish between these alternatives we need information about the rates of genetic divergence in the nonassociated sister groups of both the clades. If we find a macroevolutionary correlation between a change in the rate of genetic divergence and the origin of the association, we have strong support for a coadaptational hypothesis that the associated species have mutually modified each other's genetic structure.

The next example demonstrates the opposite situation: here there is an absence of phylogenetic association (association by colonization: host switching), coupled with a lack of similarity in degrees of genetic divergence (fig. 8.25). Once again, there are two possible explanations for this observation: the hosts and associates have had relatively independent histories of association and show no indication that their genetic structure has been affected by their relatively recent association with each other, or, the rate of genetic change has in fact been modified, but the interaction was not strong enough

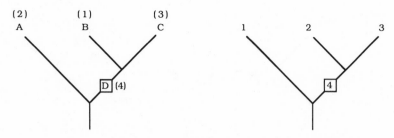

LACK OF PHYLOGENETIC CONGRUENCE BETWEEN THE TWO GROUPS

MEASURE:
distance (B, C) and distance (1, 3)
distance (A, D) and distance (2, 4)

DISCOVER:
distance (B, C) ≠ distance (1, 3)
distance (A, D) ≠ distance (2, 4)

Fig. 8.25. Interaction type IV. Little congruence between the host and associate phylogenies and no similarity between the genetic distance measurements for the appropriate host-to-host/associate-to-associate comparisons. *Letters* = host taxa; *numbers* = associate taxa. The distributions of the associates in each host are listed in parentheses above the host tree.

or has not persisted for long enough to be detected as equivalent degrees of divergence. If we find no macroevolutionary correlation between a rate change in genetic divergence and the origin of the host switch, we have a strong refutation of a hypothesis of coadaptation.

A third possibility falls between the two preceding examples. In this case, there is an absence of phylogenetic association coupled with a similarity in degree of genetic divergence in the host and associate groups (fig. 8.26). It is possible to hypothesize that the new host acquired through colonization exerted such strong directional selection pressure on the colonizing associate that the degree of genetic divergence between the two associate species approached the degree of divergence between their hosts. We have called this type of outcome **directional coadaptation.** Alternatively, what might appear to be directional coadaptation might simply reflect inherent similarities between the associated groups (i.e., the two groups possessed equivalent rates of genetic divergence before the evolutionary origin of their association). Investigation of genetic divergence rates within the clades before and after the host switch will help to distinguish between these two alternatives.

Finally, there is the possible case of hosts and associates that show cospeciation patterns yet also have different degrees of genetic divergence (fig. 8.27). Such systems would imply that it is possible for ecological associates to maintain independent genetic divergence patterns despite a long-standing association; that is, mutual descent without mutual modification, at least at

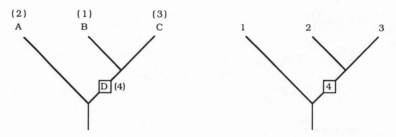

LACK OF PHYLOGENETIC CONGRUENCE BETWEEN THE TWO GROUPS

MEASURE:
distance (B, C) and distance (1, 3)
distance (A, D) and distance (2, 4)

DISCOVER:
distance (B, C) = distance (1, 3)
distance (A, D) = distance (2, 4)

Fig. 8.26. Interaction type III. Little congruence between the host and associate phylogenies and similarity between the genetic distance measurements for the appropriate host-to-host/associate-to-associate comparisons. *Letters* = host taxa; *numbers* = associate taxa. The distributions of the associates in each host are listed in parentheses above the host tree.

the genetic level. This is the case in which cospeciation is due to a casual, rather than causal, phylogenetic association.

Having discussed the theoretical possibilities, let us turn our attention to some examples of this research. Bear in mind that none of the following authors have attempted the second stage of a genetic coevolutionary analysis; tracing the macroevolutionary interactions between changes in genetic divergence rates and the evolutionary origin of the association. Since these studies are based solely upon comparisons of genetic distance data between associated species, the results are preliminary, but interesting nonetheless.

Rodents and mites (type I)

Hafner and Nadler (1988) presented a coevolutionary study of geomyid rodents and their trichodectid mite ectoparasites. They reported that (1) their host and parasite trees indicated a considerable amount of congruence (i.e., a long historical association) between the rodents and the mites, and that (2) the relative genetic distances between related hosts and between their associated parasites were roughly equal. The authors interpreted the high degree of phylogenetic association coupled with the similarity in rates of genetic divergence in host and parasite lineages to indicate a strong coevolutionary coupling of genetic change between the two clades. However, the allozyme data used to calculate the genetic distances between species were also used to construct the phylogenetic trees. Formulating evolutionary hypotheses

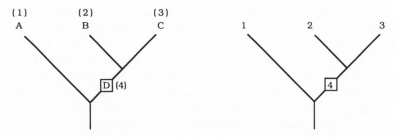

PHYLOGENETIC CONGRUENCE BETWEEN THE TWO GROUPS

MEASURE:
distance (B, C) and distance (2, 3)
distance (A, D) and distance (1, 4)

DISCOVER:
distance (B, C) ≠ distance (2, 3)
distance (A, D) ≠ distance (1, 4)

Fig. 8.27. Interaction type II. Total congruence between the host and associate phylogenies and no similarity between the genetic distance measurements for the appropriate host-to-host/associate-to-associate comparisons. *Letters* = host taxa; *numbers* = associate taxa. The distributions of the associates in each host are listed in parentheses above the host tree.

about relationships between characters using a phylogenetic reconstruction based upon those characters introduces a degree of circularity, and thus weakens the resulting evolutionary hypothesis. In addition, the authors used phenetic rather than phylogenetic methods to generate their trees. Phenetic analyses cluster taxa to maximize the fit of the data to a model of homogeneity of evolutionary rates. Since they incorporate this a priori assumption of homogeneity, phenograms are not good independent tests of similarity in coevolutionary rates of genetic divergence, even if based on data other than those used to infer the rates of divergence in the first place. What is needed here, then, is an independent phylogenetic assessment based, for example, on morphological characters, of the phylogenetic association between the rodents and the mites. Examination of Hafner and Nadler's allozyme data to determine the relative degree of genetic divergence between the hosts and parasites within this phylogenetic context would avoid the problem of circularity, thus providing a strong test of the putative coevolutionary relationships between the two groups.

Marsupials and tapeworms (type IV)

Baverstock, Adams, and Beveridge (1985) studied a group of tapeworms inhabiting Australian marsupials. In contrast to Hafner and Nadler's results, they found that (1) there was little congruence between host and parasite trees and that (2) the genetic distances between tapeworm species differed markedly from the genetic distances between their associated host species. The authors interpreted these results as indicating independent genetic histories for the host and parasite group, consistent with a limited amount of phylogenetic association between the lineages. However, their host tree (a "distance Wagner" tree constructed using genetic distances) and their parasite tree (a phenogram) were constructed using different methods of analysis, neither of which was a phylogenetic systematic approach. Based on the reanalysis of Hafner and Nadler's data, we suspect that the trees of Baverstock et al. show at least some incongruence because they were constructed by different methods.

Trematodes and frogs (type III)

Four species of parasitic flatworms, all members of the trematode genus *Glypthelmins*, inhabit the upper small intestines of North American ranid frogs. Hills, Frost, and Wright (1983), Hillis and Davis (1986), and Hillis (1988) presented a phylogenetic tree for these ranids based on a variety of morphological and molecular data. O'Grady (1987), investigating the relationships within *Glypthelmins* using morphological and ontogenetic characters, produced one tree with a consistency index of 76%. Since both host and parasite cladograms were available, O'Grady extended his study to include a cospeciation analysis using the methods described in chapter 7. He concluded that *G. intestinalis* inhabits *Rana pretiosa* as a result of cospeciation, whereas *G. californiensis* inhabits *R. aurora* as a result of a host switch. Thus the two frog hosts, *R. aurora* and *R. pretiosa*, are more closely related to each other than their associated parasites, *G. intestinalis* and *G. californiensis*, are to each other (fig. 8.28).

The first prerequisite for a coevolutionary study—the existence of phylogenetic trees, based on characters other than those used to determine genetic distances, and an analysis of the historical context of the association between the two clades—is thus met for these flatworm-frog interactions. Rannala (in press) calculated Nei's genetic distances from electrophoretic data and compared the degree of genetic divergence between the trematodes *G. californiensis* and *G. intestinalis*, and between their hosts, *R. aurora* and *R. pretiosa*. Surprisingly, the degree of genetic divergence between the frogs was statistically indistinguishable from the degree of genetic divergence between the parasites, even though the two frog species are more closely related to each other than the two parasite species are to each other. This is a possible case of directional coadaptation.

Unfortunately, few of the studies published thus far address the issue of rates of genetic divergence in associated lineages, and of those, different methods have been used to construct the phylogenetic trees. Thus the results

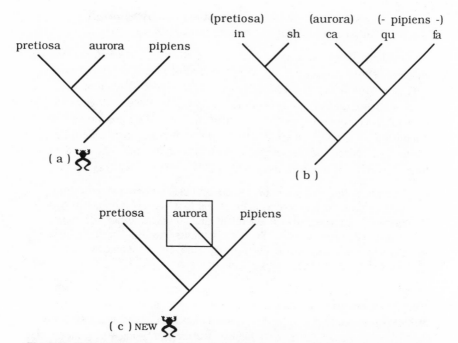

Fig. 8.28. Phylogenetic analyses of the relationships between ranid frogs and their flatworm parasites. (a) Simplified phylogenetic tree for the ranids based on rDNA restriction sequence data. Only three hosts are shown: *Rana pretiosa, R. aurora,* and *R. pipiens* (species complex). (Modified and redrawn from Hillis and Davis 1986.) (b) Simplified phylogenetic tree encompassing five species of *Glypthelmins,* based on morphological and ontogenetic characters. *in* = *G. intestinalis; sh* = *G. shastai; ca* = *G. californiensis; qu* = *G. quieta; fa* = *G. facioi.* The hosts inhabited by each species of flatworm are listed in parentheses above the species names. (Modified and redrawn from O'Grady 1987.) (c) New host cladogram derived from the phylogenetic relationships of the parasite taxa (see chapter 7 for methods). Note that *R. aurora* (*box*) is now misplaced on the cladogram; therefore, this is interpreted as a case of host switching by *G. californiensis.*

to date cannot be directly compared, and results from future investigations will continue to be incompatible until a standardized approach to coevolutionary studies is adopted. Perhaps the most important conclusion to be drawn from the three preceding examples is that similarity in degrees of genetic divergence between ecological associates does not necessarily indicate phylogenetic association, and phylogenetic congruence does not necessarily indicate similar genetic divergence between biological associates. We think this area of investigation holds exciting promise for the future, because the combination of phylogenetic, ecological, and genetic information permits both detailed reconstructions of coevolutionary pathways and an examination of the interaction between macroevolutionary patterns and microevolutionary changes.

Community Evolution: Composition and Structure of Multispecies Ecological Associations

One of the most difficult aspects of studying the evolution of communities is that biologists hold widely divergent views about just what exactly is a community. For example, some researchers view communities as associations of species so strongly tied together by their ecological and behavioral interactions (synecological attributes) that they are almost "superorganisms" (e.g., Wilson 1980, 1983). Others feel that communities are arbitrary assemblages of species that happen to be in the same place at the same time. A recent statement (Strong et al. 1984) summarizes this perspective.

> One possibility that we expect to obtain fairly commonly is the community with so few strong interactions that organization arises primarily from mutually independent autecological processes rather than synecological ones. Such communities would not be holistic entities, but rather just collections of relatively autonomous populations in the same place at the same time.

In many ways, the discussion of this question resembles the discussion among systematists about the definition of the term "species"! Ricklefs (1990:656) stated that "the term *community* has been given a variety of meanings by ecologists." As a consequence, we are not going to champion one community concept over another, nor are we going to summarize and categorize all of the literature in this area. Rather, we will assume that a variety of community types exist on this planet, and thus almost all concepts of community will be valid for particular cases.

Given this, we should be able to uncover both historical and nonhistorical components in community structure. The interesting question then becomes, Are all communities influenced to the same degree by the interaction of these two components, or has each travelled along a unique, evolutionary pathway? Phylogenetic history may confound our attempts to identify the type of community with which we are working, if species occur together due to common episodes of vicariant speciation. This historical component of the community will appear to represent a "holistic entity"; however, it does not necessarily follow that members of this "holistic entity" will exhibit strong synecological interactions. Because of this, cospeciation studies such as those outlined in chapter 7 are a useful starting point for coadaptation studies in community evolution. Needless to say, such studies are apt to be difficult because they require that we examine the members of more than one community in order to draw robust explanations for the evolution of any single community.

There is a marked similarity in many ecological associations, especially of specialized species, around the world that suggest phylogenetic influences in interaction structure as well as species composition. For example, McCoy and Heck (1976) examined the communities of corals, seagrasses, and mangroves throughout the tropics of the world, and concluded that these ecological associations all had a common origin. Hill and Smith (1984) discovered similar roosting patterns between six species of bats in a Tanzanian cave and fourteen species of bats in a cave on New Ireland Island off the northeastern coast of New Guinea. In both communities, species of the genus *Hipposideros* lived near the rear of the cave in association with species of *Rhinolophus*, while species of *Rousettus* roosted just inside the first major overhang (fig. 8.29).

On the parasitological side, Benz (cited in O'Grady 1989) examined the

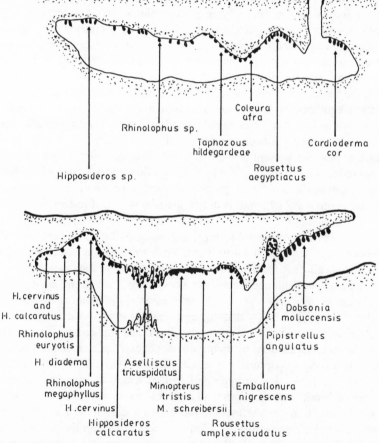

Fig. 8.29. Roosting positions of several bats species in Tanzanian (*top*) and New Ireland Island (*bottom*) caves. (From Brooks and Wiley 1988; redrawn and modified from Hill and Smith 1984.)

distribution of copepods living on sharks' gills. He discovered substantial diversification in habitat preference (site selection) within these copepod communities. Interestingly, the phylogenetic structure of these gill niches was retained even though the specific shark and copepod species varied from community to community. In almost every case, species of *Pandarus* inhabit the gill arch, *Eudactylinodes* and *Gangliopus* attach themselves to the secondary lamellae, *Nemesis* burrow into the efferent arterioles, *Phyllothyreus* inhabit the superficial portions of the interbranchial septum, *Paeon* embed in the interbranchial septum, and *Kroyeria* are found in the water channels of the secondary lamellae or embedded in the interbranchial septum (fig. 8.30).

Price (1984, 1986) outlined four models for the evolution of communities of specialists, including communities of parasites. These models, based on the differential contributions of competition, niche availability, rates of colonization and extinction, and phylogeny to molding extant patterns of community structure, may be summarized as follows: (1) **Nonasymptotic model** (Southwood 1961): the community never reaches a saturation point because its member species are too specialized to fill all the available niches. (2) **Asymptotic equilibrium model** (MacArthur and Wilson 1967; Wilson 1969): the balance between colonization and extinction rates holds the community at equilibrium. Extinctions are postulated to be driven by biological factors such as interspecific competition and predation. (3) **Asymptotic non-equilibrium model** (Connor and McCoy 1979; Lawton and Strong 1981): "vacant niches" are present in the community because colonization rates are not sufficient to fill them up. (4) **Cospeciation model** (Brooks 1980a): "niches" are duplicated by allopatric cospeciation of community members, so extant community structure is due to the persistence of historical associations.

If we use Price's "models" to represent macroevolutionary influences on the evolution of ecological associations, Price's discussion can be extended to encompass all multispecies ecological associations, be they called associations, guilds, communities, biotas, or ecosystems. Just as we believe that coevolving lineages may represent the interaction of different "models of coevolution," we think it likely that any given multispecies association may be characterized by an interaction of any, or all, of these influences. These models can be further enhanced by incorporating concepts of spatial and resource allocation among co-occurring species (Brown 1981, 1984; Brown and Maurer 1987, 1989). For example, all species within a community contain information about (1) their origin with respect to the biota (spatial allocation), and (2) the origin of traits relevant to the association, that is, traits that characterize a species' interactions with other species and with the environment (resource allocation). The occurrence of a given species in a community may be due to either phylogenetic association (its ancestor was associated with the

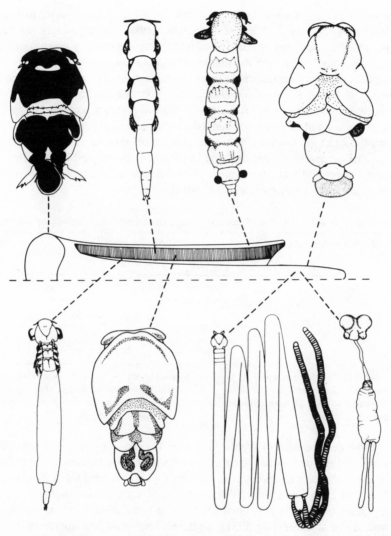

Fig. 8.30. Distribution of eight species of copepods on the gill of a shark. The copepods in this guild are (*clockwise, from the top left*) *Pandarus cranchii, Eudactylinodes uncinata, Nemesis lamna, Phyllothyreus cornutus, Paeon vaissieri, Kroyeria caseyi, Gangliopus pyriformis,* and *K. lineata.* (From O'Grady 1989.)

ancestors of other community members), in which case we refer to it as a resident species, or to colonization, in which case, not surprisingly, the taxon is termed a colonizing species. Similarly, the ecology of any given species in a community may reflect the presence of persistent ancestral traits or of recently evolved, autapomorphic traits. Starting to sound familiar? Under the guise of spatial and resource allocation, colonization, extinction, and competition, we have returned full circle to the evolutionary processes of speciation and adaptation discussed for individual clades in chapters 4 and 5. The combinations of species occurrence (speciation processes) and interactions (adaptation processes) are depicted in table 8.4. Brown and Zeng (1989) recently suggested that communities should be considered mosaics of all four of these types of historical and ecological influences.

Table 8.4 Heuristic depiction of four classes of species contributing to community structure.

Species Occurrence	Species Interactions	
	Ancestral	Derived
Ancestral	historically constrained residents I	stochastically changed residents II
Derived	noncompetitive colonizers III	competitive colonizers IV

Note: There are two components to ecological associations: Species composition: the occurrence of each species in an association is either ancestral or derived. Species interactions: the characters involved in interactions among members of the association are either ancestral or derived.

Phylogenetic history (table 8.4, type I): The conservative homeostatic portion of any community is composed of species that evolved in situ through the persistence of an ancestral association (congruent portions of phylogenies in a cospeciation analysis). Such species display the plesiomorphic condition for characters involved in interactions with other community members and with the environment (fig. 8.31). Since this section of the community is characterized by a stable relationship across evolutionary time, it may act as a stabilizing selection force on other members of the community by resisting the colonization of competing species. Macroevolutionary patterns of this nature correspond to an allopatric cospeciation model.

Ross (1986) reported a high degree of phylogenetic constraints in the structuring of contemporaneous reef-fish communities. Boucot (1982, 1983) concluded that the fossil record demonstrates the conservative nature of community structure throughout evolutionary history. He further suggested that when evolutionary changes do occur, they tend to reverberate through most

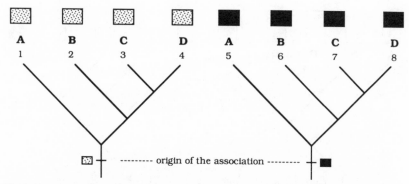

Fig. 8.31. Phylogenetic effects in community structure in area D. *Letters* = areas; *numbers* = species. Species 1 + 5 occur together in area A, species 2 + 6 occur in area B, species 3 + 7 occur in area C, and species 4 + 8 occur in area D. The state of a particular character involved in the interaction between these community members is depicted by the *boxes*. In this situation, both the origins of the traits and the origins of the association between community members are old. Extant community structure in area D thus represents *the persistence of an ancestral association, coupled with the persistence of ancestral interaction traits in both members of the community*.

of the community structure. Historical ecological methods can provide a complementary approach to this study. Consider the hypothetical case depicted in figure 8.32. Five areas (A–E) contain biotas composed of a member from each of three clades (1–5, 6–10, 11–15). Communities A, B, and C are characterized by plesiomorphic interactions among their component species (1, 6, 11; 2, 7, 12; and 3, 8, 13, respectively), whereas communities D and E are characterized by apomorphic interactions among species (4, 9, 14 and 5, 10, 15, respectively). The phylogenetic explanation for this pattern is as follows: The correlation among the plesiomorphic interaction traits is the result of historical conservatism in the evolution of these communities. Novel ecological interactions evolved in the common ancestor of species 4 + 5, the common ancestor of species 9 + 10, and the common ancestor of species 14 + 15. These clades show a pattern of historical congruence; therefore, the evolutionary changes in ecological interactions co-originated in co-occurring species in the same (ancestral) biota. We can thus hypothesize that a common cause is responsible for this suite of ecological changes.

Colonization by "preadapted species" (table 8.4, type III): This portion of the community contains species that have been added by colonization. Such species can be recognized in part because their phylogenetic history is incongruent with the histories of other community members. In addition, the term "preadapted" implies that these individuals are able to colonize the area because they already possess traits that do not conflict with the existing community structure (fig. 8.33). This scenario postulates that there is no compe-

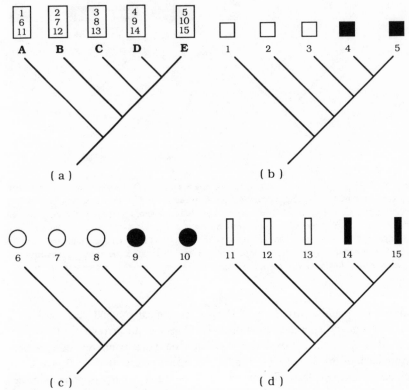

Fig. 8.32. Correlated evolutionary changes in the evolution of communities. *Letters* = five communities; *numbers* = members of three different clades that inhabit those communities. (*a*) Area cladogram of the historical relationships among the communities, based on the phylogenetic relationships of the species that occur in them. (*b–d*) The phylogenetic trees for the members of the three clades. Superimposed above each species number is a symbol indicating a particular resource utilization character. *White symbols* = plesiomorphic trait; *black symbols* = apomorphic trait. Note that the shift from plesiomorphic to apomorphic traits in each clade occurred in the common ancestor of the two most recently differentiated species, leading to the emergence of an ecological structure in communities D and E that differs from the ecological structure in communities A–C.

tition between colonizing individuals and established (resident) members of the community. On the one hand, if the appearance of these species reduces the possibilities for the subsequent addition of species into the community, then this type of macroevolutionary pattern corresponds to the asymptotic equilibrium model of MacArthur and Wilson (1967). On the other hand, if the rates of colonization are low enough, the community may persist below expected equilibrium numbers (corresponding to the asymptotic nonequilibrium model). Similarly, if the colonizers are so specialized ecologically that

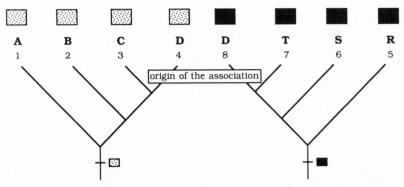

Fig. 8.33. Community structure of area D influenced by colonization of "preadapted species." *Letters* = areas; *numbers* = species. The state of a particular character involved in the interaction between these community members is depicted by the *boxes*. Species 8 has colonized area D and is now interacting with species 4. In this situation, the origins of the traits are old (plesiomorphic) while the origins of the association between community members are relatively recent (apomorphic). Extant community structure in area D thus represents *the appearance of a colonizing species, coupled with the persistence of plesiomorphic interaction traits in both members of the community.*

they do not affect other members of the community or preexisting potential niche space, community diversity may increase without approaching an apparent equilibrium (corresponding to the nonasymptotic model).

Erwin's (1985) "taxon pulse" model is an example of this kind of influence in the evolution of biotic diversity. As outlined in chapter 5, according to this model, a group of species might begin with an ancestor that displays a certain ecological propensity. As time passes, the ancestor and its descendants spread over a larger and larger geographical area, with descendant species fulfilling the same or very similar ecological roles in different locations. Subsequent to this first wave of dispersal, a new ecological trait arises in one of the descendant species in one of the localities. The species bearing this novel trait then undergoes widespread dissemination, and a new "pulse" of diversity occurs, producing a new set of descendant species, all performing similar functions in different locations. Diverse and highly structured communities could be formed in many different areas in this manner, with every community containing a member of each of the "pulses." The number of occupied "niches" within each community would thus correspond to the number of pulses represented by the species that were present. Roughgarden and Pacala (1989) presented a similar argument (using the term "taxon cycle") to explain the species composition and size structure of anoline lizard communities on Caribbean islands.

Colonization by competing species (table 8.4, type IV): All species that colonize a community will exhibit incongruence in a cospeciation analysis. However, unlike the conservative situation depicted for preadapted species (fig. 8.33), varying patterns of character evolution will be traced upon this phylogenetic framework if colonizing individuals compete with resident species. In this situation, at least one of three things must happen in order for the colonizer to become established: the colonizing species will change (fig. 8.34a), the resident species competing with the colonizer will change (fig. 8.34b), or both the resident and the colonizer will change (fig. 8.34c). This will produce a pattern in which the colonizer, the resident, or both exhibit an apomorphic condition of the traits relevant to the competitive interaction. Replacement of the resident by the colonizer would be indicated on a cospeciation analysis if (1) the extinction event is coupled with the colonization event and (2) other members of the "extinct" species' clade have similar resource requirements to the colonizer. These macroevolutionary patterns correspond most closely with the asymptotic equilibrium model of MacArthur and Wilson (1967).

Stochastic ("nonequilibrium") effects (table 8.4, type II): If there is unoccupied "niche space" in a community over extended periods of time, stochastic evolutionary processes operating on resident species may produce changes resulting in the use of some of that previously unoccupied space. In this scenario, evolutionary changes in ecological characters occur within a cospeciation framework. Species contributing to this portion of the community structure can be recognized by their historical congruence with other community members, coupled with the presence of apomorphic traits characterizing their interactions with other species and with the environment (fig. 8.35). Since these changes do not affect other community members, they may represent a type of stochastic wandering through modifications "allowed" by the existing community structure. Such species appear to diverge ecologically for no apparent reason, although care must be taken to rule out the effects of previous competition. The longer a community exists below equilibrium numbers, through any of the processes described under "colonization by preadapted species," the greater the possibility that resident species will experience these sorts of evolutionary changes.

Preliminary Examples

Communities of specialists: Neotropical stingrays and their helminth parasites

Price (1986) stated that

> one major advantage of parasite communities over others is that the habitat they live in, the host, has such a well defined structure. . . . The host microcosm is replicated through time and space much more so than habitats for most other organisms. Therefore, the study of comparative community structure is very powerful.

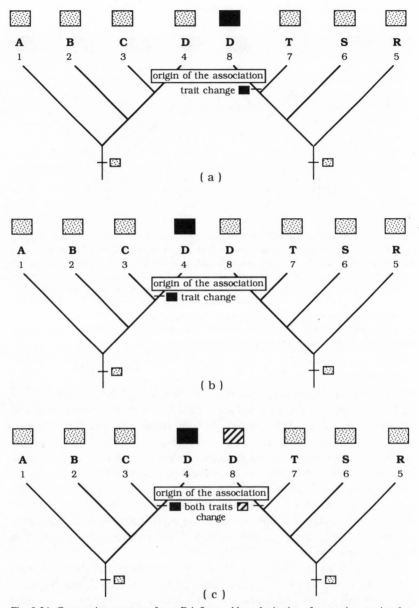

Fig. 8.34. Community structure of area D influenced by colonization of competing species. *Letters* = areas; *numbers* = species. The state of a particular character involved in the interaction between these community members is depicted by the *boxes*. Species 8 has colonized area D and is now interacting with species 4; therefore, the origins of the association between community members are recent (apomorphic). (*a*) The origin of the trait is old (plesiomorphic) in the resident species and recent (apomorphic) in the colonizer. (*b*) The origin of the trait is old (plesiomorphic) in the colonizer and recent (apomorphic) in the resident species. (*c*) The origin of the trait is recent (apomorphic) in both the colonizing and resident species. Extant community structure in area D thus represents *the appearance of a colonizing species followed by modifications in the trait(s) involved in the interaction between resident and interloper.*

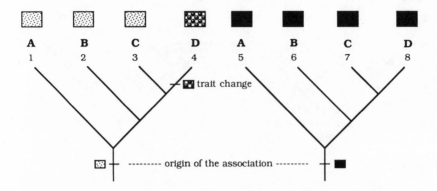

Fig. 8.35. Community structure in area D influenced by stochastic or nonequilibrium changes. *Letters* = areas; *numbers* = species. Species 1 + 5 occur together in area A, species 2 + 6 occur in area B, species 3 + 7 occur in area C, and species 4 + 8 occur in area D. The state of a particular character involved in the interaction between these community members is depicted by the *boxes*. In this situation the origin of the association between community members is old; however, while species 8 retains the ancestral interaction character (plesiomorphic), the character has been modified in species 4 (apomorphic). Extant community structure in area D thus represents *the persistence of an ancestral association, coupled with the appearance of an autapomorphic trait in one of the members of the community.*

We agree wholeheartedly and offer the relationships between neotropical freshwater stingrays and their helminth parasites as an excellent example of the historical ecological approach to studying community structure. The patterns of spatial allocation (historical biogeography) for the various host and parasite species were discussed in chapter 7. We will use the patterns of host utilization by each helminth species in each area as an indication of resource allocation. A host cladogram can be reconstructed based upon the phylogenetic relationships of the associated helminths; however, this cladogram must be considered highly preliminary because an explicit phylogenetic tree for the potamotrygonids has not yet been published. Nevertheless, it will serve as a heuristic starting point.

The area cladogram (see fig. 7.34) and the host cladogram (fig. 8.36) provide evolutionary information about the occurrence of each helminth species in a given area and in a particular host. This information, in turn, can be used to identify the portion of the community structure that is associated with each of the four influences discussed in the preceding paragraphs (table 8.5). Price (1984) suggested that communities of specialists ought to be characterized by "vacant niches," potential resources left unused due to evolutionary constraints on ecological specialization. If we consider table 8.5 to be a rough indicator of (two-dimensional) "niche space," one observation leaps out immediately: the majority of cells (171 out of 207, or approximately 83%) in the matrix are empty. Why is there so much "unused niche space"? The answer is that most of this apparent niche space is not actually accessible, due to the phylogenetic history and evolutionary specialization of the species involved. Recall that phylogenetic partitioning of ancestral biotas produces endemic areas with endemic species of hosts and parasites. As a consequence, much of the "empty niche space" depicted in the matrix results from the fact that phylogenetic history has spatially partitioned species. If we calculate the number of type I, II, III, and IV influences for each species of parasite/host/ area, and then calculate the number of "empty niches" based on the hosts and parasites that are actually known to occur in each area, there are nineteen "empty niches" out of a total of fifty-six "accessible niches." In other words, approximately 65% of the "accessible niches" are filled. Of these, nineteen (51.4%) represent type I influences, five (13.5%) represent type II influences, ten (27%) represent type III influences, and three (8.1%) represent type IV influences. Approximately 35% of the "accessible niches" are unused, again due to the effects of evolutionary specialization, as suggested by Price (1984). Thus, the apparent underutilization of "available niche space" is dramatically reduced when the distribution of such space is viewed through the filter of phylogenetic history. The phylogenetic component of community structure

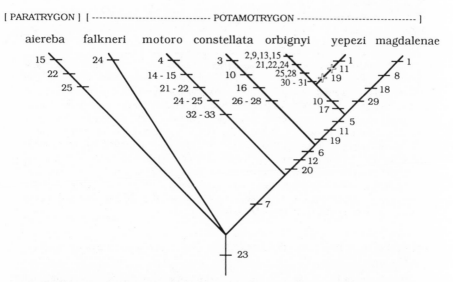

Fig. 8.36. Host cladogram for seven species of neotropical freshwater stingrays based on the phylogenetic relationships of twenty-three species of helminth parasites inhabiting them. Numbers represent parasite species and their phylogenetic relationships (see fig. 7.32 and table 8.5).

Table 8.5 Matrix listing twenty-three species of helminth parasites inhabiting neotropical freshwater stingrays, and the parasite's type of contribution to community structure in each of six communities.

Parasite	Locality[a]								
	1	2	3a	3b	4	5	6a	6b	6c
1. *Acanthobothrium quinonesi*	I	III	0	0	0	0	0	0	0
2. *Acanthobothrium regoi*	0	0	I	0	0	0	0	0	0
3. *Acanthobothrium amazonensis*	0	0	0	0	0	I	0	0	0
4. *Acanthobothrium terezae*	0	0	0	0	0	0	I	0	0
8. *Potamotrygonocestus magdalenensis*	I	0	0	0	0	0	0	0	0
9. *Potamotrygonocestus orinocoensis*	0	0	I	0	0	0	0	0	0
10. *Potamotrygonocestus amazonensis*	0	III	IV	0	0	I	0	0	0
13. *Rhinebothroides moralarai*	I	0	0	0	0	0	0	0	0
14. *Rhinebothroides venezuelensis*	0	III	I	0	0	0	0	0	0
15. *Rhinebothroides circularisi*	0	0	0	0	0	I	0	0	0
16. *Rhinebothroides scorzai*	0	0	IV	III	0	0	I	0	0
17. *Rhinebothroides freitasi*	0	0	0	0	III	0	0	0	0
18. *Rhinebothroides glandularis*	0	0	IV	0	0	0	0	0	0
24. *Eutetrarhynchus araya*	0	0	III	0	III	0	I	I	0
25. *Rhinebothrium paratrygoni*	0	0	III	0	III	0	I	I	I
26. *Paraheteronchocotyle tsalickisi*	0	0	0	0	0	I	0	0	0
27. *Potamotrygonocotyle amazonensis*	0	0	0	0	0	I	0	0	0
28. *Echinocephalus daileyi*	0	0	III	0	0	I	0	0	0
29. *Paravitellotrema overstreeti*	II	0	0	0	0	0	0	0	0
30. *Terranova edcaballeroi*	0	0	II	0	0	0	0	0	0
31. *Megapriapus ungriai*	0	0	II	0	0	0	0	0	0
32. *Leiperia gracile*	0	0	0	0	0	0	II	0	0
33. *Brevimulticaecum* sp.	0	0	0	0	0	0	II	0	0

[a]1 = Magdalena River area in host *Potamotrygon magdalenae;* 2 = Maracaibo area in host *Potamotrygon yepezi;* 3a = Orinoco Delta in host *Potamotrygon orbignyi;* 3b = Orinoco Delta in host *Paratrygon aiereba;* 4 = mid-Amazon in host *Potamotrygon motoro;* 5 = upper Amazon in host *Potamotrygon constellata;* 6a = upper Paraná in host *Potamotrygon motoro;* 6b = upper Paraná in host *Potamotrygon falkneri;* 6c = upper Paraná in host *Paratrygon aiereba.* 0 = parasite is not known to occur in that host in that area; I = phylogenetic component in terms of both spatial (biogeographic) and resource (host) allocation; II = nonequilibrium or stochastic effects (mostly colonization by parasites that normally inhabit teleost fishes or crocodilians but use intermediate hosts that are ingested by stingrays); III = colonization by "preadapted species" (i.e., geographic dispersal without host switching, or with host switching when there is no other congener in the colonized host); IV = colonization by potentially competing species (indicated by the presence of another congener in the host).

constrains the number and the topology of evolutionary pathways open to the community.

The effects of these various influences are not evenly distributed among the six communities. If we examine each community graphically (fig. 8.37), we find three general pictures of community structure. Three communities are dominated by type I influences: the upper Paraná (seven type I and two type

II influences), upper Amazon (six type I influences), and Magdalena (three type I and one type II influences). The structure of these communities is essentially determined by historical factors. Two communities are dominated by dispersal and host switching by "preadapted" species from other source areas: the mid-Amazon (two type III influences) and Maracaibo (three type III influences). In these cases, the parasite species have moved into hitherto

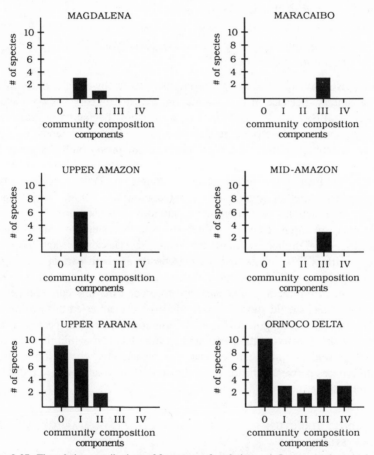

Fig. 8.37. The relative contributions of four types of evolutionary influences on the organization of six communities of helminth parasites inhabiting neotropical freshwater stingrays. O = parasite is not known to occur in that host in that area ("unoccupied niche space"); I = phylogenetic component in terms of both spatial (biogeographic) and resource (host) allocation; II = nonequilibrium or stochastic effects (mostly colonization by parasites that normally inhabit teleost fishes or crocodilians but use intermediate hosts that are ingested by the stingrays); III = colonization by "preadapted species" (i.e., geographic dispersal without host switching, or with host switching when there is no other congener in the colonized host); IV = colonization by competing species (indicated by the presence of another congener in the host).

untapped resources in colonizing the new host. No resident species are displaced; the colonizers simply increase the community's diversity. Finally, one community, in the stingrays of the Orinoco Delta, has been assembled according to variety of influences (three type I, two type II, four type III, and three type IV influences). The structure of this community is therefore the end product of a complex interaction between history, host switching into unfilled resource space, host switching by potential competitors (see legend in table 8.5), and stochastic effects.

A more familiar type of community: Some North American freshwater fishes

Biologists have been studying North American freshwater fishes for more than a century. Because of this fascination, a vast data base has been accumulated; and because of this data base, we are now beginning to catch tantalizing glimpses of historical, geographical, and ecological interactions during the evolution of these animals. Gorman (in press) built his study of freshwater fish communities upon this foundation. He was interested in uncovering the relative role of historical constraints on (1) the species composition or "structure" of communities (cospeciation) and (2) the species associations or "functions" within those communities (coadaptation).

Gorman's research centered on three rivers in the extensive Central Highland Mississippi drainage system, the White, the Gasconade, and the Wisconsin driftless. The White and the Gasconade are both part of the Ozark drainage, closer together and more similar ecologically than either is to the Wisconsin; therefore, it is not untoward to predict that the communities in these two rivers should show a closer affinity. Gorman tested this ecological prediction by examining distribution patterns for twenty-nine species of fish at two levels of analysis: the rivers and specific habitats within these rivers. We will present a step-by-step discussion of this analysis because it highlights the differences between a historical and a nonhistorical approach to the study of community structure.

COMMUNITY STRUCTURE ON A LARGE SPATIAL SCALE: COMPARISONS AMONG RIVER SYSTEMS

1. The nonhistorical approach: Map the distribution of the twenty-nine species onto an unresolved diagram for the drainages (fig. 8.38). Now, reconstruct the relationships of the river drainages by clustering according to the presence/absence of species (fig. 8.39). Clustering river systems according to the raw similarity of their community structure (presence/absence of species) places the Gasconade with the Wisconsin driftless drainage. Notice that the occurrence of *Notropis boops* and *N. greenei* in the Gasconade and White

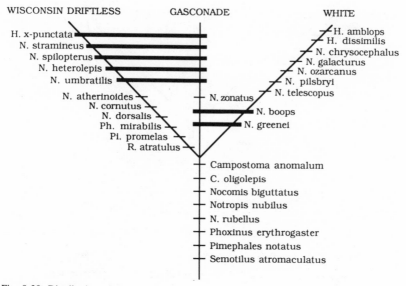

Fig. 8.38. Distribution of twenty-nine North American freshwater fish. *H.* = *Hybopsis; N.* = *Notropis; R.* = *Rhinichthys; Ph.* = *Phoxinus; Pi.* = *Pimephales; C.* = *Campostoma.*

rivers is unresolved; either these species are primitively present for all drainages and have become extinct in the Wisconsin (fig. 8.39a) or they originated in one of the Ozark rivers and dispersed into the other (fig. 8.39b). Ambiguity aside, the results of this analysis do not support the ecological expectation that the two Ozark drainages should display a more-similar community composition.

2. The phylogenetic approach: Mayden (1988) reconstructed the historical relationships among the rivers of Mississippi drainage, based upon comparing information from two sources: the geological relationships of the river drainages *and* the phylogenetic relationships of fishes living in those drainages. (Sound familiar? See chapter 7 for a discussion of this study.) According to this analysis, the three rivers of interest to us are associated in the manner shown in figure 8.40.

This phylogenetic analysis supports the intuitive expectations of community ecologists; the two Ozark drainages are more closely related to each other than either is to the Wisconsin driftless system. Now, how can we explain the anomalous results from the nonhistorical analysis? On the surface, the Wisconsin and the Gasconade drainages appear to have a more similar community structure because they share five species in common, whereas the Gasconade shares only two species with the White River (fig. 8.39). Since this conflicts with the phylogenetic hypothesis (fig. 8.40), we would have to conclude that

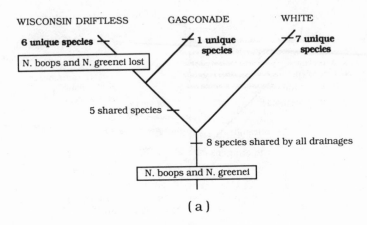

WISCONSIN DRIFTLESS

GASCONADE

WHITE

6 unique species

**1 unique
species**

**7 unique
species**

N. boops and N. greenei lost

5 shared species

8 species shared by all drainages

N. boops and N. greenei

(a)

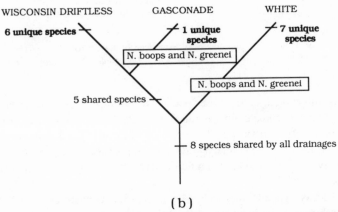

WISCONSIN DRIFTLESS

GASCONADE

WHITE

6 unique species

**1 unique
species**

**7 unique
species**

N. boops and N. greenei

N. boops and N. greenei

5 shared species

8 species shared by all drainages

(b)

Fig. 8.39. Relationships among drainages, based upon shared species. This is a **phenetic** analysis of these relationships: drainages are grouped according to raw similarity (presence/absence of species).

history is a poor predictor of community composition at this level of investigation. However, there are two types of similarities in community structure based upon the presence of shared common species and **sister species.** A nonhistorical approach paints an incomplete picture because it does not examine the entire community; only the "shared species" are investigated, the endemic or "unique species" are disregarded (fig. 8.39). By contrast, a phylogenetic analysis incorporates both historical and nonhistorical information from all species within the community.

Bearing this in mind, let us reexamine the distribution of these fishes (fig.

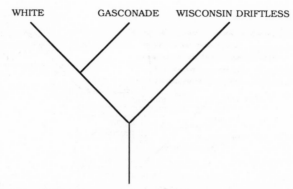

WHITE   GASCONADE   WISCONSIN DRIFTLESS

Fig. 8.40. Mayden's (1988) consensus cladogram for the three river drainages, based on an analysis of forty fish species. This cladogram indicates that *the White and Gasconade are more closely related to each other than either is to the Wisconsin.*

8.41). With an eye to history (i.e., the Ozark drainages are more closely related), we can say that, of the twenty-nine species,

1. We can predict the distribution patterns for twelve species on the basis of history—*Campostoma anomalum, C. oligolepis, Nocomis biguttatus, Notropis nubilus, Notropis rubellus, Notropis boops, Notropis greenei, Phoxinus erythrogaster, Pimephales notatus,* and *Semotilus atromaculatus. Notropis zonatus* and *Notropis pilsbryi* also represent a historical component of the community structure because they are sister species (Mayden 1988) in sister river systems.

2. We cannot predict the distribution patterns for the five species shared between the Gasconade and Wisconsin driftless systems (*Hybopsis x-punctata, Notropis stramineus, N. spilopterus, N. heterolepis, and N. umbratilis*) on the basis of the historical relationships of the drainages or their current proximity. These species are assumed to be where they are because of dispersal, but, in the absence of phylogenies for the fishes, we don't know whether they dispersed from the Wisconsin into the Gasconade or vice versa. For example, a member of both the *H. x-punctata* (*H. dissimilis:* Wiley and Mayden 1985) and *N. heterolepis* (*N. ozarcanus:* Mayden 1989) clades is found in the White River. If these pairs of relatives are sister species, their presence in the White and Gasconade rivers is explained by common history (predictable), followed by the dispersal of *H. x-punctata* and *N. heterolepis* from the Gasconade into the Wisconsin. Once we have identified the existence and direction of dispersal events, we can begin to investigate the environmental variables in common between the two river systems and the impact of dispersing species on an established community.

3. We do not have enough information about the phylogenetic relationships

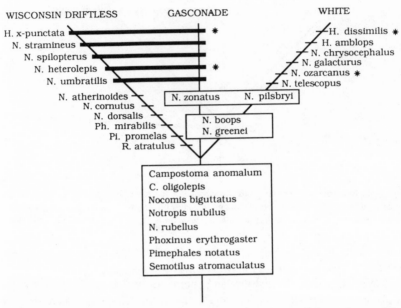

WISCONSIN DRIFTLESS

GASCONADE

WHITE

H. x-punctata
N. stramineus
N. spilopterus
N. heterolepis
N. umbratilis

*

*

H. dissimilis *
H. amblops
N. chrysocephalus
N. galacturus
N. ozarcanus *
N. telescopus

N. atherinoides
N. cornutus
N. dorsalis
Ph. mirabilis
Pi. promelas
R. atratulus

N. zonatus   N. pilsbryi

N. boops
N. greenei

Campostoma anomalum
C. oligolepis
Nocomis biguttatus
Notropis nubilus
N. rubellus
Phoxinus erythrogaster
Pimephales notatus
Semotilus atromaculatus

Fig. 8.41. Distribution of fishes (ecological data) examined from a phylogenetic perspective. Historical components of community structure are enclosed within *boxes*. * = putative members of the same clade. *H. = Hybopsis; N. = Notropis; R = Rhinichthys; Ph. = Phoxinus; Pi. = Pimephales; C. = Campostoma.*

of the remaining endemic species to determine how many are present due to dispersal (unpredictable) and how many are present due to common speciation patterns (predictable; i.e., sister species in sister drainages like *N. zonatus* and *N. pilsbryi*). In order to resolve this problem we need a phylogeny for the problematical *Notropis* group (groups?).

In summary, examination of the species presence/absence from a phylogenetic perspective has uncovered a by now familiar pattern: communities comprise both historical (cospeciation) and nonhistorical (dispersal) elements. This, in turn, highlights the need for researchers interested in investigating community structure to examine their ecological data within a phylogenetic context.

COMMUNITY STRUCTURE ON A SMALL SPATIAL SCALE: COMPARISONS AMONG HABITATS WITHIN THE RIVERS

In this section we will turn our attention to fish communities inhabiting pools and slow raceways of third- and fourth-order streams in the White, Gasconade, and Wisconsin driftless river systems. Since the presence/ab-

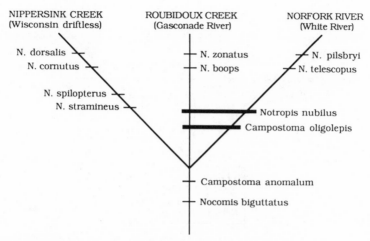

Fig. 8.42. Distribution of predominant species (> 5% of all individuals) collected in pool and slow raceway habitat of third- and fourth-order streams from three river drainages. *N.* = *Notropis*.

sence of rare fishes may be strongly affected by sampling errors, we will base this analysis upon the predominant species in these habitats.

1. The ecological data: Let us begin by mapping the distributions of predominant species onto an unresolved diagram for the creeks (fig. 8.42).

2. The phylogenetic perspective: Now, let us reexamine these distributions in light of the historical relationships of the river (creek) systems (fig. 8.43).

The distribution of species in these headwater communities indicates that Roubidoux Creek (Gasconade River drainage) and Norfork River (White River drainage) are more closely related to each other than either is to the Nippersink Creek (Wisconsin driftless drainage). Since this *agrees* with Mayden's phylogenetic analysis, it appears that **the effects of historical constraints on community structure can be detected at both the large spatial scale of river drainages and the small spatial scale of individual habitats within those drainages.** The incorporation of additional phylogenetic information into this study uncovers even more interesting aspects of the evolution of community structure. For example, the relationships depicted in figure 8.41 indicate that *Notropis nubilus* and *Campostoma oligolepis* are primitively present in all three drainages, while *N. boops* is shared between the Gasconade and White rivers. Based upon this we can see that,

1. Roubidoux Creek community structure is completely predicted by history (six out of six species): four species are primitively present in all these river drainages (*Nocomis biguttatus, C. anomalum, Notropis nubilus, C. oli-*

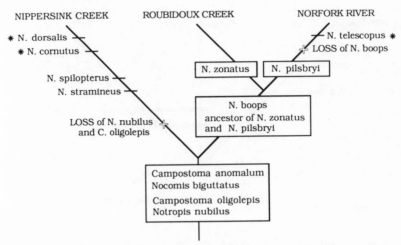

Fig. 8.43. Analysis of data within the phylogenetic framework presented by Mayden (1988) and incorporating additional information from the drainage-level analysis (fig. 8.41). * = endemic species.

golepis), one species is shared with its sister system the White River (*Notropis boops*), and one species, *Notropis zonatus*, occurs due to a shared speciation event with the White River—remember, *N. zonatus* and *N. pilsbryi* are sister species.

2. Norfork River community structure is almost completely predicted by history (five out of six species): four species are primitively present in all these river drainages (*Nocomis biguttatus, C. anomalum, Notropis nubilus, C. oligolepis*), and one species, *Notropis pilsbryi*, occurs due to a shared speciation event with the Gasconade River. The remaining species, *Notropis telescopus*, is endemic to the White River drainage so we would have expected to find it based upon our analysis at the drainage level. However, this does not tell us where that species came from originally (see discussion for endemics in the preceding section).

3. Nippersink Creek community structure is only weakly predicted by history (two out of six species): two species are primitively present (*Nocomis biguttatus, C. anomalum*), and two species are endemics, *Notropis dorsalis* and *Notropis cornutus*. The presence of *Notropis spilopterus* and *Notropis stramineus* is problematical; as discussed in the previous section, these species may have dispersed into the area.

Overall, then, thirteen out of eighteen of the predominant species in these pool and slow raceway habitats represent the presence of historical constraints on community structure. This, in turn, raises the following questions: (1) Why is *Notropis boops* absent in the Norfork River? (2) Why are *N. nubilus* and *C. oligolepis* absent in Nippersink Creek? (3) Why are *N. spilopterus* and *N. stramineus* absent in Roubidoux Creek? The answers to these questions might be found by examining the ecological interactions among species within these communities.

ANALYSIS BASED UPON CATEGORIES OF ECOLOGICAL INTERACTIONS

Let us begin by adding the ecological profiles to the information discussed in the previous section (fig. 8.44). There are some fascinating patterns here.

1. The oldest component of all these creek communities is the near-benthic forager. In the Ozark systems, the upper and midpelagic species may have been added next if the ancestor of *Notropis zonatus* and *N. pilsbryi* was a midpelagic forager. If not, the upper-pelagic forager was added to the community first, and the midpelagic species second.

2. Both the Ozark communities have the same functional structure: three near-benthic species, one lower-pelagic species, one midpelagic species, and one upper-pelagic species. This agrees with both the historical relationship of the areas and the similar environmental parameters found in the creeks.

3. Nippersink Creek has a radically different ecological composition from the Ozark creeks: one near-benthic species, three lower-pelagic species, and two midpelagic species. We need to resolve the phylogenetic relationships for the four most recent additions to this community in order to determine which, if any, of them are present because they evolved there (perhaps the endemics

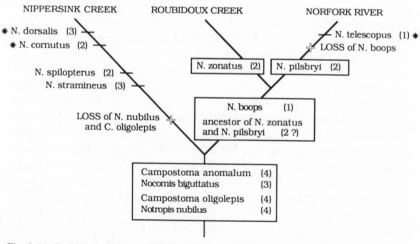

Fig. 8.44. Ecological profiles of dominant species mapped onto the cladogram for the river systems. Foraging categories: *1* = upper pelagic; *2* = midpelagic; *3* = lower pelagic; *4* = near benthic.

Notropis dorsalis and *N. cornutus*) or because they dispersed in (perhaps *N. spilopterus* and *N. stramineus*). If we can find historical evidence for dispersal, then the phylogenies can be used to determine whether the current interactions among the residents and colonizers is ancestral (type III, table 8.4) or derived (type IV). This, in turn, will tell us where to look for interspecific competition in this community.

4. "Why is *N. boops* absent in the Norfork?" The macroevolutionary patterns indicate that *N. boops* was potentially replaced by its ecological equivalent *N. telescopus*. Is there any way to move from a hypothesis based upon patterns to an experiment designed to test processes? In this case the answer to this question is definitely yes. Although rare, *N. boops* is present in the Norfork. Given this, we need to collect further information about the history of *N. telescopus* (i.e., what is its sister species, and did it evolve in the Norfork or disperse into that area?). If it dispersed into the Norfork, and if experiments reveal that it is currently out-competing *N. boops* in that river, then we have convincing evidence for competitive exclusion *without having to invoke the ghost of competition past.*

5. "Why are *N. nubilus* and *C. oligolepis* absent in Nippersink Creek?" Contrary to the *N. boops* example, these species have not been replaced with ecological equivalents. The answer to this probably lies in the structure of the substrate in Nippersink Creek (i.e., it is more sandy than in the Ozark streams, and both these species are near-benthic foragers). If so, habitat- and food-preference experiments should help resolve this problem.

ANALYSIS BASED UPON A SPECIFIC ECOLOGICAL INTERACTION

In the final section of his study, Gorman performed a series of experiments designed to investigate the role of preferred habitat use in the assemblage of the Roubidoux Creek community. The experimental protocol was simple and elegant. Preferred vertical distributions were determined for *Nocomis biguttatus, Notropis nubilus, C. anomalum, C. oligolepis, Notropis zonatus, and Notropis boops,* by observing isolated individuals in the laboratory. These patterns were then compared with distributions observed in the field and obtained in the laboratory in a mixed-species group, with stunning results. The fishes that demonstrate the same distributions in the lab (both alone and in mixed-species groups) and in the field are the four oldest members of the communities, *Nocomis biguttatus, Notropis nubilus,* and the two *Campostoma* species. Additionally, the overlap in preferred habitats among these species is low. So, as would be expected for the oldest part of an ecological association, this component has reached, and is being maintained in, a stable state. On the other hand, the more recently derived members of the community (*Notropis zonatus* and *Notropis boops*) still show evidence of interspecific interactions, that is, they are displaced from their preferred position in the field and in the mixed-species group. As discussed in the previous section, evidence of competitive interactions should be sought in two places within the community: between an older member of the community and its functional replacement (*Notropis boops* and *Notropis telescopus*) or between the most recently derived member of the community and the original residents (*Notropis zonatus*).

Overall, three generalizations about the evolution of these freshwater fish communities can be drawn from this extensive study: communities are composites of both historical and nonhistorical components; this pattern can be detected on both large and small spatial scales; and historical ecological analysis can provide a framework for experimental investigations of the impact of ecological interactions on community assemblage.

Summary

Although the preceding two studies are preliminary, we have presented them to illustrate two points. First, it is possible to combine phylogenetic, biogeographic, ecological, and behavioral data to ask questions about the relative contributions of historical and nonhistorical influences on the spatial and resource allocation patterns in biotas. Second, **patterns of evolution in communities are generally not the result of a single influence.** This is a familiar result. We discovered a similar generalization in the speciation dynamics involved in producing clades, and the coevolutionary processes involved in producing tight ecological associations. Paleontological studies (e.g., Boucot 1975a,b, 1978, 1982, 1983) tend to support this view.

Adaptive changes in community structure involve both species composition and interactions. They tend to be evolutionarily conservative today, and the paleontological evidence suggests that they have been conservative in the past. Thus the conservative nature of adaptive changes and adaptive radiations in individual clades (chapter 5) is carried through into biotic structure. Type III and IV influences in community evolution encompass the two components of the MacArthur-Wilson (1967) equilibrium model of island biogeography, colonization by preadapted species when biotas are below equilibrium numbers, and colonization by competing species when biotas are at or near equilibrium numbers. Type I and II influences are nonequilibrium phenomena, which also must be taken into account when attempting to explain biotic structure and diversity. If these nonequilibrium influences are as pervasive as we (and others—see, e.g., Wiens 1984) think, and as the examples we have presented in this book suggest, much of the structure found in communities could be due to historical association. This structure, in turn, could be misinterpreted to indicate the influence of proximal factors (e.g., strong synecological interactions) if the evolutionary interactions within the community are not investigated within a historical framework.

At the moment, it is impossible to draw any meaningful generalizations about the relative contributions of historical constraints and adaptive change to the evolution of either closely interacting species (coevolutionary studies) or interacting communities (community evolution studies). Traditionally there has been very little interaction between biologists who study these two aspects of coadaptation. We hope we have demonstrated that there is so much common ground between these research programs, at least from a macro-evolutionary perspective, that any cross-fertilization will be mutually beneficial. Indeed, some researchers are already discovering the benefits of examining complex systems within a cooperative atmosphere (see, e.g., Futuyma and Slatkin 1983 and Strong et al. 1984).

9 Prospective: Mastering the Possibilities of Historical Ecology

We hope that the analytical results and theoretical discussions from the preceding chapters have established two things about historical ecology as a research program. First, it is **feasible;** both the basic methodologies and some of the pertinent models have been developed and are awaiting application in the ecological and ethological domains. Second, it is **relevant** because *the patterns detected by this method are important components of evolutionary explanations that would not have been discovered without phylogenetic analysis.* We believe that historical ecology has a bright future, not only because so much of the data base is yet to be investigated, but because of the potential for forging connections with other research programs. In this chapter we will summarize the current data base and discuss some new areas of research that may either benefit from, or complement, historical ecological studies.

The Current Data Base

Historical ecology is concerned with uncovering and tracking the trail left by evolutionary change. This trail, although appearing to cut a uniformly straight swath through time, is in reality a complex mosaic of numerous processes. Some of these processes are directional, driving the path forward. Others seem to disrupt this progression and bend the pathway just a little with their oscillations. Some of these processes seem to be initiated by ecological phenomena, others by genealogical (especially developmental) phenomena. Regardless of the manner in which these changes are initiated, they are mediated by ecological considerations. We have concentrated our attention on two major evolutionary processes, **speciation** and **adaptation,** and on two major manifestations of these processes, **diversity** and **ecological interactions.** We have discovered that there is a paucity of rigorous phylogenetic hypotheses from which we can begin historical ecological studies, but from those that are available we can begin to perceive a glimpse of the not-so-obvious in evolution.

The Historical Ecological Perspective on Diversity: Speciation and Extinction Rates

In chapter 4 we discussed how to discriminate among the products of various speciation modes. Having delineated the patterns, we then asked how often each mode occurred and discovered that the majority of cases support an interpretation of vicariant speciation. This implies that most of the diversity component defined by species number is primarily the result of the physical disruption of gene flow by geological and geographic alterations, and not of the direct effects of adaptive changes. Allegorically speaking, speciation provides the bricks with which the evolutionary pathway is constructed, and common ancestry provides the mortar.

If speciation is affected by environmental changes, then we must search for environmental processes that, like speciation, are irreversible. Geological evolution is such a process. Although speciation can occur in the absence of geological changes, such changes will dramatically affect the rates of speciation if the rate of geological partitioning is greater than the rate of population equilibration, determined by mutation rates and gene flow and affected by short-term environmental processes. Cracraft (1982b, 1985a) suggested that the key to understanding speciation rates for any clade is the history of geological change and accompanying vicariant speciation.

Vicariant speciation may also influence our perceptions of the rates and evolutionary roles of extinctions. In every vicariant speciation event, the ancestor giving rise to the allopatric sister species becomes extinct. *If the majority of speciation events are due to vicariance, then the majority of extinctions historically have been by-products of speciation and not the result of irreplaceable losses of evolving lineages.*

Adaptation and Key Innovations

In chapter 5 we established a framework for asking four macroevolutionary questions about the second evolutionary process, adaptation. First, did the character of interest originate in the existing species or in an ancestor and, if it originated in an ancestor, how long ago and in what environmental context? Second, how often has the trait arisen in evolution, that is, is its occurrence unique or is it a repeating evolutionary theme? Third, for any adaptive scenario involving evolutionary changes in traits or the evolutionary accumulation of traits, in what sequence did the component traits appear? Are they historically independent or historically correlated, and if they are correlated, did they arise in the same or in different ancestors? Fourth, how frequent are adaptive changes in phylogeny; do ecological and behavioral changes promote or constrain phylogenetic diversification?

In every case studied so far, ecological and behavioral change is more conservative than phylogenetic diversification. This complements the discovery that the predominant speciation mode does not require adaptive responses to initiate speciation. It would appear, then, that although the production of biological diversity is the result of both speciation and adaptation processes, the two are not always tightly linked. Whenever adaptive processes, in the form of ecological or behavioral diversification, accompany speciation, they can be interpreted as contributing to the production of diversity. Conversely, whenever such adaptive changes lag behind speciation, as is usually the case, they can be interpreted as constraints, or cohesive influences, on evolutionary change. Hence, the conservative nature of evolutionary diversification in ecological correlates of adaptation plays an important role in maintaining a degree of organization and predictability in the production of diversity.

Coming as it does from a macroevolutionary perspective, the assertion that some adaptive changes function as evolutionary constraints may at first sound counterintuitive. However, if there were no underlying core of historical continuity within and between species, there would be no rationale for a comparative method. We believe that this perspective is consistent with MacArthur's proposal that we should search for evidence of competitive interactions among sympatric congeners. This proposal can be restated in historical ecological terminology: macroevolutionary conservatism in ecological and behavioral evolution leads to the potential for niche overlap in closely related species, so that when congeners are sympatric, we expect to find ecological interactions on a microevolutionary scale. Only if diversification in ecology and behavior are conservative on a macroevolutionary scale would we generally expect to find congeners with enough similarity in ecological and behavior traits to create the initial conditions for such niche overlap, competition, and habitat partitioning.

The Historical Ecological Perspective on Interactions

The conclusion that ecological diversification plays a cohesive role in evolution, based on the study of adaptation and speciation patterns within clades, was reinforced when the evolutionary framework was expanded to include interactions among clades. In chapter 7 we discussed the work of researchers interested in uncovering the extent to which different clades share common histories of speciation. Reconstruction of the speciation patterns for co-occurring groups of organisms revealed a background of phylogenetic association among clade members which often accounted for more than 50% of existing ecological associations. Superimposed upon this phylogenetic framework were independent speciation events accounting for the variation in species composition among biotas that share common historical elements. This research was extended to include the traits characterizing such ecological associations in chapter 8. Although studies of this particular evolutionary component are rare, the current data base supports an interpretation that ecological and behavioral diversification have been highly conservative phylogenetically in the evolution of biotas; hence, as we discovered in our examination of adaptational change within a clade, this process is a major cohesive force, rather than a diversifying force, in evolution.

Summary of the Current Data Base

The evidence presented in chapters 4, 5, 7, and 8 reveals a substantial degree of phylogenetic influence on both the species composition and species interactions within contemporaneous ecological associations. Geologically mediated speciation patterns (chapter 4) lead to geologically mediated cospeciation patterns (chapter 7). Conservative adaptive responses in phylogeny (chapter 5) herald a conservative interaction structure in ecological associations and biotas (chapter 8). The interchange between speciation and adaptation on all evolutionary levels produces stable assemblages of ecological associations, because the core of resident species exhibiting plesiomorphic interactions will serve as a stabilizing force with respect to each other's evolution and to the addition of colonizers. Paleontological data support the proposal that the structure and composition of ecological associations have always been highly conservative (Boucot 1975a,b, 1978, 1981, 1982, 1983). Hence, the glimpse of the evolution of global diversity that this new research program provides is one of long periods of ecological stability interrupted occasionally by adaptive shifts. At the moment the data base is limited, so these glimpses are but fleeting, tantalizing hints of future discoveries. The answer to this riddle cannot fail to be exciting, whether it is a general proposition that phylogenetic constraint rather than adaptation is the predominant force shaping species associations on the macroevolutionary scale, or the discovery that each system represents a unique outcome of the interplay between the two processes.

Are there any practical advantages for a biologist who adopts the historical ecological viewpoint? If you are a graduate student, should you invest some of your most precious commodity, time, in formulating and examining your particular question within this framework? If you are a graduate adviser, should you encourage students to master the possibilities of this program? Should you even consider leading the way by expanding the scope of your own research? We suggest that the answer to these questions is a resounding "Yes!" Historical ecology expands the spatial and temporal perspective of evolutionary biology, opening doors on more, and more varied, research possibilities. One ambitious project that is well underway is the development of the freshwater fish fauna of North America as a model system for historical ecological studies (see Mayden in press). As a result of this extensive cooperative venture, over 40% of the fauna has been analyzed phylogenetically, and a plethora of ecological, behavioral, and distributional data have been collected and examined within that phylogenetic framework. No doubt similar collaborative efforts by other groups of biologists could produce equally substantial results.

One reason for the success of the North American ichthyologists is that robust methods for studying historical effects in ecology and ethology are already available. Because of this, field and experimental biologists can concentrate on the development of the empirical data base without the distractions of constant philosophical and methodological debates. Conversely, because the program is so young, there is room for future methodological fine-tuning and innovation, so theoretically minded biologists can also participate in this time of high productivity and burgeoning ideas. Truly, this is the "best of times" for any scientific discipline! To date, the empirical research supports the hypothesis that persistent historical effects are an important component of evolutionary diversification, so there is reason to believe that embarking on similar quests will produce useful new information. In addition, since relatively few studies have been undertaken, there is little chance of embarking on a project that will not be original. We view historical ecology at this early developmental period as a form of "no-loss" research. No matter what you find, it is interesting because the field is still in its basic discovery phase. For example, what if the current data base and its implications about the importance of phylogenetic constraints on adaptive changes in the evolution of biotas is not representative of evolution in general? Using the implications of the current data base as the null hypothesis for future research, we can determine whether the generalities of today are in fact exceptions to some other general pattern, rather than the pattern itself.

It is important at this point to reemphasize the nonreductionist approach that historical ecology represents. What we seek, and what we think is more representative of the way in which evolution has occurred, are explanations that are constructed somewhat like analysis of variance (ANOVA) studies. That is, rather than asking what single factor (e.g., "history" or "ecology") accounts for the evolution of a given system, we would ask, what has been the relative contribution to, or effect of, a variety of possible processes on the evolution of the system we are studying? This is the reason we encourage closer collaboration between ecologists, ethologists, and systematists to produce an integration of microevolutionary and macroevolutionary data.

Possibilities for Future Research

Conservation

Neither of the authors of this book is a conservation biologist. Like everyone else on this planet, however, we have a vested interest in contributing to the protection of the global ecosystem. We believe that historical ecological methods can provide information that will complement current conservation/ management practices based on theories about the relationship between species number and area size or number. For example, historical biogeographical studies (cospeciation in a geographical context) can identify geographical areas that have been associated with "hot spots" of evolutionary activity in the past, and which might serve as areas of evolutionary potential in the future. Such areas of endemism are interesting because they do not always encompass the greatest number of extant species, nor are they always extremely large or centrally located. As a consequence, if, as is so often the case, a choice must be made about areas requiring protection, historical ecological data can provide information about regions that have been very important in the evolution of biological diversity.

For example, consider the neotropical freshwater stingrays and their associated parasite communities. Historical ecological studies have demonstrated that the ancestors of these organisms moved into the Amazon River at a time in the far distant past when that river flowed into the Pacific Ocean. This component of the South American freshwater fauna is thus the result of a historically unique event that can never be repeated, or recovered, if lost. Now, suppose a decidedly unusual conservationist felt that it was important to preserve the parasites of neotropical freshwater stingrays. How would we rank the six known communities of ray parasites with respect to their importance in the evolution of the fauna (see chapter 8)? The "least important" areas would be (1) the Lake Maracaibo area, because it contains only species that evolved, and currently occur, elsewhere, and (2) the mid-Amazon, because it shelters only a single endemic species. The remaining four areas all have a relatively large number of endemic species. Of these, the community in the delta of the Orinoco River has the highest species number (eleven species), due to the combination of endemics that evolved there plus species that have colonized from elsewhere. Unfortunately, it is likely that the delta of the Orinoco will face increasing pressures for industrial development in the future. The remaining three communities occur in more remote areas, the upper Paraná (six species), the upper Amazon (six species), and the Magdalena (four species). Protecting these areas would protect a high proportion of the species of stingray parasites (sixteen out of twenty-three species), including five of the eleven species of helminths that occur in the Orinoco Delta. If we examine these three areas more closely, we find that four of the six species found in the upper Amazon, three of the six species in the upper Paraná, and three of the four species in the Magdalena are not known to occur anywhere else. Because these areas are less likely to feel the encroachment of human development, they also represent good compromises between the need to protect and the need to develop.

The historical ecological perspective leads us to postulate two general cases in which traditional criteria could be detrimental to long-term conservation efforts. (1) Preserving areas based on current species numbers: If the chosen area encompasses the region of overlap between two biotas, then we may be preserving marginal habitats that have limited the expansion of members of both biotas. Confining organisms to such an area may, in turn, increase the likelihood of competitive interactions that could be detrimental to all species. It has long been postulated that the more closely related the species, the more likely they are to use the same resources (MacArthur 1958, 1972; Root 1967), something the evidence presented in this book corroborates. In undisturbed communities, congeneric species are found together less often than expected by a process of combining groups randomly from a pool of all possible species that have access to the communities (Bowers and Brown 1982). Furthermore, congeners rarely occur together often enough to allow adaptations for competitive exclusion to evolve (Maurer 1985; Graham 1986). Hence, artificially confining sets of closely related species from separate communities that evolved allopatrically could be counterproductive for the survival of any of them. (2) Preserving many small areas: In this scenario, we are asked to choose between many species with small geographic ranges and few species with large ranges. If the species exhibiting small ranges represent the outcome of peripheral isolates allopatric speciation, saving a large number of peripheral isolates areas rather than a few large central areas might destroy the central populations from which all those descendant species came. In both of these general cases, we would be trading evolutionary potential for current diversity, and possibly harming the current diversity in the process.

We are not advocating a blanket conservation policy based on phylogenetic considerations. However, we do believe that history has been a critical missing component in many conservation studies. In fact, the current historical ecological data base suggests some reasons for pessimism about global conservation efforts. Ecological structure in biotas appears to be much more conservative evolutionarily than previously thought, raising questions about the adaptability of ecosystems on time scales relevant to human activities. The greater the phylogenetic constraints on an ecosystem, the more slowly it will be able to respond to perturbations. And if it is not bad enough that current ecological structure is older and more conservative than we thought, paleontological studies indicate that this may be the way it has always been. As a consequence, our concern about the survivability of any given ecosystem should be directly proportional to the degree of phylogenetic constraint on its structure and composition. Those ecosystems that have been around for the longest time and contain the largest number of endemic species would be those least likely to survive human intervention, and thus are the most in need of immediate conservation. Similarly, those species that have been part of any given biota for the longest periods of time may be those most in need of protection against exploitation and removal.

Integrating Historical Ecology and Functional Morphology

In chapter 5 we discovered that the analysis of adaptive radiations is more complex than comparing the degree of ecological diversification with the degree of speciation in any given clade. Specifically, we suggested that evolutionary diversification in functional morphological attributes might also be involved (see also Coddington 1988). Fortunately, an explicitly phylogenetic approach to the study of functional morphology has been developing, pioneered by George Lauder (e.g., 1981, 1986, 1988, 1989; Schaefer and Lauder 1986; Lauder and Liem 1989).

Lauder's approach is based on the proposal that the realized phenotypic diversity is only a fraction of all possible phenotypic diversity, and attributing the "missing" possibilities solely to the effects of environmental selection overlooks the role for developmental dynamics and phylogenetic constraints in determining the ranges of realized phenotypes. He suggested that, since organisms are historical entities, the intrinsic phylogenetic component of organismal design may impose limitations on the way in which and the degree to which structural modifications can occur in the evolution of a lineage. Lauder (1981) proposed that the relative importance of phylogenetic constraints in the evolution of functional morphology could be assessed by examining (1) emergent structural or functional traits that have general properties within (2) an explicit phylogenetic framework to (3) test hypotheses of phylogenetic constraint by comparing general properties among clades. He further suggested that this mode of study could be integrated with studies of extrinsic (environmental) constraints on form, to generate robust explanations for the range of existing morphotypes.

Food: Lingual feeding in lizards

It has been suggested that the tongue has played a central role in the evolutionary movement from aquatic to terrestrial vertebrates (Bramble and Wake 1985; Hiiemae and Crompton 1985). However, an understanding of the evolution of vertebrate feeding patterns requires more information about the tongue's role than is presently available for many groups. Schwenk and Throckmorton (1989) began the process of producing a solid foundation for a general evolutionary study by focussing their attention on one particular group, the Sauria (lizards). They were primarily interested in examining the initial stages of prey capture and ingestion in these organisms. Independent phylogenetic analysis of the Sauria based on 148 morphological characters

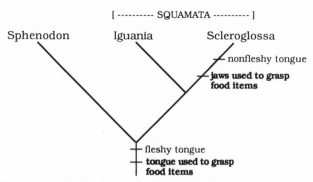

Fig. 9.1. The evolution of feeding modes in squamate lizards. Change in tongue structure and feeding behavior mapped onto a phylogeny for the squamata. (Estes, de Queiroz, and Gauthier 1988; redrawn from Schwenk and Throckmorton 1989.)

(Estes, de Queiroz, and Gauthier 1988) suggests that modern lizards comprise two basal groups, the Iguania and the Scleroglossa. Members of the fleshy-tongued Iguania capture prey items (animal or plant) by combining rapid extensions of the dorsally arched tongue with forward and downward movements of the head to pin the food item to the substrate. Food is thus both grasped and drawn into the mouth by tongue action alone. In contrast to this feeding mode, members of the "nonfleshy"-tongued Scleroglossa grasp and ingest prey items solely through the use of their jaws; the tongue is not involved in these early stages of feeding. Outgroup comparisons with *Sphenodon* (the tuatara of New Zealand) suggest that the condition found in the Iguania is plesiomorphic (fig. 9.1).

This study leads us down a number of interesting pathways. To begin with, although tongue morphology is extremely diverse (i.e., provides important taxonomic characters), there has been only one major change in feeding mode within this group of lizards. The presence of lingual feeding in all members of the Iguania represents the persistence of an ancestral condition, while the appearance of jaw feeding was an evolutionary novelty in the ancestor of the scleroglossans, which has been retained in all members of that group. The shift in feeding mode thus occurred against a background of considerable phylogenetic constraint. Interestingly, the origin of jaw feeding is associated phylogenetically with an alteration in tongue morphology, but this by itself does not establish a causal link between the two. Further research is required to uncover the origin of lingual feeding and the fleshy tongue. If they also arose at the same point in evolution, the hypothesis of a causal link between feeding mode and lingual morphology will be strengthened. Additionally, experimental studies investigating the dynamics of the derived feeding mode, using the plesiomorphic feeding mode as the control, are necessary to establish a functional connection between shifts in behavior and shifts in morphology. A cascade of new questions will naturally be produced: for example, (1) has there been an adaptive radiation of the scleroglossans and, if so, (2) can the new feeding mode be identified as the "key innovation," (3) do other behavioral modifications accompany this change, (4) have there been shifts in diet correlated with jaw feeding, and (5) what other morphological innovations appear in the ancestor of the scleroglossans? Overall then, this example is exciting because of the number of new questions generated and because it proposes that a relatively simple modification in a morphological pattern may have been responsible for a number of behavioral and ecological evolutionary changes.

Sex: Inner ears and mating calls in frogs

Auditory cues are an integral part of all anuran breeding systems. Both the structure of the male's vocalization and the female's response to the call are extremely species-specific. In fact, of the more than twenty-seven hundred frog species no two species produce identical calls (Blair 1964). These behavioral characters are thus hypothesized to be critical components in the maintenance of reproductive cohesion within a species and reproductive isolation among species. Ryan (1986) examined the hypothesis that changes in the structure of the neuroanatomy of frog ears amplified the divergence of mating-call behaviors, which in turn may have affected the rate of speciation in various lineages. The frog inner ear contains two organs, the amphibian papilla and basilar papilla, which are sensitive to airborne sound waves. Based upon an examination of the amphibian papillae in approximately eighty anuran species, Lewis (1984) divided the organ into four states characterized by an initial increase from one to two sensory patches and a subsequent increase in the complexity of the second patch. This increase in morphological complexity has behavioral consequences; frog species with the best-developed amphibian papillae respond to a wider range of call frequencies than their acoustically more restricted relatives. Ryan mapped the differences in the structure of the inner ear onto a phylogeny for the major frog families (fig. 9.2 depicts the distribution of these traits on the more recent phylogenetic tree for frog families by Duellman and Trueb 1986), using the condition in the sister group to the anurans, the urodeles (salamanders), to establish the plesiomorphic condition.

Ryan discovered that changes in the inner-ear system were highly correlated with frog phylogeny and were highly conservative. Once again, history provides the backdrop to the evolutionary play. Ryan concluded that the morphological structure of the inner ear has influenced speciation rates within the anuran clade by affecting the degree of divergence in detectable mating-call

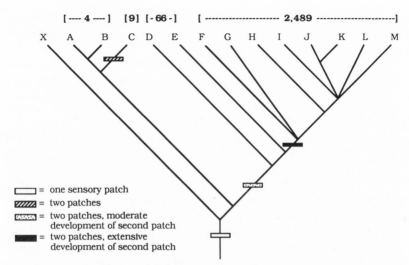

Fig. 9.2. Structure of the amphibian papilla of the inner ear, mapped onto the phylogeny for some of the anuran families. *A* = Ascaphidae; *B* = Leiopelmatidae; *C* = Discoglossidae; *D* = Pipidae; *E* = Pelobatidae; *F* = Leptodactylidae; *G* = Dendrobatidae; *H* = Hylidae; *I* = Bufonidae; *J* = Ranidae; *K* = Hyperoliidae; *L* = Rhacophoridae; *M* = Microhylidae; *X* = the outgroup, Urodela. *Numbers* = number of species associated with each character state. (Redrawn from Duellman and Trueb 1986; Ryan 1986.)

frequencies. Thus, the more frequencies you can detect, the more material is available for potential evolutionary modification and diversification. When the number of species among lineages with different inner-ear character states are examined, the expected trend towards increasing numbers of species associated with increasing complexity of the inner ear is observed. The role of breeding behavior in maintaining reproductive isolation has become a focal point of many behavioral ecological studies. Ryan's study is an important step towards integration of morphological, behavioral, and genetical information into a more complete investigation of this problem. The next step involves the collection of additional data, particularly for the families missing from this initial analysis. Additionally, it will be interesting to examine other potential correlates of speciation rates within the context of phylogeny. For example, Duellman's (1985) study suggests that other reproductive characters may be associated with anuran adaptive radiations (see also the discussion in chapter 5). The six families displaying only one reproductive mode have not undergone widespread speciation, whereas the eight most species-rich families are characterized by either widespread convergent adaptation or a combination of widespread convergence and the appearance of novel parental-care behaviors. In this case, then, the adaptive radiations of various anuran lineages may be the result of a combination of factors, including a key morphological innovation that was reinforced by a variety of ecological and behavioral innovations.

More food: Evolution of feeding modes in centrarchid fishes

Sunfishes belong to the family Centrarchidae, a group of thirty-two species of fishes endemic to North America. Thirty of the species are foraging generalists. The remaining two species, the pumpkinseed (*Lepomis gibbosus*) and the redear (*L. microlophus*) eat only snails, a dish scorned by all other self-respecting centrarchids except, occasionally, the green sunfish, *L. cyanellus*. The pumpkinseed and redear eat snails in the wild and in the laboratory in a very refined manner. Rather than simply gobbling up snails and swallowing them whole, they crush the snail shells in the pharynx, separate the shell from the body, and eject the shell detritus before swallowing the edible snail body (Lauder 1983a,b). Lauder (1986) suggested that this particular behavioral pattern is an excellent model system for studies in functional morphology and behavioral evolution for four reasons: the behavior is apomorphic and not widespread within the group; there are morphological specializations in the feeding mechanism of species that regularly exhibit the crushing behavior; the motor patterns used in the behavior can be identified experimentally by recording electrical activity in the muscles involved in snail crushing; and the crushing behavior is composed of repetitive crushing phases, and cyclically repeating motor patterns are excellent systems for studying the manner in which behavioral activities are generated by the nervous system.

Lauder's experimental studies (1983a,b) demonstrated that, in terms of muscles, bones, muscle origins, or muscle insertions, the basic musculoskeletal design of the jaw and buccal area of *L. gibbosus* and *L. microlophus* does not differ from that of any other centrarchids. There are, of course, some differences that emerge from closer examination. For example, the two snail eaters have more-robust lower pharyngeal jaws, with larger and more-rounded teeth, and the upper pharyngeal jaw teeth are hypertrophied with respect to the lower jaw teeth. There are two basic chewing patterns. First, when centrarchids are given fishes and worms to eat (the snail eaters will deign to ingest nonsnail prey in the laboratory when faced with the option of missing a free meal), all species except *L. microlophus* exhibit the same rhythmic pattern of alternating activity in the pharyngeal musculature, lasting for as much as a minute, as the prey is transported from the pharynx into the esophagus. Second, when the pumpkinseed, redear, and green sunfishes eat snails, nearly all of the pharyngeal muscles are electrically active simultaneously during the crushing process; no rhythmic alternating pattern is found. After the snail has been crushed, the shell is separated from the body by a

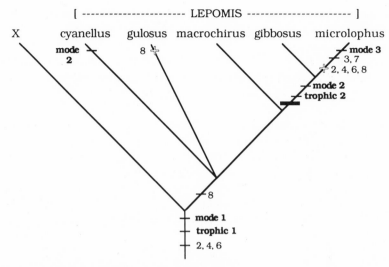

Fig. 9.3. Simplified phylogenetic tree for centrarchid fishes. X = outgroups. *Ambloplites, Pomoxis,* and *Micropterus. Numbers* = functional characters (patterns of muscle activity involved in chewing food; see Lauder 1986 for descriptions). *Mode 1* = only rhythmic pattern of chewing food; *mode 2* = rhythmic for soft foods, crushing for snails; *mode 3* = only crushing pattern. *Trophic 1* = generalist, no snails; *trophic 2* = snail specialists. *Black bar* = three morphological characters that serve as synapomorphies for *L. gibbosus* + *L. microlophus* (hypertrophy of the pharyngohyoideus muscle, hypertrophy of the levator posterior muscle, and expanded tooth areas on the upper and lower pharyngeal jaws). (Redrawn and modified from Lauder 1986.)

different motor pattern, the shell fragments are ejected from the mouth, and the snail is swallowed. This pattern is used by *L. microlophus* regardless of the prey type; unlike *L. gibbosus* and *L. cyanellus,* this species does not modulate its feeding behavior to suit its food.

Lauder's study of the morphological, functional, ecological, and behavioral characteristics involved in sunfish feeding is presented in a simplified form in figure 9.3. The phylogenetic tree suggests that the two "natural" snail eaters (*L. gibbosus* and *L. microlophus*) are sister species and are relatively highly derived members of the genus. Examination of sunfish feeding within this phylogenetic framework reveals that a new chewing pattern has appeared three times over the evolution of this group. Mode 2, the development of the crushing motor pattern, appeared convergently in the green sunfish (*L. cyanellus*) and in the common ancestor of *L. gibbosus* and *L. microlophus.* The suite of characters composing this mode was passed on to its descendants, one of which, *L. microlophus,* became even more specialized through the loss of four plesiomorphic (2, 4, 6, 8), and acquisition of two autapomorphic (3, 7), functional traits. Nevertheless, both the redear and the pumpkinseed continue to display a similar snail-eating behavioral pattern, suggesting that characters 2, 3, 4, 6, 7, and 8 are not integral components of this feeding preference.

Lepomis cyanellus is the only other sunfish that considers snails to be edible items, although it must be encouraged in this consideration. Initially, this convergent evolution of chewing mode 2 only in the snail-eating sunfishes might be offered as support for a hypothesis that the appearance of feeding mode 2 was involved in the shift to snail eating. However, closer examination reveals that, of the three changes in chewing pattern, only one was associated with a trophic shift. Since chewing pattern and food type are not tightly coupled at any other point on the tree, it appears that the origin of the crushing mode is not, of itself, sufficient to explain that species' change into a snail-eating specialist. The answer to this problem may lie in the three morphological changes that occurred in the ancestor of *L. microlophus* and *L. gibbosus* and have been transmitted to both of its descendants. To approach this from another perspective, it appears that, on their own, the functional changes in chewing pattern (from mode 1 to mode 2) explain the functional **ability** to eat snails, while, in combination with the morphological changes, they explain the trophic **specialization** on snails.

Integrating the Experimental and Phylogenetic Approaches: From Pattern to Process

The long-standing tradition of integrating experimental and evolutionary information, especially in comparative ethology, has been de-emphasized during the "eclipse of history" in ethology and ecology. The advent of historical ecological methods should reemphasize this area of research and help promote an integration of evolutionary patterns and evolutionary processes. We have already discussed examples of this type of study using ecological data (see Huey and Bennett's study of sprinting speed and thermal preferences in skinks in chapter 5 and Gorman's study of freshwater fish communities in chapter 8). In the following paragraphs we will turn our attention to the "*Drosophila*" of ethology, the stickleback fishes.

Gasterosteid fishes: Color and behavior

Sticklebacks and their relatives (family Gasterosteidae) have played a leading role in behavioral ecological research for decades. Within this family, the diminutive three-spined stickleback, *Gasterosteus aculeatus,* has commanded the most attention because breeding males undergo a wondrous transformation from inconspicuous silver-green fishes into flamboyant mosaics of intense flame-scarlet and flashing aquamarine-blue. Needless to say, such a transformation did not go unremarked. The red component of the male's breeding livery is hypothesized to play a role in (1) territory acquisition and maintenance (intrasexual selection), (2) courtship and mate acquisition (intersexual selection), or (3) paternal care (natural selection). These hypotheses are generally based upon the assumption that the origin and elaboration of male nuptial coloration occurred in an ancestral *G. aculeatus* population. As we have discovered in preceding chapters, however, studies of a single species can, at best, only address questions concerning character maintenance in that species. To investigate character origin and elaboration you need at least two other species, preferably the closest relatives to the test organism, and an outgroup.

Historical ecological methods can provide a useful way to disentangle the potential contributions of various selection pressures in the evolution of sexually dimorphic nuptial coloration in gasterosteids, by generating predictions for experimental investigation. McLennan, Brooks, and McPhail (1988) examined the macroevolutionary relationships among color and various breeding behaviors on a phylogenetic tree for the sticklebacks. They discovered the following associations.

Intersexual selection (fig. 9.4): The initial elaboration of male nuptial coloration is preceded by an increase in the complexity of courtship on the

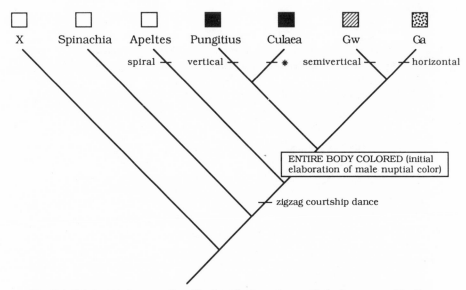

Fig. 9.4. Relationship between diversification of color and courtship in the gasterosteids. X = outgroup; Gw = *Gasterosteus wheatlandi; Ga* = *G. aculeatus*. Nuptial body colors: *white box* = no color; *black box* = black body; *striped box* = gold body; *boldly stippled box* = red body. The orientation of the zigzag courtship dance is mapped beneath each appropriate species. * indicates that the courtship dance has been replaced by a tail-flagging display in *Culaea inconstans*.

phylogenetic tree. This reinforces the intersexual selection hypothesis because an increase in the intricacy of signal exchange focusses the receiver's attention on the sender and increases the amount of information available in the interaction. This, in turn, creates the potential for the evolution of differential female response (female choice) to variability in a male character. Apparently this potential was realized in the ancestor of the *Pungitius* + *Culaea* + *Gasterosteus* clade, where nuptial coloration underwent an exaggeration (from no color to the entire body) consistent with Fisher's (1930) runaway sexual selection scenario. The phylogenetic tree supports the interpretation that initially color was not necessary in courtship (the outgroup, *Spinachia,* and *Apeltes* do not change color during the breeding season), but that it later became intimately involved with male-female interactions. Once associated, the elaboration of color and courtship are tightly coupled on the tree.

Intrasexual selection (fig. 9.5): Color may play a role in male-male interactions as part of a threat display, allowing an individual to assess the social status, experience, and motivational state of an opponent. Depending on the system employed, the information exchanged during these interactions may contain elements of both truth and bluff. However, once an encounter has escalated past threat, the emphasis should shift away from signals such as

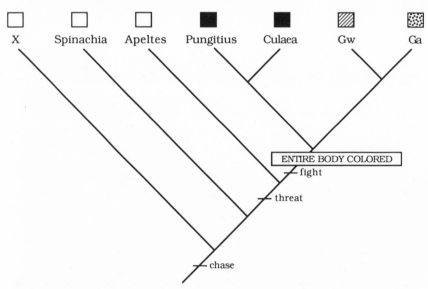

Fig. 9.5. Relationship between diversification of color and aggressive behaviors in the gasterosteids. Behaviors involved in male-male interactions have been mapped onto the tree. *X* = outgroup; *Gw* = *Gasterosteus wheatlandi; Ga* = *G. aculeatus.* Nuptial body colors: *white box* = no color; *black box* = black body; *striped box* = gold body; *boldly stippled box* = red body.

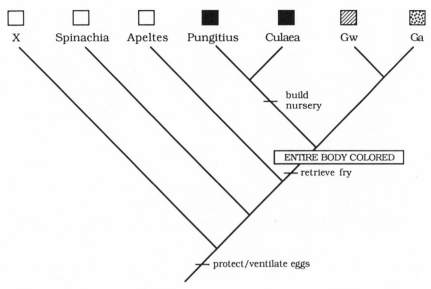

Fig. 9.6. Relationship between diversification of color and parental behaviors in the gasterosteids. Parental-care behaviors have been mapped onto the tree. *X* = outgroup; *Gw* = *Gasterosteus wheatlandi; Ga* = *G. aculeatus.* Nuptial body colors: *white box* = no color; *black box* = black body; *striped box* = gold body; *boldly stippled box* = red body.

color to factors directly involved with fighting performance (stamina, skill, strength). Examination of the phylogenetic tree indicates that, contrary to our expectations, color elaboration is initially associated with the appearance of fighting, not threat, behavior. Additionally, once threat and fight behaviors arose, they remained very conservative, whereas color continued to be elaborated. So, macroevolutionary analysis suggests that any subsequent function of color in male-male interactions is secondary to the more direct coupling of color and male-female interactions.

Natural selection (fig. 9.6): The relationship between color and parental care falls between the intra- and intersexual selection patterns. Like courtship, the initial elaboration of color is associated with an increase in parental care (prolonged fry retrieval). This supports the hypothesis that natural selection played a role in the elaboration of the color signal in this ancestor. Past this point in phylogeny, however, changes in color and parental care are not as closely associated as the macroevolutionary relationship between color and courtship.

The macroevolutionary associations between breeding color and behaviors uncovered by phylogenetic analysis provide a set of predictions about color changes across the breeding cycle and about female choice based on male color, which can be tested by laboratory studies at the population level for any of the gasterosteids. These predictions have been examined for that most handsome of fishes, the three-spined stickleback (McLennan and McPhail 1989, 1990).

The breeding cycle of *Gasterosteus aculeatus* can be divided into four behaviorally distinct stages, territory acquisition/nest building, courtship, egg guarding, and fry guarding. Fortunately, the three-spined stickleback is notoriously easy to breed in the laboratory, so McLennan and McPhail (1989) were able to take males through their paces with a maximum amount of cooperation from the fishes. The intensity of red body color was measured on a daily basis from a male's introduction into the test aquarium to the completion of fry guarding (fig. 9.7). Color changes across the breeding cycle of this species were then compared to the changes predicted from the macroevolutionary associations of color and behavior in the family Gasterosteidae.

Intrasexual selection (fig. 9.5): The nesting male in this population tends to be a lightly colored individual, displaying just a hint of flame-orange in the lower jaw and opercular areas (fig. 9.7: nest). He is, in fact, barely distinguishable from a nonterritorial male. Evidence from other microevolutionary studies on populations in Europe (Van Iersel 1953), Japan (Ikeda 1933)

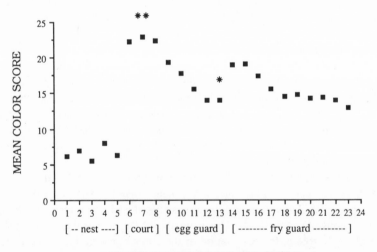

STAGE OF THE BREEDING CYCLE

Fig. 9.7. Change in total score (intensity) for red body color of *Gasterosteus aculeatus* males (N = 19) across a complete breeding cycle. The breeding cycle is broken into four distinct stages. Days 1–5, nest building/maintenance; days 6–8, courtship; days 9–13, egg guarding; days 14–23, fry guarding. All nests were completed by the end of day 2. Asterisks refer to the results of two Wilcoxon analyses: * = significant difference ($p < 0.01$) between color scores of individuals on day 13 (last day of egg guarding) and day 14 (first day of fry guarding); ** = significant difference ($p < 0.001$) between courtship peak (days 6–8) and fry-guarding peak (days 14–16).

and North America (McPhail 1969) also indicates that the development of full nuptial coloration is not a necessary prerequisite for territory establishment and maintenance. This supports the phylogenetic prediction that intrasexual selection did not play a major role in the evolutionary elaboration of the nuptial signal in the gasterosteids.

Intersexual selection (fig. 9.4): Courting males are characterized by a peak in both the intensity and the distribution of scarlet across the entire ventral and lateral surfaces of the fish (fig. 9.7: court). Since the correlation between the intensity of red body and blue eye color is also higher at this stage than at any other time in the breeding cycle, the male presents his most intense, widely dispersed, and cohesive mosaic signal during the brief courtship phase. As predicted, the strong association between the evolutionary elaboration of color and courtship behavior in the gasterosteids in general, is mirrored within *G. aculeatus* in particular.

Natural selection (fig. 9.6): The parental male presents a more somber image as a wave of dark gray sweeps forward from the caudal fin, masking all colors except the bright blue of the eyes and intense flame-scarlet in the throat region. A small decrease in mandibular color during egg guarding (fig 9.7: egg guard) is followed by a surge in intensity to courtship levels coinciding with the appearance of fry and an increase in aggression (fig. 9.7: fry guard). The second peak in overall color is lower than the level achieved during courtship, as color increases in the anterior third of the fish are countered by "losses" in the posterior and middle regions. Nevertheless, this peak suggests some role for color during the initial stages of fry guarding. As predicted by the phylogenetic analysis, color intensity is moderately tied to parental care, falling somewhere between the peaks of courtship and the valleys of territory acquisition.

The phylogenetic predictions based on the macroevolutionary relationships between the elaboration of breeding color and behaviors are thus confirmed in this microevolutionary study of color changes across the breeding cycle of *Gasterosteus aculeatus*. This, in turn, focusses our attention upon the influence of intersexual selection on the elaboration of male nuptial color in these fishes. Researchers have been investigating female mating preferences in the three-spined stickleback for over fifty years (Pelkwijk and Tinbergen 1937; McPhail 1969; Semler 1971). The results of these experiments are unequivocal: when offered a choice between red or nonred (grey or black) males, females overwhelmingly prefer red. Since red is a species-specific character (autapomorphy) for *G. aculeatus,* this research has provided ample evidence for mate recognition; that is, the female's ability to discriminate potential mates (territorial, conspecific males) from nonmates (immature, nonterritorial conspecifics and other gasterosteid males). Now, how does a female discriminate among potential mates, all of whom are red?

As discussed previously, the presence of "entire body" breeding coloration represents the persistence of an ancestral trait in three-spined sticklebacks. The actual event associated with the initial elaboration of the distribution of color occurred in the ancestor of the *Pungitius + Culaea + Gasterosteus* clade. If, as this rapid elaboration within one species suggests, this event was driven by runaway sexual selection, then it may be possible to document the effects of female choice based upon differences in intensity of body color in *G. aculeatus*. A series of female choice experiments confirmed the phylogenetic predictions: when females were offered the choice of two competing males, they displayed a significant, preferential response for the most intensely colored red male (McLennan and McPhail 1990). The results of this and the preceding experiment thus confirm the macroevolutionary prediction that intersexual selection has been the dominant force shaping the evolution of male nuptial coloration in the Gasterosteidae.

The tradition of studying behavioral evolution by combining phylogenetic and experimental information was developed by Tinbergen and Lorenz many years ago. As discussed in chapter 1, however, this tradition foundered as both ethologists and systematists became deeply enmeshed within their own theoretical revolutions. Historical ecology allows us to reestablish a dialogue between systematics and experimental ethology, to the mutual benefit of both. This section is a tribute to the original ethologists' insights, both because it refines and reemphasizes the power of a phylogenetic/experimental integration, and because it reinforces the fundamental views of stickleback behavior unveiled by Pelkwijk and Tinbergen's elegant experiments in 1937.

Adaptive Changes in Quantitative Traits: Integrating the Statistical and Phylogenetic Approaches

Over the last decade, a renewed interest in the genealogical aspects of evolution has emerged in the form of a new research program that attempts to strengthen our examination of adaptive hypotheses by considering the effects of phylogenetic constraints on quantitative traits (Stearns 1977, 1983; Gittleman 1981; Ridley 1983; Clutton-Brock and Harvey 1984; Cheverud, Dow, and Leutenegger 1985; Pagel and Harvey 1988; Gittleman and Kot 1990). These researchers have repeatedly emphasized the caveat that data from related taxa are not statistically independent, so persistent phylogenetic effects will confound statistical tests of adaptation. Most of the research in this area has centered on the evolution of life-history traits. For example, Stearns (1983) reported phylogenetic constraints on the patterns of covariation in a variety of life-history traits in mammals. Gittleman and Kot (1990) discovered significant positive correlations between (1) body weight and phylogeny within and between genera of carnivores, and (2) modal clutch size and phylogeny within and between genera for 256 western Palearctic nonpasserines. Other studies along similar lines have been published (Gittleman 1981, 1985, 1986, 1989; Ballinger 1983; Brown 1983; Ridley 1983; Stearns 1983; Bekoff, Daniels, and Gittleman 1984; Fenwick 1984; Cheverud, Dow, and Leutenegger 1985; Dunham and Miles 1985; Felsenstein 1985; Harvey and Clutton-Brock 1985; Stearns and Koella 1986; Huey and Bennett 1987; Kool 1987; Wootton 1987; Dunham, Miles, and Reznick 1988; Pagel and Harvey 1988; Fox 1989; Gittleman and Kot 1990; Faith, in press). Although a variety of different methods were used to identify and quantify phylogenetic effects, all these studies confirmed the presence of phylogenetic influences on quantitative life-history traits that had traditionally been accorded high adaptive significance. As it became apparent that phylogenetic effects may be an important component in the evolution of qualitative *and* quantitative traits, two general approaches began to emerge.

The first of these represents an attempt to examine the evolutionary diversification of quantitative traits *within a phylogenetic context* (e.g., Farris 1970; Felsenstein 1985; Huey 1987; Huey and Bennett 1987; Faith in press). This research involves methods for estimating ancestral states for quantitative traits on phylogenetic trees. Although methods differ somewhat among these authors, all agree on one point: genera, families, and other supraspecific categories do not evolve and have no status within evolutionary theory, even if they represent monophyletic groups. Therefore, the statistical study of "generic constraints" or "familial constraints" becomes an exercise in determining the consistency with which different taxonomists discern groupings, rather than an analysis of evolutionary processes. What are termed "generic-level constraints" that have evolutionary relevance correspond to attributes that originated in the common ancestor of a clade, that is, traits that originated in an ancestral **species.** Since the term "genus" is an arbitrary artifact of a classification scheme, it is likely that the degree of "generic-level constraints" will differ for each "genus," and this, in turn, might lead to mistaken conclusions about the different degrees of adaptive influences operating on different clades (fig. 9.8). Thus, *it is important to use sister-group relationships in making comparisons, rather than making comparisons across levels in a tree,* each of which might be construed as coinciding with a particular classification category (see also Felsenstein 1985).

So, let us consider the hypothetical group of "winged" species depicted in figure 9.9. An intrepid historical ecologist might first approach the problem of the evolutionary modification of the quantitative trait, wing length, in this clade by averaging across accepted taxonomic levels. Upon completion of this analysis the ecologist concludes that wing length increased at the subfamily level and decreased slightly on the generic level. There has therefore been an evolutionary trend towards increasing, followed by decreasing, wing size

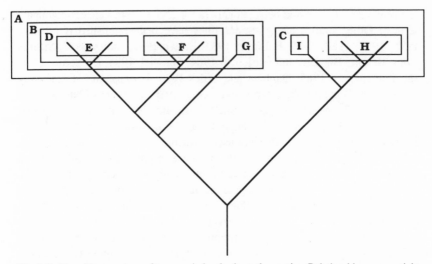

Fig. 9.8. The arbitrary nature of taxonomic levels above the species. Relationships among eight hypothetical species are depicted on a phylogenetic tree. Although the relationships among these species will remain the same, there are potentially six different classification schemes of these relationships at the genus level: (1) one genus (box A); (2) two genera (boxes B and C); (3) three genera (boxes B, I, and H); (4) three genera (boxes C, D, and G); (5) four genera (boxes C, E, F, and G); or (6) five genera (boxes E, F, G, H, and I). Researchers investigating hypotheses of adaptation based upon differences between taxonomic units larger than the species will have to choose from among these schemes; therefore, their explanations will be strongly influenced by the arbitrary and artificial nature of these supraspecific units.

in this group. In addition, it would appear that there has been some degree of evolutionary change associated with each branch on the phylogenetic tree.

During the final preparations for publication our researcher, by chance, attends a seminar concerning the pitfalls of averaging across taxonomic levels. Being of tenacious nature, the ecologist tears up the original analysis and approaches the problem by averaging the values of wing lengths between sister species (fig. 9.10). This leads to the conclusion that there has been a trend towards gradual wing reduction in the monophyletic group A + B + C + D + E, and gradual wing enlargement in the group F + G + H.

Finally, our hypothetical, and weary, historical ecologist decides to investigate wing length from yet another perspective, that of phylogenetic character optimization. This analysis produces results similar to those revealed by sister-group averaging, with some interesting differences (fig. 9.11). There is a trend towards wing reduction in one lineage and wing enlargement in the other, but this trend is superimposed upon a large amount of phylogenetic constraint in wing length within members of the entire clade. Optimization reveals that five was the ancestral wing length and that this length has been

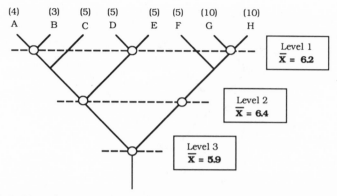

Fig. 9.9. Problems that can arise from averaging across taxonomic levels on a phylogenetic tree. *Letters* = species. Values for a character—say, wing length—are plotted above each species. The first-level value is arrived at by summing the average wing lengths of species A and B, of species D and E, and of species G and H, then dividing by 3, the number of groups. This might be considered to be a comparison across genera. The second level, perhaps corresponding to a comparison across subfamilies, is even more arbitrary. It is composed of [(the average wing length of species A, B, C, D, and E) + (the average wing length of species F, G, and H)] / 2. Finally, the last-level value represents the average wing length of species A, B, C, D, E, F, and G.

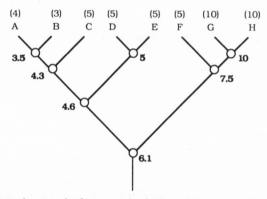

Fig. 9.10. Problems that can arise from averaging between sister groups on a phylogenetic tree. *Letters* = species. Values for the character wing length are plotted above each species. The differences in wing length demonstrated among members of this clade are postulated to be the result of a gradual progression of evolutionary modifications. Although the existing species demonstrate considerable overlap in wing length, all the ancestors in this group are hypothesized to have displayed different wing lengths.

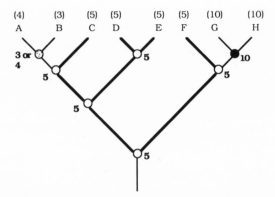

Fig. 9.11. Optimizing the values for a quantitative character on a phylogenetic tree. *Letters* = species. Values for the character wing length are plotted above each species. This analysis reveals a core of phylogenetic stasis (bold lines) and identifies the areas on the tree where substantial adaptive changes have occurred (in the ancestors of species A + B and species G + H).

retained by 50% of the clade (species C + D + E + F). There was one change from five to ten in the ancestor of species G + H, and two more changes in species A + B. The sequence of change in these latter two species cannot be resolved by optimization analysis alone. In general then, this example demonstrates that taxonomic averaging can potentially produce very misleading results, and sister-group averaging, while being the more preferable of the two methods, may obscure the historical background of adaptive change within a lineage. In this particular case it obscured the high degree of adaptive change in the ancestors of species A + B and species G + H, and overestimated the degree of adaptive change in all other lineages.

The above example uses hypothetical quantitative data that have been reduced to a single (e.g., mean or median) value for each species, which is then treated as a qualitative trait among taxa. We simplified the discussion by allowing many of the mean character values for the species to be the same. What do we do when all the values differ, and none of the character optimization procedures will work? A number of "averaging" methods have been proposed for optimizing quantitative traits (e.g., using median values, Farris 1970; using mean values, Huey and Bennett 1987), but at present there is no generally accepted approach among phylogeneticists (see Felsenstein 1988 for a review of methods). If a new spirit of cooperation and collaboration between ecologists and systematists emerges in the future, it would not surprise us to find that population biologists working on quantitative life-history traits will provide the insights necessary for phylogeneticists to develop general methods for optimizing quantitative traits.

The second general approach to studying phylogenetic effects on tests of adaptation involves attempts *to reduce spurious correlations due to phyloge-netic constraints by removing these effects from the data set*. Three statistical approaches have been developed to try to achieve this goal: an "independent contrasts" method (Felsenstein 1985; see also Sessions and Larson 1987), the use of nested analysis of variance (ANOVA) and analysis of covariance (AN-COVA; see Pagel and Harvey 1988), and the use of autocorrelation proce-dures (see Cheverud, Dow, and Leutenegger 1985; Gittleman and Kot 1990). Although we are not qualified to discuss the technical aspects of these statis-tical approaches, we can emphasize some points of similarity between this research program and historical ecology. Both of these evolutionary investi-gations require that we identify the phylogenetic components of diversity; therefore, both are subject to the same caveats with respect to phylogenetic analysis (see also Gittleman and Kot 1990; Burghardt and Gittleman 1990): (1) do not confuse similarity with relationship, (2) do not confuse plesiom-orphy with apomorphy, (3) do not confuse taxonomy with phylogeny, and (4) do not confuse averaging with optimizing.

Many of the studies we have enumerated above begin with an explicitly historical step; they examine the best current estimates of phylogenetic rela-tionships available. The second step should be an analysis of the sequence of character evolution *on the phylogenetic tree* (remember from chapter 2 that the transformation series for a given character may not correspond exactly to the phylogenetic pattern of diversification of the character). This is rarely done, however, often because explicit phylogenies are not available. As a result, there is no information available about the history of character asso-ciations (character associations that show similar correlations among extant species may have originated in a variety of different sequences), nor is there any way to distinguish phylogenetic correlations due to plesiomorphy from those due to apomorphy, or those due to homology from those due to homo-plasy. The result is often more phenetic than phylogenetic, with similarity being equated with homology and relationship.

Historical ecologists attempt to identify phylogenetic effects and incorpo-rate them into macroevolutionary explanations. The statistical research pro-gram attempts to identify and remove phylogenetic effects so they do not confound microevolutionary tests of adaptation. Ultimately, both programs are interested in producing a more rigorous set of adaptationist hypotheses, and from this a more robust theory of evolution. Fortunately there is enough overlap in these perspectives to promote the establishment of mutually bene-ficial cross-communication. For example, the first step in both programs is the identification of the phylogenetic constraint. What could be more robust than the simultaneous identification of the same qualities and quantities as the constraints? Communication between programs with different evolution-ary viewpoints will enable us to address more empirically questions about the relationship between microevolution and macroevolution. If we only discover the effects of history in a restricted class of qualitative characters, then we would be justified in assuming that such macroevolutionary patterns were of limited usefulness to discussions of evolutionary mechanisms. On the other hand, if phylogenetic effects are so pronounced even in quantitative life-history traits that they can confound purely microevolutionary studies, they might well be important aspects of evolutionary mechanisms. That is, if such effects are important enough to be partitioned out of some studies, they are important enough to be explained by general evolutionary theory. Every study that has investigated evolution within a historical framework has found evidence of the (often marked) effects of phylogenetic constraints.

How are we to explain the observation that contemporary evolutionary dynamics involve an interaction between constraints and adaptation? The extrapolationist view would suggest that today's constraints are simply yesterday's adaptations. If that is so, there must have been fewer constraints and more adaptation in the past than we now see, forcing us to conclude that phylogenetic effects should become even more pronounced in the future, at the expense of adaptive plasticity. However, it is difficult to reconcile this with the recognition that diversity of adaptations at the macroevolutionary level is conservative relative to phylogeny, while adaptive plasticity within populations of species remains relatively high. The nonextrapolationist view would suggest that evolution in the past comprised an interaction between constraints and adaptation just as it does today.

Our thought experiment with crabs, tide pools, and finches in chapter 1 demonstrated that, in some systems, we will be able to make better predictions by reference to history than to current environments, while in others, the environment will be more informative. Because of this, it is important that researchers do not feel obligated to choose between *either* history *or* adaptation when formulating evolutionary hypotheses. Since phylogenetic diversification is open-ended, *both* adaptations *and* constraints can accumulate over time. We can metaphorically view adaptations as the (ecological) tip of the iceberg of (genealogical) constraint (fig. 9.12). This metaphor gives us an important glimpse of the ecological insights that can be gained from the perspective of both the environment (adaptation) and the organism (constraint). In the next section we will address the question of how we might begin to integrate the metaphor that evolution results from an interaction between constraints and adaptation into general evolutionary theory.

Integrating Historical Ecology with General Evolutionary Theory

Evolutionary theory is experiencing a period of intensive reexamination. Some researchers assert that the synthetic theory of evolution, or neo-

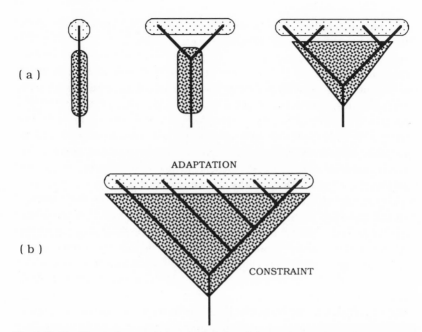

Fig. 9.12. The historical accumulation of adaptation (*light stippling*) and constraint (*dark stippling*) in the phylogenetic diversification of a group of species (*a*). Each species lineage is a mixture of ancestral traits (historical constraints) and adaptive changes. Speciation increases the number of species present, hence increasing both the constrained and the adaptive components of biological diversity. Adaptation is thus metaphorically the tip of the iceberg of constraints (*b*).

Darwinism, is essentially a complete theory (e.g., Stebbins and Ayala 1981; Charlesworth, Lande, and Slatkin 1982; Buss 1987). Others argue that there are gaps in the theory that can only be addressed by integrating traditional principles and research programs with new ideas. This is exemplified by recent texts and articles that promote attempts to "expand" (Gould 1980), "finish" (Eldredge 1985), "extend" (Wicken 1987), or "unify" (Brooks and Wiley 1988; Brooks, Collier, Maurer, Smith, and Wiley 1989) evolutionary biology. The mechanisms of evolutionary constraint are being examined on many levels, from phylogenetic to developmental (Alberch et al. 1979; Lauder 1981, 1986, 1988, 1989; Alberch 1982; Fink 1982; Charlesworth, Lande, and Slatkin 1982; Kauffman 1983; Buss 1987), to genetic (e.g., Kauffman 1974; Zuckerkandl 1976; Dover 1982), to those derived from basic physicochemical laws (e.g., Brooks and Wiley 1986, 1988; Kauffman 1986; Wicken 1987; Brooks, Collier, Maurer, Smith, and Wiley 1989; Weber et al. 1989). We believe that a richer theory, based upon the incorporation of microevolutionary and macroevolutionary information, is emerging from these debates. Historical ecology may play a role in this emergence because it provides us with a way to integrate the study of macroevolution with the study of microevolution in a nonreductionist manner.

Throughout this book we have concentrated on the reconstruction and interpretation of macroevolutionary patterns without directly addressing the issue of underlying mechanisms. It is now time to ask, Is there any reason to believe that macroevolution is more than just "microevolution writ large"? Like many questions in science, researchers who have an opinion about the answer to this question find themselves in one of two philosophical camps. The "extrapolationists" believe that macroevolutionary patterns simply reflect the accumulation of a series of independent microevolutionary events over time (Eldredge 1985). The patterns may be real, that is, they may give us an accurate picture of part of the past, but they do not indicate the actions of macroevolutionary processes. Schoch (1986) suggested that three predictions could be drawn from this perspective: grades (paraphyletic groups) are real and discrete evolutionary units, speciation patterns are not correlated geographically, and ecology and behavior are better correlated with the local environment than with phylogenetic history. He then suggested that phylogenetic systematic research had largely refuted all three of these predictions, and the evidence presented in this book corroborates his assertion. The current data base thus supports the nonextrapolationist view that "something else" beside microevolutionary processes is going on in evolution.

Two general features of evolution that might conceivably fall under the heading of "something else" have emerged from our considerations of historical ecology. First, evolution results from an interaction of ecological and genealogical phenomena; ecological processes do not "cause" genealogy, nor is genealogy independent of the environment. Second, viewing biological systems on different temporal and spatial scales may produce very different pictures of evolutionary patterns and processes. In the remaining pages we will present a more detailed discussion of these two themes as our contribution to the ongoing study of macroevolutionary processes.

The Two Biological Hierarchies

Environmental and genealogical phenomena are intimately connected; therefore, it is difficult to disentangle "environmental effects" from "genetic (genealogical) effects" in evolutionary studies. This difficulty has prompted some authors to propose that two forms of hierarchically organized systems coexist in biology (Eldredge and Salthe 1984; Salthe 1985; Eldredge 1985, 1986; Brooks 1988b; Brooks and Wiley 1988; Brooks, Collier, Maurer, Smith, and Wiley 1989). The **ecological hierarchy** is an energy-flow system, manifested by patterns of energy and matter exchange between the organism and its environment. It is the hierarchy of biological classes, such as trophic levels or ecological associations (e.g., herbivore-crop, predator-prey, host-parasite). Since any plant and any herbivore are sufficient to define an herbivore-crop association, the particular identities of the species involved is unimportant. The **genealogical hierarchy** is an information-flow system, manifested by genealogical relationships over short (genetic) and long (phylogenetic) time scales. It is the hierarchy of individuals. From the genealogical perspective identity is paramount. It does not matter what each species does so long as its members find resources sufficient for survival and perpetuation.

The relationship between the two hierarchies can be illustrated with the following sports metaphor: the ecological hierarchy establishes the dimensions of the playing field, while the genealogical hierarchy establishes the rules of the game being played. In other words, biological systems obey rules of self-organization transmitted genealogically (historically) and played out within environmentally defined boundaries. However, the genealogical processes that characterize life and evolution are autonomous enough from environmental conditions to be capable of overrunning available required resources and of changing the environment substantially. Because of this, the game may redefine the boundaries of the playing field and may be subsequently constrained by these self-imposed changes. For example, the evolution of photosynthetic prokaryotes from anaerobic ancestors resulted in increased oxygen content in the atmosphere. This increase, in turn, ultimately altered the diversity and distribution of anaerobic organisms, limiting them to relatively rare environments.

The ecological hierarchy is the means by which two different genealogies, or two different generations in one genealogy, can causally influence one another. We believe that this is the reason historical ecology is important to the development of evolutionary theory in general. The conservative nature of ecological diversification uncovered so far by historical ecological methods implies that adaptive processes act as cohesive rather than as diversifying forces in evolution. This, of course, begs the question of just what is the diversifying force. At the moment, the answer to this question remains widely and passionately disputed (see, e.g., Brooks and Wiley 1986, 1988; Brooks, Collier, Maurer, Smith, and Wiley 1989; Weber *et al.* 1989 and references therein).

Spatial and Temporal Scaling Effects

In the preceding chapters we have uncovered numerous examples of the ways in which the effects of temporal and spatial scaling shape our evolutionary perspective. One such example was considered without its temporal implications when we discussed methodology in chapter 2. The designation of plesiomorphic or apomorphic status to character states is a relative, not absolute, statement. All characters begin as evolutionary novelties (autapomor-

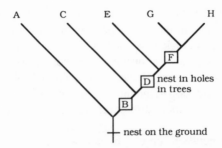

Fig. 9.13. The plesiomorphic or apomorphic status of a character is relative to the temporal scale of the investigation. There has been an evolutionary change from nesting on the ground to nesting in trees in this hypothetical clade of birds. Nesting in tree holes is an autapomorphy for species D, a synapomorphy for species D + E + F, and a plesiomorphy for species D + E + F + G + H.

phies) in a new species. If the species undergoes another speciation event before the character changes again, the character becomes a synapomorphy uniting the descendant sister species. If speciation continues in this lineage while the character remains unchanged, the character will come to be considered plesiomorphic for the group (fig. 9.13). This reemphasizes the basic phylogenetic assertion that only synapomorphies (homologies on an intermediate temporal scale) are useful for reconstructing phylogenetic relationships. Homologies on a small temporal scale (autapomorphies) or homologies on a large temporal scale (plesiomorphies) do not contain information useful to this reconstruction. In a similar vein, the difference between convergent and divergent adaptation is also dependent upon the temporal scale of the investigation. All convergent adaptation is the accumulation of parallel independent episodes of divergent adaptation between sister species (see *Montanoa* example in chapter 5; fig. 9.14).

We have discussed several instances in which the evolution of a particular association or portion of a biota might be explained by reference to a specific model of coevolution or community evolution, whereas the explanation for the evolution of associated clades or of entire biotas would invoke the influences of several such models. This implies that the kind of ecological associations, as well as the kinds of explanations relevant to studies of the evolution of those associations, depends on the temporal and spatial scale chosen by the researcher. The original concern of many evolutionary ecologists was to partition out the effects of different scales by adopting what Wiens (1984) termed a Goldilocks approach. That is, do not choose a scale so large that patterns will be influenced by phylogeny and other historical factors, but also do not choose a scale that is so small that no regularities of interest emerge.

Ricklefs (1987, 1990; see chapter 1) and Brown and Maurer (1989) have recently discussed the significance of scaling effects in explaining the struc-

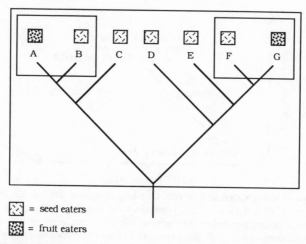

= seed eaters

= fruit eaters

Fig. 9.14. Identification of convergent and divergent adaptation is dependent upon the temporal scale of the investigation. The change from eating seeds to eating fruit represents independent cases of divergent evolution between sister species A + B and sister species F + G. The temporal scale must be increased to include all the members of this clade before the convergent evolution of fruit eating in species A and G can be identified.

ture and evolution of biotas. Their approach differs from the original attitude adopted by evolutionary ecologists by virtue of fact that they wish to take such scaling effects into account and integrate them into ecological studies. On the systematic side, Brooks (1988b) recently suggested that phylogenetic studies in biogeography could be related to the biological relationships between phylogenetic (temporal) and spatial scaling effects. On the smallest spatial scales, biogeographic patterns result from microhabitat distribution and vagility of the organisms being studied. Increasing the power of the spatial telescope increases the number of species involved in the biogeographic tapestry until an intermediate level is reached where particular communities, ecosystems, or biotas form the threads of the pattern. The processes dominating these spatial scale patterns are species-composition phenomena such as immigration-emigration dynamics and biotic expansion and contraction. Finally, the entire tapestry is revealed on large spatial scales where groups of biotas compose the units of study. Since most species occur allopatrically from their closest relatives (see chapter 4), expanding the study area increases the likelihood that the relevant biogeographical patterns will involve groups of phylogenetically related species. Or, to put this another way, increasing the spatial scale chosen for observation of a biological system increases the influence of historical constraints (effects of the temporal scale) on the evolution of observed diversity and distribution patterns. Figure 9.15 presents a

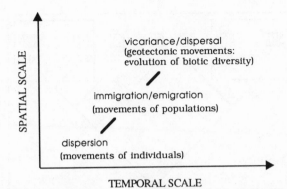

Fig. 9.15. Spatial and temporal scaling effects on biogeographical patterns. On the smallest scales, movements of individuals within populations are revealed, whereas on intermediate scales, movements among populations are important. It is only on the largest spatial and temporal scale that the evolution of biological diversity, resulting from the interaction between biotic dispersal and geographical changes, is observed. (Redrawn and modified from Brooks 1988b.)

heuristic view of biogeography unified by spatial and phylogenetic (temporal) scaling effects (Brooks 1988b). Analogous reasoning about the significance of scaling effects can be applied to studies of coevolution in an ecological context by substituting "hosts" or "other associated species" for "areas."

Although heuristic, this diagram of the interaction between space and time in evolutionary processes produces two testable predictions: increasing the analytical spatial scale increases the likelihood of finding vicariant components in any patterns of ecological association; and if ecological associations reflect organized patterns of energy use, then the larger the space occupied by an association, the greater the proportion of phylogenetically determined energy-use patterns (Brown and Maurer 1987, 1989).

Both the phylogenetic and geographical scales determine what macroevolutionary regularities you will see. In chapter 4 we discussed a variety of modes of speciation. Every speciation event is the result of a single mode, but the evolution of a clade might well involve a variety of speciation modes. Hence, at the macroevolutionary level, we would ask questions about the relative frequency of modes of speciation. In addition, enlarging the phylogenetic and geographic scale of speciation studies allows us to ask questions about the relative frequency of cospeciation among clades (chapter 7). Similarly, there are a variety of adaptive modes (chapter 5) and models of coevolution (chapter 8), any number of which might be involved in the evolution and coevolution of a diverse group of species. Again, at the macroevolutionary level, we would ask questions about the relative frequency of adaptive changes or of different coevolutionary effects. Finally, individual species occurring in communities may correspond to as many as four different classes of evolutionary patterns (chapter 8). Communities may differ macroscopically as a result of the differential degree of influence by these various effects (see also Emig 1985).

The preceding discussion highlights the most important aspect of scaling considerations: *there is no objective level of organization, time interval, or spatial interval for biological evolution.* A variety of evolutionary processes operate on all levels and at all scales; however, they do not all play equally important roles at all levels. Therefore, the macroscopic manifestations of evolutionary principles will differ depending on the window of observation (see also Salthe 1985). Gould (1982b) summarized this elegantly in his defense of the hierarchical approach to evolution.

> We do not wish to advocate the despairing conclusion that micro- and macroevolution are absolutely separate in principle and that nothing about one illuminates the other. . . . Rather, the same processes of variation and selection operate throughout the hierarchy. But they work differently upon the varying materials (individuals) of ascending levels in a discontinuous hierarchy. . . . Important ties of feedback unite all levels, but new modes emerge at higher levels and reduction to natural selection upon organisms will not render all evolution. Nothing about microevolutionary population genetics, or any other aspect of microevolutionary theory, is wrong or inadequate at its level. Little of it is irrelevant to students of macroevolution. But it is not everything.

We agree with Gould's conclusion that microevolutionary processes, although important, are not the sole forces of evolution. They are currently high-profile processes because they dominate evolution on limited temporal and spatial scales, scales that are the most easily accessible windows of study for organisms with our biological and career life-span constraints. Processes relevant to evolution are either generative (originating or diversifying) in their effects, or conservative (maintaining or cohesive) in their effects. The interplay of diversity-promoting and diversity-limiting processes through time produces historically constrained order. Many processes affect biological systems, at all levels of organization and at all times, but their effects are often manifested on different time scales. Changes occurring on time scales shorter than speciation rates will appear as microevolutionary patterns; those occurring on time scales longer than speciation rates will appear as macroevolutionary patterns. **In this sense, macroevolutionary processes are neither reducible to, nor autonomous from, microevolutionary processes.**

The Hierarchy of Evolution: Looking through Windows of Time

The relationships among various components of evolutionary biology are depicted heuristically in figure 9.16. This figure summarizes our perspective

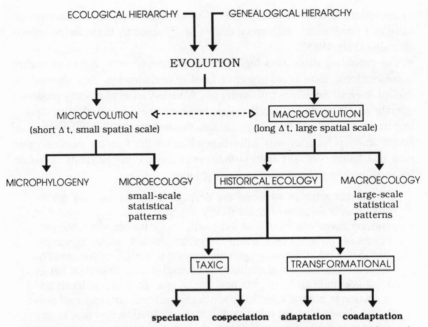

Fig. 9.16. Heuristic flowchart depicting the relationship of historical ecological research to various areas of evolutionary biology.

on the role and place of historical ecology within the evolutionary framework. Microevolutionary research emphasizes short time-scale and small spatial-scale evolutionary processes. It comprises two elements, one dealing with ecological factors, microecology, and the other with genealogical factors, microphylogeny. Microecology incorporates population ecology, ecological aspects of population genetics, and the statistical analysis of phylogenetic constraints on population-level adaptive processes. The results of microecological research are generally presented as statistical patterns on short time scales and small spatial scales. Studies of microphylogeny use historical reconstructions rather than statistical summaries to answer evolutionary questions at the population level. This category therefore includes population genetical studies of speciation and phylogenetic analysis based on population-level information (Avise 1989).

Macroevolutionary research is concerned with long time-scale and large spatial-scale evolutionary patterns and processes. Like microevolution, it incorporates both ecological and genealogical influences; however, the distinction between the two is somewhat blurred at this level. Students of macroecology (Brown and Maurer 1989) are primarily concerned with documenting the **statistical patterns** of spatial and resource allocation of energy use (e.g.,

Maurer and Brown 1988). Students of historical ecology are interested in uncovering the **phylogenetic patterns** of spatial and resource allocation of biological information. There are two classes of research programs within historical ecology. The taxic approach concentrates upon explanations of speciation and extinction rates and patterns. The transformational approach focusses upon adaptive changes in evolution. Investigations at the macroevolutionary level thus highlight the mutually dependent nature of interactions between ecological and genealogical factors.

Evolution is not a process, it is a result. When we document evolutionary change within any group of organisms, what we are really describing is the outcome of an interaction between a variety of processes. Darwin (1872:403) himself detailed the existence of several evolutionary "laws," including growth, the "correlation of parts," reproduction, variability in traits, competition, natural selection, character divergence, and extinction. These processes, in turn, are characterized by both the unique properties of the ecological and genealogical hierarchies and the properties arising from their interactions. Because of this, evolutionary explanations that do not consider all of these components are inherently incomplete.

Modern ecology, ethology, and systematics were founded upon the Darwinian tradition of integrating genealogical and ecological information into explanations of descent with modification. All these disciplines have drifted from this pathway, moving along a star burst of increasingly specialized and independent trajectories. However, evolution is more than the observations, results, and theories of any one discipline. In the past decade, we have finally come to the realization that questions of global diversity and global ecology are the concern of all people, so a reconciliation between these research programs is now critically important. This reconciliation can be accomplished by incorporating historical and nonhistorical, biological and nonbiological processes into our evolutionary perspective. Evolution binds all organisms in both a common hierarchy of life on this planet and a common hierarchy of processes shaping the universe. This "web of existence" is strongly influenced by the irreversible and indelible effects of time. Because of this, the potential of the future is hidden within the constraints of the past in biological systems. Uncovering, understanding, and preserving this potential will be the task of a coalition of evolutionary biologists working on all spatial and temporal scales.

References

Adams, E. N. 1972. Consensus techniques and the comparison of taxonomic trees. *Syst. Zool.* 21:390–97.

Åhman, I. 1981. Några kålväxters potential som värdväxter för skidgallmyggen. *Ent. Tidskr.* 102:111–19.

————. 1985. Larval feeding period and growth of *Dasineura brassicae* Winn. (Dipt., Cecidomyiidae) on brassica host plants. *Oikos* 44:191–94.

Alberch, P. 1982. Developmental constraints in evolutionary processes. In *Evolution and development*, ed. J. T. Bonner, 313–32. New York: Springer Verlag.

Alberch, P., S. J. Gould, G. F. Oster, and D. B. Wake. 1979. Size and shape in ontogeny and phylogeny. *Paleobiology* 5:296–315.

Albrecht, H. 1966. Zur Stammesgeschichte einiger Bewegungsweisen bei Fischen, untersucht am Verhalten von *Haplochromis* (Pisces, Cichlidae). *Z. Tierpsychol.* 23:270–302.

Alcock, J. 1984. *Animal behavior.* 3d ed. Sunderland, Mass.: Sinauer Assoc.

Alexander, R. D. 1962. The role of behavioral study in cricket classification. *Syst. Zool.* 11:53–72.

Andersen, N. M. 1982. The semi-aquatic bugs (Hemiptera, Gerromorpha): Phylogeny, adaptations, biogeography, and classification. *Entomograph* 3:1–455.

Armbruster, W. S. 1988. Multilevel comparative analysis of the morphology, function, and evolution of *Dalechampia* blossoms. *Ecology* 69:1746–61.

Ashlock, P. D. 1974. The uses of cladistics. *Ann. Rev. Ecol. Syst.* 5:81–99.

Atz, J. W. 1970. The application of the idea of homology to behavior. In *Development and evolution of behavior,* ed. L. R. Aronson, E. Tobach, D. S. Lehrman, and J. S. Rosenblatt, 53–74. San Francisco: W. H. Freeman & Co.

Avise, J. C. 1983. Protein variation and phylogenetic reconstruction. In *Protein polymorphism: Adaptive and taxonomic significance,* ed. G. S. Oxford and D. Rollinson, 103–30. New York: Academic Press.

Avise, J. C. 1989. Gene trees and organismal histories: A phylogenetic approach to population biology. *Evolution* 43:1192–1208.

Avise, J. C., and C. F. Aquadro. 1982. A comparative summary of genetic distances in vertebrates. *Evol. Biol.* 15:151–85.

Ayala, F. J. 1982a. Beyond Darwinism? The challenge of macroevolution to the synthetic theory of evolution. *Philos. Sci. Assoc.* 2:275–91.

————. 1982b. The genetic structure of species. In *Perspectives on evolution,* ed. R. Milkman, 3–81. Sunderland, Mass.: Sinauer Assoc.

Baerends, G. P. 1958. Comparative methods and the concept of homology in the study of behaviour. *Arch. Neerl. Zool.* 13 (suppl. 1): 401–17.

Baker, F. C. 1927. Molluscan associations of White Lake, Michigan: A study of a small inland lake from an ecological and systematic viewpoint. *Ecology* 8:353–70.

Bakker, R. T. 1983. The deer flees, the wolf pursues: Incongruencies in predator-prey coevolution. In *Coevolution,* ed. D. J. Futuyma and M. Slatkin, 350–82. Sunderland, Mass.: Sinauer Assoc.

Ballinger, R. E. 1983. Life-history variations. In *Lizard ecology: Studies of a model vertebrate,* ed. R. B. Huey, E. R. Pianka, and T. W. Schoener, 241–60. Cambridge: Harvard Univ. Press.

Bandoni, S. M., and D. R. Brooks, 1987a. Revision and phylogenetic analysis of the Amphilinidea Poche, 1922 (Platyhelminthes: Cercomeria: Cercomeromorpha). *Can. J. Zool.* 65:1110–28.

———. 1987b. Revision and phylogenetic analysis of the Gyrocotylidea Poche, 1926 (Platyhelminthes: Cercomeria: Cercomeromorpha). *Can. J. Zool.* 65:2369–89.

Barrett, S. C. H. 1989. Mating system evolution and speciation in heterostylous plants. In *Speciation and its consequences,* ed. D. Otte and J. A. Endler, 257–83. Sunderland, Mass.: Sinauer Assoc.

Barton, N. H. 1989. Founder effect speciation. In *Speciation and its consequences,* ed. D. Otte and J. A. Endler, 229–56. Sunderland, Mass.: Sinauer Assoc.

Barton, N. H., and B. Charlesworth. 1984. Genetic revolutions, founder effects, and speciation. *Ann. Rev. Ecol. Syst.* 15:133–64.

Bateson, P., ed. 1983. *Mate choice.* Cambridge: Cambridge Univ. Press.

Baverstock, P. R., M. Adams, and I. Beveridge. 1985. Biochemical differentiation in bile duct cestodes and their marsupial hosts. *Mol. Biol. Evol.* 2:321–37.

Bekoff, M., T. J. Daniels, and J. L. Gittleman. 1984. Life history patterns and the comparative social ecology of carnivores. *Ann. Rev. Ecol. Syst.* 15:191–232.

Bell, G. 1989. A comparative method. *Amer. Nat.* 133:553–71.

Benz, G. W., and G. B. Deets. 1988. Fifty-one years later: An update on *Entepherus,* with a phylogenetic analysis of Cecropidae Dana, 1849 (Copepoda: Siphonostomatoidea). *Can. J. Zool.* 66:856–65.

Berenbaum, M. R. 1983. Coumarins and caterpillars: A case for coevolution. *Evolution* 37:163–79.

Berlocher, S. H., and G. L. Bush. 1982. An electrophoretic analysis of *Rhagoletis* (Diptera: Tephritidae) phylogeny. *Syst. Zool.* 31:136–55.

Blair, W. F. 1964. Isolating mechanisms and interspecies interactions in anuran amphibians. *Quart. Rev. Biol.* 39:334–44.

Bock, W. J. 1979. The synthetic explanation of macroevolutionary change: A reductionistic approach. *Bull. Carnegie Mus. Nat. Hist.* 13:20–69.

Boeger, W. A., and D. C. Kritsky. 1989. Phylogeny, coevolution, and revision of the Hexabothriidae Price, 1942 (Monogenea). *Int. J. Parasitol.* 19:425–40.

Boucot, A. J. 1975a. *Evolution and extinction rate controls.* New York: Elsevier.

———. 1975b. Standing diversity of fossil groups in successive intervals of geologic time viewed in the light of changing levels of provincialism. *J. Paleontol.* 49:1105–11.

———. 1978. Community evolution and rates of cladogenesis. In *Evolutionary biol-*

ogy, ed. M. K. Hecht, W. C. Steere, and B. Wallace, 11:545–655. New York: Plenum.

————. 1981. *Principles of benthic marine paleoecology.* New York: Academic Press.

————. 1982. *Paleobiologic evidence of behavioral evolution and coevolution.* Corvallis, Oreg.: Author.

————. 1983. Does evolution take place in an ecological vacuum? II. *J. Paleontol.* 57:1–30.

Bowers, M. A., and J. H. Brown. 1982. Body size and coexistence in desert rodents: Chance or community structure? *Ecology* 63:391–400.

Boyden, A. 1947. Homology and analogy: A critical review of the meanings and implication of these concepts in biology. *Amer. Midl. Nat.* 37:648–60.

Bradbury, J. W., and M. B. Andersson, eds. 1987. *Sexual selection: Testing the alternatives.* London: John Wiley & Sons.

Bragg, A. N., and C. C. Smith. 1943. Observations on the ecology and natural history of anura. IV. The ecological distribution of toads in Oklahoma. *Ecology* 24:285–309.

Bramble, D. M., and D. B. Wake. 1985. Feeding mechanisms of lower tetrapods. In *Functional vertebrate morphology,* ed. M. Hildebrand, D. M. Bramble, K. F. Liem, and D. B. Wake, 230–61. Cambridge: Harvard Univ. Press.

Brooks, D. R. 1977. Evolutionary history of some plagiorchioid trematodes of anurans. *Syst. Zool.* 26:277–89.

————. 1978. Systematic status of proteocephalid cestodes from reptiles and amphibians in North America with descriptions of three new species. *Proc. Helminthol. Soc. Wash.* 45:1–28.

————. 1979a. Testing hypotheses of evolutionary relationships among parasitic helminths: The digeneans of crocodilians. *Amer. Zool.* 12:1225–38.

————. 1979b. Testing the context and extent of host-parasite coevolution. *Syst. Zool.* 28:299–307.

————. 1980a. Allopatric speciation and non-interactive parasite community structure. *Syst. Zool.* 29:192–203.

————. 1980b. Revision of the Acanthostominae (Digenea: Cryptogonimidae). *Zool. J. Linn. Soc.* 70:313–82.

————. 1981. Hennig's parasitological method: A proposed solution. *Syst. Zool.* 30:229–49.

————. 1982. Higher level classification of parasitic platyhelminths and fundamentals of cestode classification. In *Parasites: Their world and ours,* ed. D. F. Mettrick and S. S. Desser, 189–93. Amsterdam: Elsevier Biomedical.

————. 1985. Historical ecology: A new approach to studying the evolution of ecological associations. *Ann. Missouri Bot. Garden* 72:660–80.

————. 1988a. Macroevolutionary comparisons of host and parasite phylogenies. *Ann. Rev. Ecol. Syst.* 19:235–59.

————. 1988b. Scaling effects in historical biogeography: A new view of space, time, and form. *Syst. Zool.* 38:237–44.

————. 1989a. The phylogeny of the Cercomeria (Platyhelminthes: Rhabdocoela) and general evolutionary principles. *J. Parasitol.* 75:606–16.

————. 1989b. A summary of the database pertaining to the phylogeny of the major groups of parasitic platyhelminths, with a revised classification. *Can. J. Zool.* 67:714–20.

———. 1990. Parsimony analysis in historical biogeography and coevolution: Methodological and theoretical update. *Syst. Zool.* 39. (In press.)

Brooks, D. R., and S. M. Bandoni. 1988. Coevolution and relicts. *Syst. Zool.* 37:19–33.

Brooks, D. R., S. M. Bandoni, C. M. Macdonald, and R. T. O'Grady. 1989. Aspects of the phylogeny of the Trematoda Rudolphi, 1808 (Platyhelminthes: Cercomeria). *Can. J. Zool.* 67:2609–24.

Brooks, D. R., J. Collier, B. A. Maurer, J. D. H. Smith, and E. O. Wiley. 1989. Entropy and information in evolving biological systems. *Biol. Philos.* 4:407–32.

Brooks, D. R., and T. L. Deardorff. 1988. *Rhinebothrium devaneyi* n. sp. (Eucestoda: Tetraphyllidea) and *Echinocephalus overstreeti* Deardorff and Ko, 1983 (Nematoda: Gnathostomidae) in a thorny back ray, *Urogymnus asperrimus,* from Enewetak Atoll, with phylogenetic analysis of both species groups. *J. Parasitol.* 74:459–65.

Brooks, D. R., and D. R. Glen. 1982. Pinworms and primates: A case study in coevolution. *Proc. Helminthol. Soc. Wash.* 49:76–85.

Brooks, D. R., and C. A. Macdonald. 1986. A new species of *Phyllodistomum* Braun, 1899 (Digenea: Gorgoderidae) in a neotropical catfish, with discussion of the generic relationships of the Gorgoderidae. *Can. J. Zool.* 64:1326–30.

Brooks, D. R., and C. Mitter. 1984. Analytical basis of coevolution. In *Fungus-insect relationships: Perspectives in ecology and evolution,* ed. Q. Wheeler and M. Blackwell, 42–53. New York: Columbia Univ. Press.

Brooks, D. R., and R. T. O'Grady. 1989. Crocodilians and their helminth parasites: Macroevolutionary considerations. *Amer. Zool.* 29:873–83.

Brooks, D. R., R. T. O'Grady, and D. R. Glen. 1985a. Phylogenetic analysis of the Digenea (Platyhelminthes: Cercomeria) with comments on their adaptive radiation. *Can. J. Zool.* 63:411–43.

———. 1985b. The phylogeny of the Cercomeria Brooks, 1982 (Platyhelminthes). *Proc. Helminthol. Soc. Wash.* 52:1–20.

Brooks, D. R., and R. M. Overstreet. 1978. The family Liolopidae (Digenea) including a new genus and two new species from crocodilians. *Int. J. Parasitol.* 8:267–73.

Brooks, D. R., and G. Rasmussen. 1984. Proteocephalidean cestodes from Venezuelan siluriform fishes, with a revised classification of the Monticelliidae. *Proc. Biol. Soc. Wash.* 97:748–60.

Brooks, D. R., T. B. Thorson, and M. A. Mayes. 1981. Freshwater stingrays (Potamotrygonidae) and their helminth parasites: Testing hypotheses of evolution and coevolution. In *Advances in cladistics: Proceedings of the first meeting of the Willi Hennig society,* ed. V. A. Funk and D. R. Brooks, 147–75. New York: New York Botanical Garden.

Brooks, D. R., and E. O. Wiley. 1985. Theories and methods in different approaches to systematics. *Cladistics* 1:1–14.

———. 1986. *Evolution as entropy: Toward a unified theory of biology.* Chicago: Univ. of Chicago Press.

————. 1988. *Evolution as entropy: Toward a unified theory of biology.* 2d ed. Chicago: Univ. of Chicago Press.

Brown, J. H. 1971. Mechanisms of competitive exclusion between two species of chipmunks. *Ecology* 52:305–11.

————. 1973. Species diversity of seed-eating desert rodents in sand dune habitats. *Ecology* 54:775–87.

————. 1981. Two decades of homage to Santa Rosalia: Toward a general theory of diversity. *Amer. Zool.* 21:877–88.

————. 1984. On the relationship between abundance and distribution of species. *Amer. Nat.* 124:255–79.

Brown, J. H., and A. C. Gibson. 1984. *Biogeography.* St. Louis: Mosby.

Brown, J. H., and B. A. Maurer. 1987. Evolution of species assemblages: Effects of energetic constraints and species dynamics on the diversification of the North American avifauna. *Amer. Nat.* 130:1–17.

————. 1989. Macroecology: The division of food and space among species on continents. *Science* 243:1145–50.

Brown, J. H., and Z. Zeng. 1989. Comparative population ecology of eleven species of rodents in the Chihuahuan desert. *Ecology* 70:1507–25.

Brown, K. 1983. Do life history tactics exist at the intraspecific level? Data from freshwater snails. *Amer. Nat.* 121:871–79.

Brues, C. T. 1920. The selection of food-plants by insects, with special reference to lepidopterous larvae. *Amer. Nat.* 54:313–32.

Brundin, L. 1966. Transantarctic relationships and their significance, as evidenced by chironomid midges. *Kungl. Svenska Vetenskap. Handl.* 11:1–472.

————. 1972. Evolution, causal biology, and classification. *Zool. Scripta* 1:107–20.

Burghardt, G. M. and J. L. Gittleman. 1990. Comparative behavior and phylogenetic analysis. In *Interpretation and explanation in the study of behavior: Comparative perspectives,* ed. M. Bekoff and D. Jamieson. Boulder, Colo.: Westview Press.

Bush, G. L. 1966. The taxonomy, cytology, and evolution of the genus *Rhagoletis* in North America (Diptera, Tephrytidae). *Bull. Mus. Comp. Zool.* 134:431–562.

————. 1969. Sympatric host race formation and speciation in frugivorous flies of the genus *Rhagoletis* (Diptera, Tephrytidae). *Evolution* 23:237–51.

————. 1974. The mechanism of sympatric host race formation in the true fruit flies (Tephrytidae). In *Genetic mechanisms of speciation in insects,* ed. M. J. D. White, 3–23. Sydney: Australia & New Zealand Book Co.

————. 1975a. Modes of animal speciation. *Ann. Rev. Ecol. Syst.* 6:339–64.

————. 1975b. Sympatric speciation in phytophagous parasitic insects. In *Evolutionary strategies of parasitic insects and mites,* ed. P. W. Price, 187–206. New York: Plenum.

————. 1982. What do we really know about speciation? In *Perspectives on evolution,* ed. R. Milkman, 119–28. Sunderland, Mass.: Sinauer Assoc.

Buss, L. 1987. *The evolution of individuality.* Princeton: Princeton Univ. Press.

Buth, D. G. 1984. The application of electrophoretic data in systematics. *Ann. Rev. Ecol. Syst.* 15:501–22.

Butlin, R. K. 1987. Speciation by reinforcement. *Trends Ecol. Evol.* 2:8–13.

————. 1989. Reinforcement of premating isolation. In *Speciation and its conse-*

quences, ed. D. Otte and J. A. Endler, 158–79. Sunderland, Mass.: Sinauer Assoc.

Cade, T. J. 1963. Observations of torpidity in captive chipmunks of the genus *Eutamias. Ecology* 44:255–61.

Cain, S. A. 1944. *Foundations of plant biogeography.* New York: Harper & Row.

Caisse, M., and J. Antonovics. 1978. Evolution in closely adjacent plant populations. IX. Evolution of reproductive isolation in clinal populations. *Heredity* 40:371–84.

Camp, W. H. 1947. Distributional patterns in modern plants and the problems of ancient dispersals. *Ecol. Mono.* 17:159–83.

Carothers, J. H. 1984. Sexual selection and sexual dimorphism in some herbivorous lizards. *Amer. Nat.* 124:244–54.

Carpenter, C. C. 1956. Body temperatures of three species of *Thamnophis. Ecology* 37:732–35.

Carson, H. L. 1975. The genetics of speciation at the diploid level. *Amer. Nat.* 109:83–92.

———. 1982. Speciation as a major reorganization of polygenic balances. In *Mechanisms of speciation,* ed. C. Barigozzi, 411–33. New York: Alan R. Liss.

Carson, H. L., and K. Y. Kaneshiro. 1976. *Drosophila* of Hawaii: Systematics and evolutionary genetics. *Ann. Rev. Ecol. Syst.* 7:311–46.

Carson, H. L., and A. R. Templeton. 1984. Genetic revolutions in relation to speciation phenomena: The founding of new populations. *Ann Rev. Ecol. Syst.* 15:97–131.

Chapin, J. P. 1917. The classification of the weaver-birds. *Bull. Amer. Mus. Nat. Hist.* 37:243–80.

Charlesworth, B., R. Lande, and M. Slatkin. 1982. A neo-Darwinian commentary on macroevolution. *Evolution* 36:474–98.

Charlesworth, B., and S. Rouhani. 1988. The probability of peak shifts in a founder population. II. An additive polygenic trait. *Evolution* 42:1129–45.

Cheverud, J. M., M. M. Dow, and W. Leutenegger. 1985. The quantitative assessment of phylogenetic constraints in comparative analyses: Sexual dimorphism in body weight among primates. *Evolution* 39:1335–51.

Christie, P., and M. R. MacNair. 1987. The distribution of post-mating reproductive isolating genes in populations of the yellow monkey flower, *Mimulus guttatus. Evolution* 41:571–78.

Clutton-Brock, T. H., and P. H. Harvey. 1984. Comparative approaches to investigating adaptation. In *Behavioural ecology: An evolutionary approach,* 2d ed., ed. J. R. Krebs and N. B. Davies, 7–29. Sunderland, Mass.: Sinauer Assoc.

Coddington, J. A. 1988. Cladistic tests of adaptational hypotheses. *Cladistics* 4:3–22.

Collette, B. B. 1982. South American freshwater needlefishes of the genus *Potamorrhaphis* (Beloniformes: Belonidae). *Proc. Biol. Soc. Wash.* 95:714–47.

Collette, B. B., and J. L. Russo. 1985. Interrelationships of the Spanish mackerels (Pisces: Scombridae: *Scomberomorus*) and their copepod parasites. *Cladistics* 1:141–58.

Collins, J. P. 1986. Evolutionary ecology and the use of natural selection in ecological theory. *J. Hist. Biol.* 19:257–88.

Compagno, L. J. V. 1977. Phyletic relationships of living sharks and rays. *Amer. Zool.* 17:303–22.

Conant, R. 1975. *A field guide to reptiles and amphibians of eastern and central North America*. Boston: Houghton Mifflin.

Connell, J. H. 1978. Diversity in tropical rain forests and coral reefs. *Science* 199:1302–9.

Connell, J. H., and E. Orias. 1964. The ecological regulation of species diversity. *Amer. Nat.* 98:399–414.

Connor, E. F., and E. D. McCoy. 1979. The statistics and biology of the species-area relationship. *Amer. Nat.* 113:791–833.

Constance, L. 1953. The role of plant ecology in biosystematics. *Ecology* 34:642–49.

Coyne, J. A., and H. A. Orr. 1989. Two rules of speciation. In *Speciation and its consequences,* ed. D. Otte and J. A. Endler, 180–207. Sunderland, Mass.: Sinauer Assoc.

Coyne, J. A., and M. Kreitman. 1986. Evolutionary genetics of two sibling species, *Drosophila simulans* and *D. sechella. Evolution* 40:673–91.

Cracraft, J. 1982a. Geographic differentiation, cladistics, and vicariance biogeography: Reconstructing the tempo and mode of evolution. *Amer. Zool.* 22:411–24.

———. 1982b. A nonequilibrium theory for the rate-control of speciation and extinction and the origin of macroevolutionary patterns. *Syst. Zool.* 31:348–65.

———. 1983a. Cladistic analysis and vicariance biogeography. *Amer. Sci.* 7:273–81.

———. 1983b. Species concepts and speciation analysis. In *Current ornithology,* ed. R. F. Johnston, 1:159–87. New York: Plenum.

———. 1984. Conceptual and methodological aspects of the study of evolutionary rates, with some comments on bradytely in birds. In *Living fossils,* ed. N. Eldredge and S. M. Stanley, 95–104. New York: Springer Verlag.

———. 1985a. Biological diversification and its causes. *Ann. Missouri Bot. Garden* 72:794–822.

———. 1985b. Species selection, macroevolutionary analysis, and the "hierarchical theory of evolution." *Syst. Zool.* 34:222–29.

———. 1986. Origin and evolution of continental biotas: Speciation and historical congruence within the Australian avifauna. *Evolution* 40:977–96.

———. 1988. Deep-history biogeography: Retrieving the historical pattern of evolving continental biotas. *Syst. Zool.* 37:221–36.

———. 1989. Speciation and its ontology: The empirical consequences of alternative species concepts for understanding patterns and processes of speciation. In *Speciation and its consequences,* ed. D. Otte and J. A. Endler, 28–59. Sunderland, Mass.: Sinauer Assoc.

Cracraft, J., and R. O. Prum. 1988. Patterns and processes of diversification: Speciation and historical congruence in some neotropical birds. *Evolution* 42:603–20.

Cressey, R. F., B. B. Collette, and J. Russo. 1983. Copepods and scombrid fishes: A study in host-parasite relationships. *Fish. Bull.* 81:227–65.

Crisci, J. V., and T. F. Stuessy. 1980. Determining primitive character states for phylogenetic reconstructions. *Syst. Bot.* 5:112–35.

Croizat, L., G. Nelson, and D. E. Rosen. 1974. Centers of origin and related concepts. *Syst. Zool.* 23:265–87.

Cronquist, A. 1981. *Integrated system of classification of flowering plants*. New York: Columbia Univ. Press.

Crother, B. I., M. M. Miyamoto, and W. F. Presch. 1986. Phylogeny and biogeography of the lizard family Xantusiidae. *Syst. Zool.* 35:37–45.

Dahlgren, R., and K. Bremer. 1985. Major clades of the angiosperms. *Cladistics* 1:349–68.

Darwin, C. 1859. *The origin of species by means of natural selection.* London: John Murray.

———. 1872. *The origin of species by means of natural selection.* 6th ed. London: John Murray.

———. 1877. *The different forms of flowers on plants of the same species.* London: John Murray.

Davidson, D. W., and S. R. Morton. 1984. Dispersal adaptations of some *Acacia* species in the Australian arid zone. *Ecology* 65:1038–51.

Davidson, J. F. 1952. The use of taxonomy in ecology. *Ecology* 33:297–99.

Deets, G. B. 1987. Phylogenetic analysis and revision of *Kroeyerina* Wilson, 1932 (Siphonostomatoida: Kroeyeriidae), copepods parasitic on Chondrichthyans, with descriptions of four new species and the erection of a new genus, *Prokroyeria. Can. J. Zool.* 65:2121–48.

Deets, G. B., and J.-S. Ho. 1988. Phylogenetic analysis of the Eudactylinidae (Crustacea: Copepoda: Siphonostomatoida), with descriptions of two new genera. *Proc. Biol. Soc. Wash.* 101:317–39.

Dickinson, H., and J. Antonovics. 1973. Theoretical consideration of sympatric divergence. *Amer. Nat.* 107:256–74.

Diehl, S. R., and G. L. Bush. 1989. The role of habitat preference in adaptation and speciation. In *Speciation and its consequences,* ed. D. Otte and J. A. Endler, 345–65. Sunderland, Mass.: Sinauer Assoc.

Dietz, R. S., and J. C. Holden. 1966. Miogeoclines (Miogeosynclines) in space and time. *J. Geol.* 74:566–83.

Dobson, F. S. 1985. The use of phylogeny in behavior and ecology. *Evolution* 39:1384–88.

Dobzhansky, T. 1937. *Genetics and the origin of species.* New York: Columbia Univ. Press.

———. 1940. Speciation as a stage in evolutionary divergence. *Amer. Nat.* 74:312–21.

———. 1951. *Genetics and the origin of species.* 3d ed. New York: Columbia Univ. Press.

———. 1970. *Genetics of the evolutionary process.* New York: Columbia Univ. Press.

Dobzhansky, T., and A. B. da Cunha. 1955. Differentiation of nutritional preferences in Brazilian species of *Drosophila. Ecology* 36:34–39.

Dojiri, M., and G. B. Deets. 1988. *Norkus cladocephalus,* new genus, new species (Siphonostomatoida: Sphyriidae), a copepod parasitic on an elasmobranch from southern California waters, with a phylogenetic analysis of the Sphyriidae. *J. Crust. Biol.* 8:679–87.

Donoghue, M. J. 1985. A critique of the biological species concept, and recommendation for a phylogenetic alternative. *Bryologist* 88:172–81.

———. 1989. Phylogenies and the analysis of evolutionary sequences, with examples from seed plants. *Evolution* 43:1137–56.

Donoghue, M. J., and P. D. Cantino. 1984. The logic and limitations of the outgroup substitution approach to cladistic analysis. *Syst. Bot.* 9:192–202.

Donoghue, M. J., J. A. Doyle, J. Gauthier, A. G. Kluge, and T. Rowe. 1989. The importance of fossils in phylogeny reconstruction. *Ann. Rev. Ecol. Syst.* 20:431–60.

Dover, G. A. 1982. Molecular drive: A cohesive mode of species evolution. *Nature* 229:111–17.

Doyle, J. A., and M. J. Donoghue. 1986. Seed plant phylogeny and the origin of angiosperms: An experimental cladistic approach. *Bot. Rev.* 52:321–431.

Ducke, A. 1913. Über Phylogenie und Klassification der sozialen Vespiden. *Zool. Jahrb. Abt. Syst. Oekol. Geogr. Tiere* 36:303–30.

Duellman, W. E. 1985. Reproductive modes in anuran amphibians: Phylogenetic significance of adaptive strategies. *S. Afr. J. Sci.* 81:174–78.

Duellman, W. E., and L. Trueb. 1986. *Biology of amphibians.* New York: McGraw-Hill.

Duffels, J. P. 1986. Biogeography of Indo-Pacific Cicadoidea: A tentative recognition of areas of endemism. *Cladistics* 2:318–36.

Dunbar, R. I. M. 1982. Adaptation, fitness, and the evolutionary tautology. In *Current problems in sociobiology,* ed. King's College Sociobiology Group, 9–28. Cambridge: Cambridge Univ. Press.

Dunford, C., and R. Davis. 1975. Cliff chipmunk vocalizations and their relevance to the taxonomy of coastal Sonoran chipmunks. *J. Mammal.* 56:207–12.

Dunham, A. E., and D. B. Miles. 1985. Patterns of covariation in the life history traits of squamate reptiles: The effects of size and phylogeny reconsidered. *Amer. Nat.* 126:231–57.

Dunham, A. E., D. B. Miles, and D. N. Reznick. 1988. Life history patterns in squamate reptiles. In *Biology of the Reptilia,* ed. C. Gans and R. B. Huey, 16B:441–522. New York: Alan R. Liss.

Ehrlich, P. R., and P. H. Raven. 1964. Butterflies and plants: A study in coevolution. *Evolution* 18:586–608.

Eibl-Eibesfeldt, I. 1975. *Ethology: The biology of behavior.* 2d ed. New York: Holt, Rinehart & Winston.

Eldredge, N. 1976. Differential evolutionary rates. *Paleobiology* 2:174–77.

———. 1979. Alternative approaches to evolutionary theory. *Bull. Carnegie Mus. Nat. Hist.* 13:7–19.

———. 1985. *Unfinished synthesis: Biological hierarchies and modern evolutionary thought.* New York: Columbia Univ. Press.

———. 1986. Information, economics, and evolution. *Ann. Rev. Ecol. Syst.* 17:351–69.

———. 1989. *Macroevolutionary dynamics: Species, niches, and adaptive peaks.* New York: McGraw-Hill.

Eldredge, N., and J. Cracraft. 1980. *Phylogenetic patterns and the evolutionary process.* New York: Columbia Univ. Press.

Eldredge, N., and S. J. Gould. 1972. Punctuated equilibria: An alternative to phyletic gradualism. In *Models in paleobiology,* ed. T. J. M. Schopf, 82–115. San Francisco: W. H. Freeman & Co.

Eldredge, N., and S. N. Salthe. 1984. Hierarchy and evolution. In *Oxford surveys in evolutionary biology*, ed. R. Dawkins and M. Ridley, 1:182–206. Oxford: Oxford Univ. Press.

Emerson, A. E. 1938. Termite nests: A study of the phylogeny of behavior. *Ecol. Mono.* 8:247–84.

Emig, C. C. 1985. Relations entre l'espèce, structure dissapatrice biologique, et l'écosytème, structure dissapatrice écologique: Contribution à la théorie de l'évolution des systèmes non–en équilibre. *C. R. Acad. Sci. Paris* 300:323–26.

Endler, J. A. 1977. *Geographic variation, speciation, and clines.* Princeton, N.J.: Princeton Univ. Press.

———. 1982. Problems in distinguishing historical from ecological factors in biogeography. *Amer. Zool.* 22:441–52.

———. 1989. Conceptual and other problems in speciation. In *Speciation and its consequences*, ed. D. Otte and J. A. Endler, 625–48. Sunderland, Mass.: Sinauer Assoc.

Endler, J. A., and T. McLellan. 1988. The processes of evolution: Toward a newer synthesis. *Ann. Rev. Ecol. Syst.* 19:395–421.

Erwin, T. 1985. The taxon pulse: A general pattern of lineage radiation and extinction among carabid beetles. In *Taxonomy, phylogeny, and zoogeography of beetles and ants*, ed. G. E. Ball, 437–72. Netherlands: Dordrecht, W. Junk.

Estes, R., and G. Pregill, eds. 1988. *Phylogenetic relationships of the lizard families.* Palo Alto, Calif.: Stanford Univ. Press.

Estes, R., K. de Queiroz, and J. Gauthier. 1988. Phylogenetic relationships within Squamata. In *Phylogenetic relationships of the lizard families*, ed. R. Estes and G. Pregill, 119–281. Palo Alto, Calif.: Stanford Univ. Press.

Fahrenholz, H. 1913. Ectoparasiten und Abstammungslehre. *Zool. Anz.* (Leipzig) 41:371–74.

Faith, D. P. In press. Homoplasy as pattern: Multivariate analysis of morphological convergence in Anseriformes. *Cladistics.*

Farris, J. S. 1970. Methods for computing Wagner trees. *Syst. Zool.* 19:83–92.

———. 1981. Distance data in phylogenetic analysis. In *Advances in cladistics: Proceedings of the first meeting of the Willi Hennig Society*, ed. V. A. Funk and D. R. Brooks, 3–23. New York: New York Botanical Garden.

———. 1982. Outgroups and parsimony. *Syst. Zool.* 31:328–34.

———. 1985. Distance data revisited. *Cladistics* 1:67–85.

Farris, J. S., A. G. Kluge, and M. J. Eckardt. 1970. A numerical approach to phylogenetic systematics. *Syst. Zool.* 19:172–89.

Feeny, P. P. 1976. Plant apparency and chemical defense. In *Recent advances in phytochemistry: Biochemical interactions between plants and insects*, ed. J. Wallace and R. Mansell, 10:1–40. New York: Plenum.

Felsenstein, J. 1981. Skepticism towards Santa Rosalia, or why are there so few kinds of animals? *Evolution* 35:124–38.

———. 1982. Numerical methods for inferring phylogenetic trees. *Quart. Rev. Biol.* 57:379–404.

———. 1985. Phylogenies and the comparative method. *Amer. Nat.* 125:1–15.

———. 1988. Phylogenies and quantitative characters. *Ann. Rev. Ecol. Syst.* 19:445–71.

Fenwick, G. D. 1984. Life-history tactics of brooding crustacea. *J. Exp. Mar. Biol. Ecol.* 84:247–64.

Ferris, V. R., and J. M. Ferris. 1989. Why ecologists need systematists: Importance of systematics to ecological research. *J. Nematol.* 21:308–14.

Fink, W. L. 1982. The conceptual relationship between ontogeny and phylogeny. *Paleobiology* 8:254–64.

Fisher, D. C. 1982. Phylogenetic and macroevolutionary patterns within the Xiphosurida. *Proc. 3d N. Amer. Paelontol. Conv.* 1:175–80.

Fisher, R. A. 1930. *The genetical theory of natural selection.* Oxford: Clarendon Press.

———. 1958. *The genetical theory of natural selection.* 2d ed. New York: Dover.

Fitzpatrick, J. W. 1988. Why so many passerine birds? A response to Raikow. *Syst. Zool.* 37:71–76.

Fox, G. A. 1989. Consequences of flowering-time variation in a desert annual: Adaptation and history. *Ecology* 70:1294–1306.

Fraser, D. F. 1976. Coexistence of salamanders in the genus *Plethodon:* A variation of the Santa Rosalia theme. *Ecology* 57:238–51.

Friedmann, H. 1929. *The cowbirds.* Springfield, Ill.: Charles C. Thomas.

Funk, V. A. 1982. Systematics of *Montanoa* (Asteraceae: Heliantheae). *Mem. N. Y. Bot. Garden* 36:1–135.

———. 1985. Phylogenetic patterns and hybridization. *Ann. Missouri Bot. Garden* 72:681–715.

Funk, V. A., and D. R. Brooks. 1990. Phylogenetic systematics as the basis of comparative biology. *Smithsonian Contr. Bot.* 73:1–45.

Funk, V. A., and P. H. Raven. 1980. Ploidy in *Montanoa* (Cerv.) (Compositae, Heliantheae). *Taxon* 29:417–19.

Futuyma, D. J. 1986. *Evolutionary biology.* 2d ed. Sunderland, Mass.: Sinauer Assoc.

———. 1989. Speciational trends and the role of species in macroevolution. *Amer. Nat.* 134:318–21.

Futuyma, D. J., and J. Kim. 1987. Phylogeny and coevolution. *Science* 237:441–47.

Futuyma, D. J., and G. C. Mayer. 1980. Non-allopatric speciation in animals. *Syst. Zool.* 29:254–71.

Futuyma, D. J., and G. Moreno. 1988. The evolution of ecological specialization. *Ann. Rev. Ecol. Syst.* 19:207–33.

Futuyma, D. J., and M. Slatkin, eds. 1983. *Coevolution.* Sunderland, Mass.: Sinauer Assoc.

Gauld, I. D. 1983. The classification, evolution, and distribution of the Labeninae, an ancient southern group of Ichneumonidae (Hymenoptera). *Syst. Entomol.* 8:167–78.

Geel, B. van, and T. van der Hammen. 1973. Upper Quaternary vegetation and climatic sequence of the Fuquene area (eastern cordillera, Colombia). *Palaeogeogr. Palaeoclimatol. Palaeoecol.* 14:9–92.

Geesink, R., and D. J. Kornet. 1989. Speciation and Malesian Leguminoseae. In *Tropical forests: Botanical dynamics, speciation, and diversity,* ed. L. B. Holm-Nielsen, I. C. Nielsen, and H. Balsely, 135–51. London: Academic Press.

Ghiselin, M. 1974. A radical solution to the species problem. *Syst. Zool.* 23:536–44.

Gilbert, C. R. 1976. Composition and derivation of the North American freshwater fish fauna. *Fla. Sci.* 39:104–11.

Gilinsky, N. L. 1981. Stabilizing selection in the Archaeogastropoda. *Paleobiology* 7:316–31.

Gittenberger, E. 1988. Sympatric speciation in snails: A largely neglected model. *Evolution* 42:826–28.

Gittleman, J. L. 1981. The phylogeny of parental care in fishes. *Anim. Behav.* 29:936–41.

———. 1985. Carnivore body size: Ecological and taxonomic correlations. *Oecologia* 67:540–54.

———. 1986. Carnivore brain size, behavioral ecology, and phylogeny. *J. Mammal.* 67:23–36.

———. 1989. The comparative approach in ethology: Aims and limitations. In *Perspectives in ethology,* ed. P. P. G. Bateson and P. H. Klopfer, 8:55–83. London: Plenum.

Gittleman, J. L., and M. Kot. 1990. Adaptation: Statistics and a null model for estimating phylogenetic effects. *Amer. Nat.* (In press.)

Glazier, D. S. 1987. Energetics and taxonomic patterns of species diversity. *Syst. Zool.* 36:62–71.

Glen, D. R., and D. R. Brooks. 1985. Phylogenetic relationships of some strongylate nematodes of primates. *Proc. Helminthol. Soc. Wash.* 52:227–36.

———. 1986. Parasitological evidence pertaining to the phylogeny of the hominoid primates. *Biol. J. Linn. Soc.* 27:331–54.

Goldschmidt, R. 1940. *The material basis of evolution.* New Haven, Conn.: Yale Univ. Press.

Goodnight, C. J. 1987. On the effect of founder events on epistatic genetic variance. *Evolution* 41:80–91.

Gorman, O. T. In press. Evolutionary ecology versus historical ecology: Assembly, structure, and organization of stream fish communities. In *Systematics, historical ecology, and North American freshwater fishes,* ed. R. L. Mayden. Palo Alto, Calif.: Stanford Univ. Press.

Gould, S. J. 1980. Is a new and general theory of evolution emerging? *Paleobiology* 6:119–30.

———. 1981. G. G. Simpson, paleontology, and the modern synthesis. In *The evolutionary synthesis: Perspectives on the unification of biology.* ed. E. Mayr and W. B. Provine, 153–72. Cambridge: Harvard Univ. Press.

———. 1982a. Change in developmental timing as a mechanism of macroevolution. In *Evolution and development,* ed. J. T. Bonner, 333–46. New York: Springer Verlag.

———. 1982b. The meaning of punctuated equilibrium and its role in validating a hierarchical approach to macroevolution. In *Perspectives on evolution,* ed. R. Milkman, 83–104. Sunderland, Mass.: Sinauer Assoc.

———. 1989. A developmental constraint in *Cerion,* with comments on the definition and interpretation of constraint in evolution. *Evolution* 43:516–39.

Gould, S. J., and N. Eldredge. 1977. Punctuated equilibria: The tempo and mode of evolution reconsidered. *Paleobiology* 3:115–51.

Gould, S. J., and R. C. Lewontin. 1979. The spandrels of San Marco and the Pan-glossian paradigm: A critique of the adaptationist programme. *Proc. R. Soc. London* B205:581–98.

Gould, S. J., D. M. Raup, J. J. Sepkoski, T. J. M. Schopf, and D. S. Simberloff. 1977. The shape of evolution: A comparison of real and random clades. *Paleobiology* 3:23–40.

Graham, R. W. 1986. Response of mammalian communities to environmental changes during the late Quaternary. In *Community ecology,* ed. J. Diamond and T. J. Case, 300–313. New York: Harper & Row.

Grant, P. R., and B. R. Grant. 1989. Sympatric speciation and Darwin's finches. In *Speciation and its consequences,* ed. D. Otte and J. A. Endler, 433–57. Sunderland, Mass.: Sinauer Assoc.

Grant, V. 1981. *Plant speciation.* 2d ed. New York: Columbia Univ. Press.

Griffith, H. 1987. Phylogenetic relationships and evolution in the genus *Dissodactylus* Smith, 1870 (Crustacea: Brachyura: Pinnotheridae). *Can. J. Zool.* 65:2292–2310.

Groves, C. P. 1983. Phylogeny of the living species of rhinos. *Z. zool. Syst. Evolut.-forsch.* 21:293–313.

Haffer, J. 1969. Speciation in Amazonian forest birds. *Science* 165:131–37.

Hafner, M. S., and S. A. Nadler. 1988. Phylogenetic trees support the coevolution of parasites and their hosts. *Nature* 332:258–60.

Hairston, N. G. 1951. Interspecies competition and its probable influence upon the vertical distribution of Appalachian salamanders of the genus *Plethodon. Ecology* 32:266–74.

———. 1981. An experimental test of a guild: Salamander competition. *Ecology* 62:65–72.

Haldane, J. B. S. 1932. *The causes of evolution.* London: Longman.

Hammen, T. van der. 1972. Changes in vegetation and climate in the Amazon Basin and surrounding areas during the Pleistocene. *Geol. & Mijnb.* 51:641–43.

Hammen, T. van der, and E. Gonzales. 1960. Upper Pleistocene and Holocene climate and vegetation of the "Sabana de Bogota" (Colombia, South America). *Leidse Geol. Meded.* 25:261–315.

Hansen, T. A. 1983. Models of larval development and rates of speciation in Early Tertiary neogastropods. *Science* 220:501–2.

Harlan, J. R., and J. M. J. DeWet. 1963. The compilospecies concept. *Evolution* 17:497–501.

Harold, A. S., and M. Telford. 1990. Systematics, phylogeny, and biogeography of the genus *Mellita* (Echinoidea: Clyperasteroida). *J. Nat. Hist.* (In press.)

Harrison, L. 1914. The Mallophaga as a possible clue to bird phylogeny. *Austral. Zool.* 1:7–11.

———. 1915a. Mallophaga from *Apteryx,* and their significance: With a note on *Rallicola. Parasitology* 8:88–100.

———. 1915b. The relationship of the phylogeny of the parasite to that of the host. *Rep. British Assoc. Adv. Sci.* 85:476–77.

———. 1916. Bird-parasites and bird-phylogeny. *Ibis* 10:254–63.

———. 1922. On the Mallophagan family Trimenoponidae, with a description of a new genus and species from an Australian marsupial. *Austral. Zool.* 2:154–59.

————. 1924. The migration route of the Australian marsupial fauna. *Austral. Zool.* 3:247–63.

————. 1926. Crucial evidence for antarctic radiation. *Amer. Nat.* 60:374–83.

————. 1928a. Host and parasite. *Proc. Linn. Soc. New South Wales* 53:ix–xxxi.

————. 1928b. On the genus *Stratiodrilus* (Archiannelida: Histriobdellidae), with a description of a new species from Madagascar. *Rec. Austral. Mus.* 16:116–21.

————. 1929. The composition and origins of the Australian fauna, with special reference to the Wegener hypothesis. *Rep. Mtgs. Australas. Assoc. Adv. Sci.* (Perth) 1926:332–96.

Harrison, R. G., and D. M. Rand. 1989. Mosaic hybrid zones and the nature of species boundaries. In *Speciation and its consequences,* ed. D. Otte and J. A. Endler, 111–33. Sunderland, Mass.: Sinauer Assoc.

Hart, J. A. 1985a. Peripheral isolation and the origin of diversity in *Lepechinia* sect. *Parviflorae* (Lamiaceae). *Syst. Bot.* 10:134–46.

————. 1985b. Evolution of dioecism in *Lepechinia* Willd. sect. *Parviflorae* (Lamiaceae). *Syst. Bot.* 10:147–54.

Harvey, P., J. J. Bull, M. Pemberton, and R. J. Paxton. 1982. The evolution of aposematic coloration in distasteful prey: A family model. *Amer. Nat.* 119:710–19.

Harvey, P. H., and T. Clutton-Brock, 1985. Life history variation in primates. *Evolution* 39:559–81.

Hastenrath, S. 1971a. On snowline depression and atmospheric circulation in the tropical Americas during Pleistocene. *S. Afr. Geog. J.* 53:53–69.

————. 1971b. On the Pleistocene snowline depression in the arid regions of the South American Andes. *J. Glaciol.* 10:255–67.

Heinroth, O. 1911. Beiträge zur Biologie, namentlich Ethologie und Psychologie der Anatiden. *Verh. Ver. Int. Ornithol. Kongr.* (Berlin) 1910:589–702.

Henderson, R. W., T. A. Noeske-Hallin, B. I. Crother, and A. Schwartz. 1988. The diets of Hispaniolan colubrid snakes. II. Prey species, prey size, and phylogeny. *Herpetologica* 44:55–70.

Hennig, W. 1950. *Grundzüge einer Theorie der phylogenetischen Systematik.* Berlin: Deutscher Zentralverlag.

————. 1966. *Phylogenetic systematics.* Urbana: Univ. of Illinois Press.

Herrick, F. H. 1911. Nest and nest-building in birds. *J. Anim. Behav.* 1:159–92, 244–77, 336–73.

Hewitt, G. M. 1989. The subdivision of species by hybrid zones. In *Speciation and its consequences.* ed. D. Otte and J. A. Endler, 85–110. Sunderland, Mass.: Sinauer Assoc.

Hickey, L. J., and J. A. Wolfe, 1975. The bases of angiosperm phylogeny: Vegetative morphology. *Ann. Missouri Bot. Garden* 62:583–89.

Hiiemae, K. M., and A. W. Crompton. 1985. Mastication, food transport, and swallowing. In *Functional vertebrate anatomy,* ed. M. Hildebrand, D. M. Bramble, K. F. Liem, and D. B. Wake, 262–90. Cambridge: Harvard Univ. Press.

Hill, J. E., and J. D. Smith. 1984. *Bats: A natural history.* London: British Museum (Natural History).

Hillis, D. M. 1985. Evolutionary genetics of the Andean lizard genus *Pholidobolus* (Sauria: Gymnopthalmidae): Phylogeny, biogeography, and a comparison of tree construction techniques. *Syst. Zool.* 34:109–26.

———. 1988. Systematics of the *Rana pipiens* complex: Puzzle and paradigm. *Ann. Rev. Ecol. Syst.* 19:39–63.

Hillis, D. M., and S. K. Davis. 1986. Evolution of ribosomal DNA: Fifty million years of recorded history in the frog genus *Rana*. *Evolution* 40:1275–88.

Hillis, D. M., J. S. Frost, and D. A. Wright. 1983. Phylogeny and biogeography of the *Rana pipiens* complex: A biochemical evaluation. *Syst. Zool.* 32:132–43.

Hixon, M. A. 1980. Competitive interactions between California reef fishes of the genus *Embiotoca*. *Ecology* 61:918–31.

Ho, J.-S. 1988. Cladistics of *Sunaristes*, a genus of harpacticoid copepods associated with hermit crabs. *Hydrobiologia* 167/168:555–60.

Ho, J.-S., and T. T. Do. 1985. Copepods of the family Lernanthropidae parasitic on Japanese marine fishes, with a phylogenetic analysis of the Lernanthropid genera. *Rep. Sado Mar. Biol. Stat. Niigata Univ.* 15:31–76.

Hoberg, E. P. 1986. Evolution and historical biogeography of a host-parasite assemblage: *Alcataenia* spp. (Cyclophyllidea: Dilepididae) in Alcidae (Charadriiformes). *Can. J. Zool.* 64:2576–89.

Holmes, J. C., and P. W. Price. 1980. Parasite communities: The roles of phylogeny and ecology. *Syst. Zool.* 29:203–13.

Horton, C. C., and D. H. Wise. 1983. The experimental analysis of competition between two syntopic species of orb-web spiders (Araneae: Araneidae). *Ecology* 64:929–44.

Howes, G. J., and G. G. Teugels. 1989. New bariliin cyprinid fishes from West Africa, with a consideration of their biogeography. *J. Nat. Hist.* 23:873–902.

Hubbell, S. P., and L. K. Johnson. 1978. Comparative foraging behavior of six stingless bee species exploiting a standardized resource. *Ecology* 59:1123–36.

Huey, R. B. 1987. Phylogeny, history, and the comparative method. In *New directions in ecological physiology*, ed. M. E. Feder, A. F. Bennett, W. Burggren, and R. B. Huey, 76–98. Cambridge: Cambridge Univ. Press.

Huey, R. B., and A. F. Bennett. 1987. Phylogenetic studies of coadaptation: Preferred temperatures versus optimal performance temperatures of lizards. *Evolution* 41:1098–1115.

Huey, R. B., and E. R. Pianka. 1977. Patterns of niche overlap among broadly sympatric versus narrowly sympatric Kalahari lizards (Scincidae: *Mabuya*). *Ecology* 58:119–28.

Huey, R. B., and T. P. Webster. 1976. Thermal biology of *Anolis* lizards in a complex fauna: The *Cristatellus* group on Puerto Rico. *Ecology* 57:985–94.

Hull, D. L. 1976. Are species really individuals? *Syst. Zool.* 25:174–91.

———. 1978. A matter of individuality. *Philos. Sci.* 45:335–60.

———. 1980. Individuality and selection. *Ann. Rev. Ecol. Syst.* 11:311–32.

———. 1988. *Science as a process*. Chicago: Univ. of Chicago Press.

Humphery-Smith, I. 1989. The evolution of phylogenetic specificity among parasitic organisms. *Parasitol. Today* 5:385–87.

Humphries, C. J. 1981. Biogeographical methods and the southern beeches (Fagaceae: *Nothofagus*). In *Advances in cladistics: Proceedings of the first meeting of the Willi Hennig Society*, ed. V. A. Funk and D. R. Brooks, 177–207. New York: New York Botanical Garden.

Humphries, C. J., J. M. Cox, and E. S. Nielson. 1986. *Nothofagus* and its parasites:

A cladistic approach to coevolution. In *Coevolution and systematics,* ed. A. R. Stone and D. L. Hawksworth, 55–76. Oxford: Clarendon Press.

Humphries, C. J., and L. Parenti. 1986. *Cladistic biogeography.* London: Academic Press.

Hutchinson, G. E. 1957. Concluding remarks. *Cold Spring Harbor Symp. Quant. Biol.* 22:415–27.

————. 1959. Homage to Santa Rosalia or why are there so many kinds of animals? *Amer. Nat.* 93:145–59.

Iersel, J. J. A. van. 1953. An analysis of the parental behaviour of the male three-spined stickleback (*Gasterosteus aculeatus* L.). *Behav. Suppl.* 3:1–159.

Ihering, H. von. 1891. On the ancient relations between New Zealand and South America. *Trans. & Proc. New Zealand Inst.* 24:431–45.

————. 1902. Die Helminthen als Hilfsmittel der zoogeographischen Forschung. *Zool. Anz.* (Leipzig) 26:42–51.

Ikeda, K. 1933. Effect of castration on the secondary sexual characters of anadromous three-spined stickleback, *Gasterosteus aculeatus. Jap. J. Zool.* 5:135–57.

Jablonski, D. 1982. Evolutionary rates and modes in Late Cretaceous gastropods: Role of larval ecology. *Proc. 3d N. Amer. Paleontol. Conv.* 1:257–62.

Jackson, J. B. C. 1974. Biogeographic consequences of eurytypy and stenotypy among marine bivalves and their evolutionary significance. *Amer. Nat.* 108:541–60.

Jacobson, H. R., D. H. Kistner, and J. M. Pasteels. 1986. Generic revision, phylogenetic classification, and phylogeny of the termitophilous tribe Corotocini (Coleoptera: Staphylinidae). *Sociobiology* 12:1–245.

Jermy, T. 1976. Insect-host plant relationships: Coevolution or sequential evolution? *Symp. Biol. Hung.* 16:109–13.

————. 1984. Evolution of insect/host plant relationships. *Amer. Nat.* 124:609–30.

Johnston, S. J. 1912. On some trematode parasites of Australian frogs. *Proc. Linn. Soc. New South Wales* 37:285–362.

————. 1914. Australian trematodes and cestodes. *Med. J. Austral.* 1:243–44.

————. 1916. On the trematodes of Australian birds. *J. & Proc. R. Soc. New South Wales* 50:187–261.

Jong, R. de. 1980. Some tools for evolutionary and phylogenetic studies. *Z. zool. Syst. Evolut.-forsch.* 18:1–23.

Kaneshiro, K. Y. 1980. Sexual isolation, speciation, and the direction of evolution. *Evolution* 34:437–44.

Kauffman, S. A. 1974. The large scale structure and dynamics of gene control circuits: An ensemble approach. *J. Theor. Biol.* 44:167–90.

————. 1983. Developmental constraints: Internal factors in evolution. In *Development and evolution,* ed. B. C. Goodwin, N. Holder, and C. G. Wylie, 195–225. Cambridge: Cambridge Univ. Press.

————. 1986. Autocatalytic sets of proteins. *J. Theor. Biol.* 119:1–24.

Keen, W. H. 1982. Habitat selection and interspecific competition in two species of plethodontid salamanders. *Ecology* 63:94–102.

Kellogg, V. L. 1896. New Mallophaga, 1: With special reference to a collection from maritime birds of the bay of Monterey, California. *Proc. Cal. Acad. Sci.* 6:31–168.

————. 1913. Distribution and species-forming of ectoparasites. *Amer. Nat.* 47:129–58.

Kethley, J. B., and D. E. Johnston. 1975. Resource tracking patterns in bird and mammal ectoparasites. *Misc. Publ. Entomol. Soc. Amer.* 9:231–36.

Key, K. H. L. 1968. The concept of stasipatric speciation. *Syst. Zool.* 17:14–22.

Kiester, A. R., R. Lande, and D. W. Schemske. 1984. Models of coevolution and speciation in plants and their pollinators. *Amer. Nat.* 124:220–43.

Kim, K. C., ed. 1985. *Coevolution of parasitic arthropods and mammals.* New York: Wiley-Intersci.

King, C. E. 1964. Relative abundance of species and MacArthur's model. *Ecology* 45:716–27.

Kingsland, S. 1985. *Modelling nature.* Chicago: Univ. of Chicago Press.

Kingsolver, J. G. 1983. Thermoregulation and flight in *Colias* butterflies: Elevational patterns and mechanistic limitations. *Ecology* 64:534–45.

Klassen, G. J., and M. Beverly-Burton. 1987. Phylogenetic relationships of *Ligictaluridus* spp. (Monogenea: Ancyrocephalidae) and their ictalurid (Siluriformes) hosts: An hypothesis. *Proc. Helminthol. Soc. Wash.* 54:84–90.

————. 1988. North American fresh water Ancyrocephalids (Monogenea) with articulating haptoral bars: Host-parasite coevolution. *Syst. Zool.* 37:179–89.

Kluge, A. G. 1985. Ontogeny and phylogenetic systematics. *Cladistics* 1:13–27.

————. 1988. Parsimony in vicariance biogeography: A quantitative method and a Greater Antillean example. *Syst. Zool.* 37:315–28.

Kluge, A. G., and J. S. Farris. 1969. Quantitative phyletics and the evolution of anurans. *Syst. Zool.* 18:1–32.

Kochmer, J. P., and R. H. Wagner. 1988. Why are there so many kinds of passerine birds? Because they are so small. *Syst. Zool.* 37:68–69.

Kohn, A. J. 1959. The ecology of *Conus* in Hawaii. *Ecol. Mono.* 29:47–90.

Kool, S. 1987. Significance of radular characters in reconstruction of thaidad phylogeny (Neogastropoda: Muricacea). *Nautilus* 101:117–32.

Ladiges, P. Y., and C. J. Humphries. 1986. Relationships in the stringybarks, *Eucalyptus* L'Herit., informal subgenus *Monocalyptus,* series Capitellatae and Olsenianae: Phylogenetic hypotheses, biogeography, and classification. *Austral. J. Bot.* 34:603–31.

Ladiges, P. Y., C. J. Humphries, and M. I. H. Brooker, 1987. Cladistic and biogeographic analysis of western Australian species of *Eucalyptus* L'Herit., informal subgenus *Moncalyptus* Pryor & Johnson. *Austral. J. Bot.* 35:251–81.

Laerm, J. 1974. A functional analysis of morphological variation and differential niche utilization in basilisk lizards. *Ecology* 55:404–11.

Lande, R. 1980a. Genetic variation and phenotypic evolution during allopatric speciation. *Amer. Nat.* 116:463–79.

————. 1980b. Microevolution in relation to macroevolution. *Paleobiology* 6:235–38.

————. 1981. Models of speciation by sexual selection on polygenic traits. *Proc. Nat. Acad. Sci. USA* 78:3721–25.

————. 1982. Rapid origin of sexual isolation and character divergence in a cline. *Evolution* 36:213–23.

Larson, A., D. B. Wake, L. R. Maxson, and R. Highton. 1981. A molecular phylogenetic perspective on the origins of morphological novelties in the salamanders of the tribe Plethodontini (Amphibia, Plethodontidae). *Evolution* 35:405–22.

Lauder, G. V. 1981. Form and function: Structural analysis in evolutionary biology. *Paleobiology* 7:430–42.

———. 1982. Historical biology and the problem of design. *J. Theor. Biol.* 97:57–67.

———. 1983a. Functional and morphological bases of trophic specialization in sunfishes (Teleostei: Centrarchidae). *J. Morphol.* 178:1–21.

———. 1983b. Neuromuscular patterns and the origin of trophic specialization in fishes. *Science* 219:1235–37.

———. 1986. Homology, analogy, and the evolution of behavior. In *Evolution of animal behavior,* ed. M. H. Nitecki and J. A. Kitchell, 9–40. Oxford: Oxford Univ. Press.

———. 1988. Phylogeny and physiology. *Evolution* 42:1113–14.

———. 1989. Caudal fin locomotion in ray-finned fishes: Historical and functional analyses. *Amer. Zool.* 29:85–102.

Lauder, G. V., and K. F. Liem. 1989. The role of historical factors in the evolution of complex organismal functions. In *Complex organismal functions: Integration and evolution in vertebrates,* ed. D. B. Wake and G. Roth, 63–78. London: John Wiley and Sons.

Lawton, J. H., and D. R. Strong. 1981. Community patterns and competition in folivorous insects. *Amer. Nat.* 118:317–38.

Levinton, J. S. 1983. Stasis in progress: The empirical basis of macroevolution. *Ann. Rev. Ecol. Syst.* 14:103–37.

Lewis, D. 1942. The evolution of sex in flowering plants. *Biol. Rev.* 17:46–67.

Lewis, E. R. 1984. On the frog amphibian papilla. *Scanning Electron Micros.* 4:1899–1913.

Lewontin, R. C. 1978. Adaptation. *Sci. Amer.* 239:212–30.

Lichtenfels, J. R., and P. A. Pilitt. 1983. Cuticular ridge patterns of *Nematodirella* (Nematoda: Trichostrongyloidea) of North American ruminants, with a key to species. *Syst. Parasitol.* 5:271–85.

Liebherr, J. K. 1988. General patterns in West Indian insects, and graphical biogeographic analysis of some circum-Caribbean *Platynus* beetles (Carabidae). *Syst. Zool.* 38:385–409.

Liem, K. F. 1973. Evolutionary strategies and morphological innovations: Cichlid pharyngeal jaws. *Syst. Zool.* 22:424–41.

Liem, K. F., and D. B. Wake. 1985. Morphology: Current approaches and concepts. In *Functional vertebrate morphology,* ed. M. Hildebrand, D. M. Bramble, K. F. Liem, and D. B. Wake, 366–77. Cambridge: Harvard Univ. Press.

Linardi, P. M. 1984. Relacoes taxonomicas e filogeneticas entre os generos de sifonapteros ropalopsilinos obtidos do estudo das relacoes hospedeiro/parasito. *Rev. Brasil. Biol.* 44:329–34.

Lindeman, R. L. 1942. The trophic-dynamic aspect of ecology. *Ecology* 23:399–418.

Lorenz, K. 1941. Vergleichende Bewegungsstudien an Anatien. *J. Ornithol.* 89:194–294.

————. 1950. The comparative method in studying innate behaviour patterns. *Symp. Soc. Exp. Biol.* 4:221–68.

————. 1958. The evolution of behavior. *Sci. Amer.* 199:67–78.

Lundberg, J. G. 1970. The evolutionary history of the North American catfishes, family Ictaluridae. Ph.D. diss., Univ. of Michigan, Ann Arbor.

Lynch, J. D. 1975. A review of the Andean Leptodactylid frog genus *Phrynopus*. *Occ. Pap. Mus. Nat. Hist. Univ. Kansas* 35:1–51.

————. 1982. Relationships of the frogs of the genus *Ceratophrys* (Leptodactylidae) and their bearing on hypotheses of Pleistocene forest refugia in South America and punctuated equilibria. *Syst. Zool.* 31:166–79.

————. 1986. Origins of the high Andean herpetological fauna. In *High altitude tropical biogeography,* ed. F. Vuilleumier and M. Monasterio, 478–99. London: Oxford Univ. Press.

————. 1988. Refugia. In *Analytical biogeography: An integrated approach to the study of animal and plant distributions,* ed. A. A. Myers and P. S. Giller, 311–42. London and New York: Chapman & Hall.

————. 1989. The gauge of speciation: On the frequencies of modes of speciation. In *Speciation and its consequences,* ed. D. Otte and J. A. Endler, 527–53. Sunderland, Mass.: Sinauer Assoc.

MacArthur, R. H. 1958. Population ecology of some warblers of northeastern coniferous forests. *Ecology* 39:599–619.

————. 1965. Patterns of species diversity. *Biol. Rev.* 40:510–33.

————. 1972. *Geographical ecology.* New York: Harper & Row.

MacArthur, R. H., and E. O. Wilson. 1967. *The theory of island biogeography.* Princeton, N.J.: Princeton Univ. Press.

McClearn, G. E., and J. C. DeFries. 1973. *Introduction to behavioral genetics.* San Francisco: W. H. Freeman & Co.

McClure, M. S., and P. W. Price. 1975. Competition among sympatric *Erythroneura* leaf-hoppers (Homoptera: Cicadellidae) on American sycamore. *Ecology* 56:1388–97.

McCoy, E. D., and K. L. Heck, Jr. 1976. Biogeography of corals, seagrasses, and mangroves: An alternative to the center of origin concept. *Syst. Zool.* 25:201–10.

Macdonald, C. A., and D. R. Brooks. 1989. Revision and phylogenetic analysis of the North American species of *Telorchis* Luehe, 1899 (Cercomeria: Trematoda: Digenea: Telorchiidae). *Can. J. Zool.* 67:2301–20.

McIntosh, R. P. 1985. *The background of ecology: Concept and theory.* Cambridge: Cambridge Univ. Press.

————. 1987. Pluralism in ecology. *Ann. Rev. Ecol. Syst.* 18:321–41.

McKitrick, M. C., and R. M. Zink. 1988. Species concepts in ornithology. *Condor* 90:1–14.

McLain, D. K., and K. S. Rain. 1986. Reinforcement for ethological isolation: The Southeast Asian *Aedes albipictus* subgroup (Diptera: Culicidae). *Evolution* 40:1346–50.

McLennan, D. A., D. R. Brooks, and J. D. McPhail. 1988. The benefits of communication between comparative ethology and phylogenetic systematics: A case study using gasterosteid fishes. *Can. J. Zool.* 66:2177–90.

McLennan, D. A., and J. D. McPhail. 1989. Experimental investigations of the evolutionary significance of sexually dimorphic nuptial colouration in *Gasterosteus aculeatus* (L.): Temporal changes in the structure of the male mosaic signal. *Can. J. Zool.* 67:1767–77.

―――. 1990. Experimental investigations of the evolutionary significance of sexually dimorphic nuptial colouration in *Gasterosteus aculeatus* (L.): The relationships between male colour and female behaviour. *Can. J. Zool.* 68:482–92.

McMillan, C. 1954. Parallelisms between ecology and systematics. *Ecology* 35:92–94.

McPhail, J. D. 1969. Predation and the evolution of a stickleback (*Gasterosteus*). *J. Fish. Res. Bd. Can.* 26:3183–3208.

Maddison, W. P., M. J. Donoghue, and D. R. Maddison. 1984. Outgroup analysis and parsimony. *Syst. Zool.* 33:83–103.

Maderson, P. F. A. 1982. The role of development in macroevolutionary change: Group report. In *Evolution and development,* ed. J. T. Bonner, 279–312. New York: Springer Verlag.

Maurer, B. A. 1985. On the ecological and evolutionary roles of interspecific competition. *Oikos* 45:300–302.

―――. 1989. Diversity dependent species dynamics: Incorporating the effects of population level processes on species dynamics. *Paleobiology* 15:133–46.

Maurer, B. A., and J. H. Brown. 1988. Distribution of energy use and biomass among species of North American terrestrial birds. *Ecology* 69:1923–32.

May, M. L. 1977. Thermoregulation and reproductive activity in tropical dragonflies of the genus *Micrathyria. Ecology* 58:787–98.

Mayden, R. L. 1985. Biogeography of the Ouachita Highland fishes. *Southwest. Nat.* 30:195–211.

―――. 1986. Speciose and depauperate phylads and tests of punctuated and gradual evolution: Fact or artifact? *Syst. Zool.* 35:591–602.

―――. 1987a. Historical ecology and North American highland fishes: A research program in community ecology. In *Community and evolutionary ecology of North American stream fishes,* ed. W. J. Matthews and D. C. Heins, 210–22. Norman: Univ. of Oklahoma Press.

―――. 1987b. Pleistocene glaciation and historical biogeography of North American Central Highland fishes. In *Quaternary environments of Kansas,* ed. W. C. Johnson, 141–51. Guidebook Series 5. Lawrence: Kansas Geological Survey.

―――. 1988. Biogeography, parsimony, and evolution in North American freshwater fishes. *Syst. Zool.* 37:329–55.

―――. 1989. Phylogenetic studies of North American minnows, with emphasis on the genus *Cyprinella* (Teleostei: Cypriniformes). *Misc. Publ. Univ. Kansas Mus. Nat. Hist.* 80:1–189.

―――, ed. In press. *Systematics, historical ecology, and North American freshwater fishes.* Palo Alto, Calif.: Stanford Univ. Press.

Maynard Smith, J. 1966. Sympatric speciation. *Amer. Nat.* 100:637–50.

Mayr, E. 1942. *Systematics and the origin of species.* New York: Columbia Univ. Press.

————. 1954. Change of genetic environment and evolution. In *Evolution as a process*, ed. J. Huxley, A. C. Hardy, and E. B. Ford, 157–80. London: Allen & Unwin.

————. 1958. Behavior and systematics. In *Behavior and Evolution*, ed. A. Roe and G. G. Simpson, 341–66. New Haven, Conn.: Yale Univ. Press.

————. 1960. The emergence of evolutionary novelties. In *Evolution after Darwin*, ed. S. Tax, 349–80. Chicago: Univ. of Chicago Press.

————. 1963. *Animal species and evolution.* Cambridge: Harvard Univ. Press.

————. 1970. *Populations, species, and evolution.* Cambridge, Mass.: Belknap Press.

————. 1982. Processes of speciation in animals. In *Mechanisms of speciation*, ed. C. Barigozzi, 1–19. New York: Alan R. Liss.

————. 1988. *Toward a new philosophy of biology: Observations of an evolutionist.* Cambridge, Mass.: Belknap Press.

Metcalf, M. M. 1920. Upon an important method of studying problems of relationship and geographical distribution. *Proc. Nat. Acad. Sci. USA* 6:432–33.

————. 1922. The host parasite method of investigation and some problems to which it gives approach. *Anat. Rec.* 23:117.

————. 1923a. The opalinid ciliate infusorians. *Bull. US Nat. Mus.* 120:1–484.

————. 1923b. The origin and distribution of the Anura. *Amer. Nat.* 57:385–411.

————. 1929. Parasites and the aid they give in problems of taxonomy, geographic distribution, and paleogeography. *Smithsonian Misc. Coll.* 81:1–36.

————. 1940. Further studies on the opalinid ciliate infusorians. *Proc. US Nat. Mus.* 87:465–634.

Michener, C. D. 1953. Life-history studies in insect systematics. *Syst. Zool.* 2:112–18.

————. 1967. Diverse approaches to systematics. In *Evolutionary biology*, ed. T. Dobzhansky, 1–38. New York: Appleton-Century-Crofts.

Miller, J. S. 1987. Host-plant relationships in the Papilionidae (Lepidoptera): Parallel cladogenesis or colonization? *Cladistics* 3:105–20.

Milne, M. J., and L. J. Milne. 1939. Evolutionary trends in caddis-worm case construction. *Ann. Entomol. Soc. Amer.* 32:533–42.

Mishler, B. D, and S. P. Churchill. 1984. A cladistic approach to the phylogeny of the "Bryophytes." *Brittonia* 36:406–24.

Mishler, B. D., and M. J. Donoghue. 1982. Species concepts: A case for pluralism. *Syst. Zool.* 31:491–503.

Mitter, C., and D. R. Brooks. 1983. Phylogenetic aspects of coevolution. In *Coevolution*, ed. D. J. Futuyma and M. Slatkin, 65–98. Sunderland, Mass.: Sinauer Assoc.

Mitter, C., B. Farrel, and B. Wiegemann. 1988. The phylogenetic study of adaptive zones: Has phytophagy promoted insect diversification? *Amer. Nat.* 132:107–28.

Miyamoto, M. M. 1985. Consensus cladograms and general classifications. *Cladistics* 1:186–89.

Mode, C. J. 1958. A mathematical model for the co-evolution of obligate parasites and their hosts. *Evolution* 12:158–65.

Möhn, E. 1961. Gallmücken (Diptera, Itonidae) aus El Salvador. 4. Zur Phylogenie der neotropischen und holarktischen Region. *Senck. Biol.* 42:131–330.

Moore, B. 1920. The scope of ecology. *Ecology* 1:3–5.

Moore, J., and D. R. Brooks. 1987. Asexual reproduction in cestodes (Cyclophyllidea: Taeniidae): Ecological and phylogenetic influences. *Evolution* 41:882–91.

Moore, W. S. 1987. Random mating in the northern flicker hybrid zone: Implications for the evolution of bright and contrasting plumage patterns in birds. *Evolution* 41:539–46.

Moran, N. A. 1989. A 48-million-year-old aphid-plant host association and complex life cycle: Biogeographic evidence. *Science* 245:173–75.

Morse, J. C. 1977. Evolution of the caddisfly genus *Ceraclea* in Africa: Implications for the age of Leptoceridae (Trichoptera). In *Proc. 2d Int. Symp. Trichoptera,* 199–205. The Hague: W. Junk.

Morse, J. C., and D. F. White, Jr. 1979. A technique for analysis of historical biogeography and other characters in comparative biology. *Syst. Zool.* 28:356–65.

Moyle, P. B., and J. J. Cech, Jr. 1982. *Fishes: An introduction to ichthyology.* Englewood Cliffs, N.J.: Prentice Hall.

Muller, H. J. 1942. Isolating mechanisms, evolution, and temperature. *Biol. Symp.* 6:71–125.

Murphy, R. W. 1988. The problematic phylogenetic analysis of interlocus heteropolymer isozyme characters: A case study from sea snakes and cobras. *Can. J. Zool.* 66:2628–33.

Nelson, G. 1969. Infraorbital bones and their bearing on the phylogeny and geography of osteoglossomorph fishes. *Amer. Mus. Nat. Hist. Nov.* 2394:1–37.

———. 1972. Cephalic sensory canals, pitlines, and the classification of esocoid fishes, with notes on galaxiids and other teleosts. *Amer. Mus. Nat. Hist. Nov.* 2492:1–49.

———. 1978. Ontogeny, phylogeny, paleontology, and the biogenetic law. *Syst. Zool.* 21:324–45.

———. 1979. Cladistic analysis and synthesis: Principles and definitions, with a historical note on Adanson's *Familles des plantes* (1763–1767). *Syst. Zool.* 28:1–21.

———. 1983. Reticulation in cladograms. In *Advances in cladistics: Proceedings of the second meeting of the Willi Hennig Society,* ed. N. I. Platnick and V. A. Funk, 105–11. New York: Columbia Univ. Press.

———. 1984. Identity of the anchovy *Hildebrandichthys setiger* with notes on relationships and biogeography of the genera *Engraulis* and *Cetengraulis. Copeia* 1984:422–27.

———. 1989. Species and taxa: Systematics and evolution. In *Speciation and its consequences,* ed. D. Otte and J. A. Endler, 60–81. Sunderland, Mass.: Sinauer Assoc.

Nelson, G., and N. I. Platnick. 1981. *Systematics and biogeography: Cladistics and vicariance.* New York: Columbia Univ. Press.

Nitecki, M. H., ed. 1983. *Coevolution.* Chicago: Univ. of Chicago Press.

Noonan, G. R. 1988. Biogeography of North American and Mexican insects, and a critique of vicariance biogeography. *Syst. Zool.* 37:366–84.

Novacek, M. J. 1984. Evolutionary stasis in the elephant-shrew, *Rhynchocyon*. In *Living fossils*, ed. N. Eldredge and S. M. Stanley, 4–22. New York: Springer Verlag.

Novacek, M. J., and L. G. Marshall. 1976. Early biogeographic history of Ostariophysan fishes. *Copeia* 1976:1–12.

O'Connor, B. M. 1984. Co-evolutionary patterns between astigmatid mites and primates. In *Acarology VI*, ed. D. E. Griffiths and C. E. Bowman, 1:19–28. Chicester, England: Ellis Horwood.

————. 1985. Hypoderatid mites (Acari) associated with cormorants (Aves: Phalacrocoracidae), with description of a new species. *J. Med. Entomol.* 22:324–31.

————. 1988. Host associations and coevolutionary relationships of astigmatid mite parasites of New World primates. I. Families Psoroptidae and Audycoptidae. *Fieldiana* 39:245–60.

Odum, E. P. 1969. The strategy of ecosystem development. *Science* 164:262–70.

O'Grady, R. T. 1987. Phylogenetic systematics and the evolutionary history of some intestinal flatworm parasites (Trematoda: Digenea: Plagiorchioidea) of anurans. Ph.D. diss., Univ. of British Columbia, Vancouver.

————. 1989. Parasite-host specificity. In *Parasitology: The biology of animal parasites*, 6th ed., ed. E. R. Noble and G. A. Noble, 495–511. Philadelphia: Lea and Febiger.

Ohta, A. T. 1978. Ethological isolation and phylogeny in the *grimshawi* species complex of Hawaiian *Drosophila*. *Evolution* 32:485–92.

Otte, D., and J. A. Endler, eds. 1989. *Speciation and its consequences*. Sunderland, Mass.: Sinauer Assoc.

Padian, K. 1982. Macroevolution and the origin of major adaptations: Vertebrate flight as a paradigm for the analysis of patterns. *Proc. 3d N. Amer. Paleontol. Conv.* 2:387–92.

Page, R. D. M. 1987. Graphs and generalized tracks: Quantifying Croizat's panbiogeography. *Syst. Zool.* 36:1–17.

————. 1988. Quantitative cladistic biogeography: Constructing and comparing area cladograms. *Syst. Zool.* 37:254–70.

Pagel, M. D., and P. H. Harvey. 1988. Recent developments in the analysis of comparative data. *Quart. Rev. Biol.* 63:413–40.

Paine, R. T. 1966. Food web complexity and species diversity. *Amer. Nat.* 100:65–75.

Parenti, L. R. 1981. A phylogenetic and biogeographic analysis of cyprinodontiform fishes (Teleostei, Atherinomorpha). *Bull. Amer. Mus. Nat. Hist.* 168:341–557.

Park, O. 1945. Observations concerning the future of ecology. *Ecology* 26:1–9.

Park, T., and M. B. Frank. 1948. The fecundity and development of the flour beetles, *Tribolium confusum* and *Tribolium castaneum*, at three constant temperatures. *Ecology* 29:368–74.

Parker, J. S. 1930. Some effects of temperature and moisture upon *Melanoplus mexicanus* Saussure and *Camnula pellucida* Scudder (Orthoptera). *Univ. Montana Agric. Exp. Sta. Bull.* no. 223.

Parrish, J. D., and S. B. Saila. 1970. Interspecific competition, predation, and species diversity. *J. Theor. Biol.* 27:207–20.

Parsons, P. A. 1972. Genetic determination of behaviour (mice and men). In *Genetics, environment, and behaviour,* ed. L. Ehrman, S. Omenn, and E. Caspari, 75–103. New York: Academic Press.

Paterson, H. E. H. 1985. The recognition concept of species. In *Species and speciation,* ed. E. S. Vrba, 21–29. Transvaal Mus. Monogr. 4. Pretoria: Transvaal Museum.

Patten, B. C. 1978. Systems approach to the concept of environment. *Ohio J. Sci.* 78:206–22.

————. 1982. Environs: Relativistic elementary particles for ecology. *Amer. Nat.* 119:179–219.

Patterson, C. 1981. Methods of paleobiogeography. In *Vicariance biogeography: A critique,* ed. G. Nelson and D. E. Rosen, 446–89. New York: Columbia Univ. Press.

————. 1982. Morphological characters and homology. In *Problems of phylogeny reconstruction,* ed. K. A. Joysey and A. E. Friday, 21–74. London: Academic Press.

Patton, J. L., and S. W. Sherwood. 1983. Chromosome evolution and speciation in rodents. *Ann. Rev. Ecol. Syst.* 14:139–58.

Patton, J. L., and M. F. Smith. 1989. Population structure and the genetic and morphologic divergence among pocket gopher species (genus *Thonomys*). In *Speciation and its consequences,* ed. D. Otte and J. A. Endler, 284–304. Sunderland, Mass.: Sinauer Assoc.

Pavan, C., T. Dobzhansky, and H. Burla. 1950. Diurnal behavior of some neotropical species of *Drosophila. Ecology* 31:36–43.

Pelkwijk, J. J. ter, and N. Tinbergen. 1937. Eine reizbiologische Analyse einiger Verhaltensweisen von *Gasterosteus aculeatus.* L. *Z. Tierpsychol.* 1:193–200.

Petrunkevitch, A. 1926. The value of instinct as a taxonomic character in spiders. *Biol. Bull.* 50:427–32.

Pitelka, L. F. 1977. Energy allocation in annual and perennial lupines (*Lupinus:* Leguminosae). *Ecology* 58:1055–65.

Plath, O. E. 1934. *Bumblebees and their ways.* New York: Macmillan.

Platnick, N. I., and G. Nelson. 1978. A method of analysis for historical biogeography. *Syst. Zool.* 27:1–16.

Platt, T. R. 1984. Evolution of the Elaphostrongylinae (Nematoda: Metastrongyloidea: Protostrongylidae) parasites of cervids. *Proc. Helminthol. Soc. Wash.* 51:196–204.

————. 1988. Phylogenetic analysis of the North American species of the genus *Hapalorhynchus* Stunkard, 1922 (Trematoda: Spirorchiidae), blood flukes of freshwater turtles. *J. Parasitol.* 74:870–74.

Powers, J. H. 1909. Are species realities or concepts only? *Amer. Nat.* 43:598–610.

Presch, W. 1980. Evolutionary history of the South American microteiid lizards (Teiidae: Gymnophthalminae). *Copeia* 1980:36–56.

————. 1989. Systematics and science: A comment. *Syst. Zool.* 38:181–89.

Preston, E. M. 1973. A computer simulation of competition among five sympatric congeneric species of xanthid crabs. *Ecology* 54:469–83.

Price, P. W. 1980. *Evolutionary biology of parasites*. Princeton, N.J.: Princeton Univ. Press.

————. 1984. Communities of specialists: Vacant niches in ecological and evolutionary time. In *Ecological communities: Conceptual issues and the evidence*, ed. D. R. Strong, Jr., D. Simberloff, L. G. Abele, and A. B. Thistle, 510–23. Princeton, N.J.: Princeton Univ. Press.

————. 1986. Evolution in parasite communities. *Int. J. Parasitol.* 17:209–14.

Queiroz, K. de. 1985. The ontogenetic method for determining character polarity and its relevance to phylogenetic systematics. *Syst. Zool.* 34:280–99.

Raikow, R. J. 1986. Why are there so many kinds of passerine birds? *Syst. Zool.* 35:255–59.

Ramirez, W. 1974. Specificity of Agaonidae: The coevolution of *Ficus* and its pollinators. Ph.D. diss. Univ. of Kansas, Lawrence.

Rand, A. S. 1964. Ecological distribution in anoline lizards of Puerto Rico. *Ecology* 45:745–57.

Rannala, B. H. In press. Electrophoretic evidence concerning the relationship between *Haplometrana* Lucker, 1931 and *Glypthelmins* Stafford, 1905 (Digenea: Plagiorchiiformes). *J. Parasitol.*

Rau, P. 1929. The habitat and dissemination of four species of *Polistes* wasps. *Ecology* 10:191–200.

————. 1931. Polistes wasps and their use of water. *Ecology* 12:690–93.

Raup, D. M., and S. J. Gould. 1974. Stochastic simulation and evolution of morphology: Towards a nomothetic paleontology. *Syst. Zool.* 23:305–22.

Raup, D. M., S. J. Gould, T. J. M. Schopf, and D. S. Simberloff. 1973. Stochastic models of phylogeny and the evolution of diversity. *J. Geol.* 81:525–42.

Reduker, D. W., D. W. Duszynski, and T. L. Yates. 1987. Evolutionary relationships among *Eimeria* spp. (Apicomplexa) infecting cricetid rodents. *Can. J. Zool.* 65:722–35.

Remane, A. 1956. *Die Grundlagen des natürlichen System der vergleichenden Anatomie und Phylogenetik*. 2d ed. Leipzig: Geest & Portig.

————. 1961. Gedanken zum Problem: Homologie und Analogie, Preadaptation und Parallelität. *Zool. Anz.* 166:447–70.

Resh, V. H., J. C. Morse, and I. D. Wallace. 1976. The evolution of the sponge feeding habit in the caddisfly genus *Ceraclea* (Trichoptera: Leptoceridae). *Ann. Entomol. Soc. Amer.* 69:937–41.

Ricklefs, R. E. 1987. Community diversity: Relative roles of local and regional processes. *Science* 235:167–71.

————. 1989. Speciation and diversity: The integration of local and regional processes. In *Speciation and its consequences*, ed. D. Otte and J. A. Endler, 599–622. Sunderland, Mass.: Sinauer Assoc.

————. 1990. *Ecology*. 3d ed. San Francisco: W. H. Freeman & Co.

Ridley, M. 1983. *The Explanation of organic diversity: The comparative method and adaptations for mating*. Oxford: Clarendon Press.

Riedl, R. 1978. *Order in living organisms.* New York: John Wiley & Sons.

Ringo, J. M. 1977. Why 300 species of Hawaiian *Drosophilia? Evolution* 31:694–96.

Root, R. B. 1967. The niche exploitation theory of the blue-grey gnatcatcher. *Ecol. Mono.* 37:317–50.

Rosen, D. E. 1975. A vicariance model of Caribbean biogeography. *Syst. Zool.* 24:431–64.

———. 1978. Vicariant patterns and historical explanation in biogeography. *Syst. Zool.* 27:159–88.

———. 1979. Fishes from the uplands and intermontane basins of Guatemala: Revisionary studies and comparative biogeography. *Bull. Amer. Mus. Nat. Hist.* 162:267–376.

———. 1985. Geological hierarchies and biogeographic congruence in the Caribbean. *Ann. Missouri Bot. Garden* 72:636–59.

Roskam, J. C. 1985. Evolutionary patterns in gall midge-host plant associations (Diptera, Cecidiomyiidae). *Tijdschr. Entomol.* 128:193–213.

Roskam, J. C., and G. van Uffelen. 1981. Biosystematics of insects living in female birch catkins. III. Plant-insect relations between white birches, *Betula* L., section *Excelsae* (Koch), and gall midges of the genus *Semudobia* Kieffer (Diptera, Cecidiomyiidae). *Neth. J. Zool.* 31:533–53.

Ross, H. H. 1972a. The origin of species diversity in ecological communities. *Taxon* 21:253–59.

———. 1972b. An uncertainty principle in ecological evolution. In *A symposium on ecosystematics,* ed. R. T. Allen and F. C. James, 133–57. Occasional paper 4. Fayetteville: Univ. of Arkansas Museum.

———. 1974. *Biological systematics.* Reading, Mass.: Addison-Wesley.

Ross, M. D. 1978. The evolution of gynodioecy and subdioecy. *Evolution* 32:174–88.

Ross, S. T. 1986. Resource partitioning in fish assemblages: A review of field studies. *Copeia* 1986:352–88.

Roughgarden, J. 1974. Niche width: Biogeographic patterns among *Anolis* lizard populations. *Amer. Nat.* 108:429–42.

Roughgarden, J., and S. Pacala. 1989. Taxon cycle among *Anolis* lizard populations: Review of evidence. In *Speciation and its consequences,* ed. D. Otte and J. A. Endler, 403–31. Sunderland, Mass.: Sinauer Assoc.

Ryan, M. J. 1986. Neuroanatomy influences speciation rates among anurans. *Proc. Nat. Acad. Sci. USA* 83:1379–82.

Rymer, L. 1979. The epistemology of historical ecology. 1. Documentary evidence. *Env. Conserv.* 6:278.

Sabrosky, C. W. 1950. Taxonomy and ecology. *Ecology* 31:151–52.

Saigusa, T., A. Nakanishi, H. Shima, and O. Yata. 1982. Phylogeny and geographical distribution of the swallow-tail subgenus *Graphium* (Lepidoptera: Papilionidae). *Entomol. Gen.* 8:59–69.

Salthe, S. N. 1985. *Evolving hierarchical systems: Their structure and representation.* New York: Columbia Univ. Press.

Savage, J. M. 1982. The enigma of the Central American herpetofauna: Dispersals or vicariance? *Ann. Missouri Bot. Garden* 69:464–547.

Schaefer, S. A., and G. V. Lauder. 1986. Historical transformation of functional design: Evolutionary morphology of feeding mechanisms in loricarioid catfishes. *Syst. Zool.* 35:489–508.

Schmidt, G. D., and L. S. Roberts. 1988. *Foundations of parasitology.* 4th ed. St. Louis: Times Mirror/Mosby College Publishing.

Schmidt-Nielsen, K. 1984. *Scaling: Why is animal size so important?* Cambridge: Cambridge Univ. Press.

Schneirla, T. C. 1952. A consideration of some conceptual trends in comparative psychology. *Psychol. Bull.* 49:559–97.

Schoch, R. M. 1986. *Phylogeny reconstruction in paleontology.* New York: Van Nostrand Reinhold.

Schoener, T. W. 1965. The evolution of bill size differences among sympatric congeneric species of birds. *Evolution* 19:189–213.

———. 1968. Sizes of feeding territories among birds. *Ecology* 49:123–41.

Schoener, T. W., and G. C. Gorman. 1968. Some niche differences in three Lesser Antillean lizards of the genus *Anolis*. *Ecology* 49:819–30.

Schroder, G. D. 1987. Mechanisms for coexistence among three species of *Dipodomys:* Habitat selection and an alternative. *Ecology* 68:1071–83.

Schuh, R. T., and G. M. Stonedahl. 1986. Historical biogeography in the Indo-Pacific: A cladistic approach. *Cladistics* 2:337–55.

Schwenk, K., and G. S. Throckmorton. 1989. Functional and evolutionary morphology of lingual feeding in squamate reptiles: Phylogenetics and kinematics. *J. Zool.* 219:153–75.

Seaman, F. C., and V. A. Funk. 1983. Cladistic analysis of complex natural products: Developing transformation series from sesquiterpene lactone data. *Taxon* 32:1–27.

Semler, D. E. 1971. Some aspects of adaptation in polymorphism for breeding colours in the threespine stickleback (*Gasterosteus aculeatus*). *J. Zool. London* 165:291–302.

Sessions, S. K., and A. Larson. 1987. Developmental correlates of genome size in plethodontid salamanders and their implications for genome evolution. *Evolution* 41:1239–51.

Sillen-Tullberg, B. 1988. Evolution of gregariousness in aposematic butterfly larvae: A phylogenetic analysis. *Evolution* 42:293–305.

Simberloff, D. S. 1971. Population sizes of congeneric bird species on islands. *Amer. Nat.* 105:190–93.

———. 1987. Calculating probabilities that cladograms match: A method of biogeographical inference. *Syst. Zool.* 36:175–95.

———. 1988. Effects of drift and selection on detecting similarities between large cladograms. *Syst. Zool.* 37:56–59.

Simpson, B. B. 1975. Pleistocene changes in the flora of the high tropical Andes. *Paleobiology* 1:273–94.

Simpson, G. G. 1944. *Tempo and mode in evolution.* New York: Columbia Univ. Press.

———. 1953. *The major features of evolution.* New York: Columbia Univ. Press.

Skuhravy, V., M. Skuhrava, and J. W. Brewer. 1983. Ecology of the saddle gall midge *Haplodiplosis marginata* (Von Roser) (Diptera, Cecidomyiidae). *Z. Angew. Entomol.* 96:476–90.

Slobodkin, L. B., and H. L. Sanders. 1969. On the contribution of environmental predictability to species diversity. In *Diversity and stability in ecological systems,* ed. G. M. Woodwell and H. H. Smith, 82–95. Upton, N.Y.: Brookhaven National Laboratory.

Smiley, J. 1978. Plant chemistry and the evolution of host specificity: New evidence from *Heliconius* and *Passiflora. Science* 201:745–46.

Smith, C. C., and A. N. Bragg. 1949. Observations on the ecology and natural history of anura. VII. Food and feeding habits of the common species of toads in Oklahoma. *Ecology* 30:333–49.

Sober, E. 1988. The conceptual relationship of cladistic phylogenetics and vicariance biogeography. *Syst. Zool.* 37:245–53.

Sokal, R. R., and P. H. A. Sneath. 1963. *The principles of numerical taxonomy.* San Francisco: W. H. Freeman & Co.

Southwood, T. R. E. 1961. The number of species of insects associated with various trees. *J. Anim. Ecol.* 30:1–8.

Spieth, H. T. 1974. Mating behavior and evolution of the Hawaiian *Drosophila.* In *Genetic mechanisms of speciation in insects,* ed. M. J. D. White, 94–101. Sydney: Australia & New Zealand Book Co.

Stanley, S. M. 1979. *Macroevolution: Pattern and process.* San Francisco: W. H. Freeman & Co.

Stanley, S. M., P. W. Signor, S. Lidgard, and A. F. Karr. 1981. Natural clades differ from "random" clades: Simulations and analysis. *Paleobiology* 7:115–27.

Stearns, S. C. 1977. The evolution of life history traits: A critique of the theory and a review of the data. *Ann. Rev. Ecol. Syst.* 8:145–71.

———. 1983. The influence of size and phylogeny on patterns of covariation among life-history traits in the mammals. *Oikos* 41:173–87.

Stearns, S. C., and J. Koella. 1986. The evolution of phenotypic plasticity in life-history traits: Predictions of reaction norms for age and size at maturity. *Evolution* 40:893–913.

Stebbins, G. L., and F. J. Ayala. 1981. Is a new evolutionary synthesis necessary? *Science* 213:967–71.

Stenseth, N. C. 1984. Why mathematical models in evolutionary ecology? In *Trends for ecological research for the 1980's,* ed. J. H. Cooley and F. B. Golley, 239–87. New York: Plenum.

Stevens, P. F. 1980. Evolutionary polarity of character states. *Ann. Rev. Ecol. Syst.* 11:333–58.

Stiassny, M. L. J., and J. Jensen. 1987. Labroid interrelationships revisited: Morphological complexity, key innovations, and the study of comparative diversity. *Bull. Mus. Comp. Zool.* 151:269–319.

Stonor, C. R. 1936. The evolution and mutual relationships of some members of the Paradiseidae. *Proc. Zool. Soc. London* 1936:1177–85.

Stratton, G. E., and G. W. Uetz. 1986. The inheritance of courtship behavior and its role as a reproductive isolating mechanism in two species of *Schizocosa* wolf spider (Araneae: Lycosidae). *Evolution* 40:129–41.

Strauch, J. G. 1985. The phylogeny of Alcidae. *Auk* 102:520–39.

Strong, D. R., Jr., D. Simberloff, L. G. Abele, and A. B. Thistle, eds. 1984. *Eco-*

logical communities: Conceptual issues and the evidence. Princeton, N.J.: Princeton Univ. Press.

Sturgeon, K. B., and J. B. Mitton. 1986. Allozyme and morphological differentiation of mountain pine beetles *Dendroctonus ponderosae* Hopkins (Coleoptera: Scolytidae) associated with a host tree. *Evolution* 40:290–302.

Sussman, R. W., and P. H. Raven. 1978. Pollination by lemurs and marsupials: An archaic coevolutionary system. *Science* 200:731–36.

Swofford, D. L., and S. H. Berlocher. 1987. Inferring evolutionary trees from gene frequency data under the principle of maximum parsimony. *Syst. Zool.* 36:293–325.

Swofford, D. L., and W. P. Maddison. 1987. Reconstructing ancestral character states under Wagner parsimony. *Math. Biosci.* 87: 199–229.

Talbot, M. 1934. Distribution of ant species in the Chicago region with reference to ecological factors and physiological toleration. *Ecology* 15:416–39.

———. 1945. A comparison of flights of four species of ants. *Amer. Midl. Nat.* 34:504–10.

———. 1948. A comparison of two ants of the genus *Formica*. *Ecology* 29:316–25.

Tauber, C. A., and M. J. Tauber. 1977a. A genetic model for sympatric speciation through habitat diversification and seasonal isolation. *Nature* 268:702–5.

———. 1977b. Sympatric speciation based on allelic changes at three loci: Evidence from natural populations in two habitats. *Science* 197:1298–99.

———. 1989. Sympatric speciation in insects: Perception and perspective. In *Speciation and its consequences,* ed. D. Otte and J. A. Endler, 307–44. Sunderland, Mass.: Sinauer Assoc.

Taylor, W. R. 1969. A revision of the catfish genus *Noturus* Rafinesque with an analysis of higher groups in the Ictaluridae. *Bull. US Nat. Mus.* 282:1–315.

Telford, M. 1982. Echinoderm spine structure, feeding, and host relationships of four species of *Dissodactylus* (Brachyura: Pinnotheridae). *Bull. Mar. Sci.* 32:584–94.

Templeton, A. R. 1979. Once again, why 300 species of Hawaiian *Drosophila?* *Evolution* 33:513–17.

———. 1980. The theory of speciation by the founder principle. *Genetics* 92:1011–38.

———. 1981. Mechanisms of speciation: A population genetic approach. *Ann. Rev. Ecol. Syst.* 12:23–48.

———. 1982. Genetic architectures of speciation. In *Mechanisms of speciation,* ed. C. Barigozzi, 105–21. New York: Alan R. Liss.

———. 1989. The meaning of species and speciation: A genetic perspective. In *Speciation and its consequences,* ed. D. Otte and J. A. Endler, 3–27. Sunderland, Mass.: Sinauer Assoc.

Thompson, J. N. 1988. Evolutionary genetics of oviposition preference in swallowtail butterflies. *Evolution* 42:1223–34.

Thompson, P., and J. W. Sites, Jr. 1986. Comparison of population structure in chromosomally polytypic and monotypic species of *Sceloporus* (Sauria: Iguanidae) in relation to chromosomally-mediated speciation. *Evolution* 40:303–14.

Thornhill, R., and J. Alcock. 1983. *The evolution of insect mating systems.* Cambridge: Harvard Univ. Press.

Thorson, T. B., D. R. Brooks, and M. A. Mayes. 1983. The evolution of freshwater adaptation in stingrays. *Nat. Geog. Soc. Res. Rep.* 15:663–94.

Tinbergen, N. 1951. *The study of instinct.* London: Oxford Univ. Press.

———. 1953. *Social behaviour in animals: With special reference to vertebrates.* London: Methuen & Co.

———. 1964. On aims and methods of ethology. *Z. Tierpsychol.* 20:410–33.

Turner, J. R. G. 1971. Studies of Müllerian mimicry and its evolution in burnet moths and heliconid butterflies. In *Ecological genetics and evolution,* ed. R. Creed, 224–60. Oxford: Blackwell.

Valentine, J. W., and D. Jablonski. 1983. Larval adaptations and patterns of brachiopod diversity in space and time. *Evolution* 37:1052–61.

Vehrenkamp, S. L., and J. W. Bradbury. 1984. Mating systems and ecology. In *Behavioural ecology: An evolutionary approach,* 2d ed., ed. J. R. Krebs and N. B. Davies, 251–78. Sunderland, Mass.: Sinauer Assoc.

Vermeij, G. J. 1988. The evolutionary success of passerines: A question of semantics? *Syst. Zool.* 37:69–71.

Verschaffelt, E. 1910. The cause determining the selection of food in some herbivorous insects. *Proc. Acad. Sci. Amsterdam* 13:536–42.

Vigneault, G., and E. Zouros. 1986. The genetics of asymmetrical male sterility in *Drosophila mojavensis* and *Drosophila arizonensis* hybrids: Interaction between the Y chromosome and autosomes. *Evolution* 40:1160–70.

Voorzanger, B., and W. J. van der Steen. 1982. New perspectives on the biogenetic law? *Syst. Zool.* 31:202–5.

Vrba, E. S. 1980. Evolution, species, and fossils: How does life evolve? *S. Afr. J. Sci.* 76:61–84.

———. 1983. Macroevolutionary trends: New perspectives on the roles of adaptation and incidental effect. *Science* 221:387–89.

———. 1984a. Evolutionary pattern and process in the sister-group Alcelaphini-Aepycerotini (Mammalia: Bovidae). In *Living fossils,* ed. N. Eldredge and S. M. Stanley, 62–79. New York: Springer Verlag.

———. 1984b. What is species selection? *Syst. Zool.* 33:318–28.

Vrijenhoek, R. C. 1989. Genotypic diversity and coexistence among sexual and clonal lineages of *Peociliopsis.* In *Speciation and its consequences,* ed. D. Otte and J. A. Endler, 386–400. Sunderland, Mass.: Sinauer Assoc.

Wake, D. B., K. P. Yanev, and M. M. Frelow. 1989. Sympatry and hybridization in a "ring species": The plethodontid salamander *Ensatina eschscholtzii.* In *Speciation and its consequences,* ed. D. Otte and J. A. Endler, 134–57. Sunderland, Mass.: Sinauer Assoc.

Wallace, A. R. 1878. *Tropical nature and other essays.* London and New York: Macmillan.

Wallace, B. 1955. Inter-population hybrids in *Drosophila melanogaster. Evolution* 9:302–16.

Wanntorp, H.-E. 1983. Historical constraints in adaptation theory: Traits and nontraits. *Oikos* 41:157–60.

Wanntorp, H.-E., D. R. Brooks, T. Nilsson, S. Nylin, F. Ronqvist, S. C. Stearns, and N. Weddell. 1990. Phylogenetic approaches in ecology. *Oikos* 57:119–32.

Watrous, L. E., and Q. Wheeler. 1981. The outgroup comparison method of character analysis. *Syst. Zool.* 30:1–11.

Wcislo, T. 1989. Behavioral environments and evolutionary change. *Ann. Rev. Ecol. Syst.* 20:137–69.

Weber, B. H., D. J. Depew, C. Dyke, S. N. Salthe, E. D. Schneider, R. E. Ulanowicz, and J. S. Wicken. 1989. Evolution in thermodynamic perspective: An ecological approach. *Biol. Philos.* 4:374–406.

Weitzmann, S. H., and S. V. Fink. 1985. Xenurobryconin phylogeny and putative pheromone pumps in glandulocaudine fishes (Teleostei: Characidae). *Smithsonian Contr. Zool.* 421:1–121.

Weitzmann, S. H., and W. L. Fink. 1983. Relationships of the neon tetras, a group of South American freshwater fishes (teleostei, Characidae), with comments on the phylogeny of New World characiforms. *Bull. Mus. Comp. Zool.* 150:339–95.

West-Eberhard, M. J. 1983. Sexual selection, social competition, and speciation. *Quart. Rev. Biol.* 58:155–83.

————. 1989. Phenotypic plasticity and the origins of diversity. *Ann. Rev. Ecol. Syst.* 20:249–78.

Wheeler, Q., and M. Blackwell, eds. 1984. *Fungus-insect relationships: Perspectives in ecology and evolution.* New York: Columbia Univ. Press.

Wheeler, W. M. 1919. The parasitic *Aculeata*, a study in evolution. *Proc. Amer. Philos. Soc.* 58:1–40.

————. 1928. *The social insects: Their origin and evolution.* New York: Harcourt, Brace & Co.

White, M. J. D. 1978. *Modes of speciation.* San Francisco: W. H. Freeman & Co.

Whitehead, D. R. 1972. Classification, phylogeny, and zoogeography of *Schizogenius* Putzeys (Coleoptera: Carabidae: Scaritini). *Quaest. Entomol.* 8:131–348.

————. 1976. Classification and evolution of *Rhinochenus* Lucas (Coleoptera: Curculionidae: Cryptorhynchinae), and Quaternary Middle American zoogeography. *Quaest. Entomol.* 12:118–201.

Whitman, C. O. 1899. Animal behavior. In *Biological lectures, Woods Hole,* ed. C. O. Whitman, 285–338. Boston: Ginn & Co.

Wicken, J. S. 1987. *Evolution, thermodynamics, and information: Extending the Darwinian paradigm.* New York: Oxford Univ. Press.

Wickler, W. 1961. Ökologie und Stammesgeshichte von Verhaltensweisen. *Fortschr. Zool.* 13:303–65.

Wiens, J. A. 1984. On understanding a non-equilibrium world: Myth and reality in community patterns and processes. In *Ecological communities: Conceptual issues and the evidence,* ed. D. R. Strong, Jr., D. Simberloff, L. G. Abele, and A. B. Thistle, 439–57. Princeton, N.J.: Princeton Univ. Press.

Wiley, E. O. 1976. The phylogeny and biogeography of fossil and recent gars (Actinopterygii: Lepisosteidae). *Misc. Publ. Univ. Kansas Mus. Nat. Hist.* 64:1–111.

————. 1978. The evolutionary species concept revisited. *Syst. Zool.* 27:17–26.

————. 1980a. Is the evolutionary species concept fiction? A consideration of classes, individuals, and historical entities. *Syst. Zool.* 29:76–80.

————. 1980b. The metaphysics of individuality and its consequences for systematic biology. *Brain & Behav. Sci.* 4:302–3.

————. 1981. *Phylogenetics: The theory and practice of phylogenetic systematics.* New York: Wiley-Intersci.

————. 1986a. Approaches to outgroup comparison. In *Systematics and evolution: A matter of diversity,* ed. P. Hovenkamp, 173–91. Utrecht: Univ. of Utrecht Press.

————. 1986b. The evolutionary basis for phylogenetic classification. In *Systematics and evolution: A matter of diversity,* ed. P. Hovenkamp, 55–64. Utrecht: Univ. of Utrecht Press.

————. 1986c. Historical ecology and coevolution. In *Systematics and evolution: A matter of diversity,* ed. P. Hovenkamp, 331–41. Utrecht: Univ. of Utrecht Press.

————. 1986d. Methods in vicariance biogeography. In *Systematics and evolution: A matter of diversity,* ed. P. Hovenkamp, 283–306. Utrecht: Univ. of Utrecht Press.

————. 1986e. Process and pattern: Cladograms and trees. In *Systematics and evolution: A matter of diversity,* ed. P. Hovenkamp, 233–47. Utrecht: Univ. of Utrecht Press.

————. 1986f. La sistemática en la revolución Darwiniana. *An Mus. Hist. Nat. Valparaiso* 17:25–31.

————. 1988a. Parsimony analysis and vicariance biogeography. *Syst. Zool.* 37:271–90.

————. 1988b. Vicariance biogeography. *Ann. Rev. Ecol. Syst.* 19:513–42.

————. 1989. Kinds, individuals, and theories. In *What biology is,* ed. M. Ruse, 289–300. Dordrecht, Netherlands: Kluwer Academic.

Wiley, E. O., and D. R. Brooks. 1982. Victims of history: A nonequilibrium approach to evolution. *Syst. Zool.* 31:1–24.

Wiley, E. O., and R. L. Mayden. 1985. Species and speciation in phylogenetic systematics, with examples from the North American fish fauna. *Ann. Missouri Bot. Garden* 72:596–635.

Wiley, E. O., D. J. Siegel-Causey, D. R. Brooks, and V. A. Funk. 1990. *The compleat cladist: A primer of phylogenetic procedures.* Lawrence: Museum of Natural History, Univ. of Kansas.

Williams, C. B. 1964. *Patterns in the balance of nature.* London: Academic Press.

Wilson, A. C., G. L. Bush, S. M. Case, and M. C. King. 1975. Social structuring of mammalian populations and rate of chromosomal evolution. *Proc. Nat. Acad. Sci. USA* 72:5061–65.

Wilson, D. S. 1980. *The natural selection of populations and communities.* Menlo Park, Calif.: Benjamin Cummings.

————. 1983. The group selection controversy: History and current status. *Ann. Rev. Ecol. Syst.* 14:159–89.

Wilson, E. O. 1969. The species equilibrium. In *Diversity and stability in ecological systems,* ed. G. M. Woodwell and H. H. Smith, 38–47. Upton, N.Y.: Brookhaven National Laboratory.

————. 1971. The plight of taxonomy. *Ecology* 52:741.

Wilz, K. J. 1971. Comparative aspects of courtship behavior in the ten-spined stickleback, *Pygosteus pungitius* (L.). *Z. Tierpsychol.* 29:1–10.

Wood, T. K., and S. I. Guttman. 1983. *Enchenopa binotata* complex: Sympatric speciation? *Science* 220:310–12.

Woodruff, D. S. 1972. The evolutionary significance of hybrid zones in *Pseudophryne* (Anura: Leptodactylidae). Ph.D. thesis, Univ. of Melbourne, Australia.

Wootton, J. T. 1987. The effects of body mass, phylogeny, habitat, and trophic level on mammalian age at first reproduction. *Evolution* 41:732–49.

Wootton, R. J. 1976. *The biology of the sticklebacks.* London: Academic Press.

Wright, D. H. 1983. Species-energy theory: An extension of species-area theory. *Oikos* 41:496–506.

Wright, S. 1931. Evolution in Mendelian populations. *Genetics* 16:97–159.

———. 1940. Breeding structure of populations in relation to speciation. *Amer. Nat.* 74:232–48.

———. 1978a. *Evolution and the genetics of populations.* Vol. 4. *Variability within and among populations.* Chicago: Univ. of Chicago Press.

———. 1978b. Modes of speciation. *Paleobiology* 4:373–79.

Wulff, E. V. 1950. *An introduction to historical plant biogeography.* Waltham, Mass.: Chronica Botanica.

Zandee, M., and M. C. Roos. 1987. Component-compatibility in historical biogeography. *Cladistics* 3:305–32.

Zouros, E. 1986. A model for the evolution of asymmetrical male hybrid sterility and its implications for speciation. *Evolution* 40:1171–84.

Zuckerkandl, E. 1976. Programs of gene action and progressive evolution. In *Molecular anthropology,* ed. M. Goodman, R. E. Tashian, and J. E. Tashian, 387–447. New York: Plenum.

Author Index

Subject Index

adaptation, 3, 15, 19–22, 20, 72–73, 80–87,
133–85, 190, 322, 343–45
coadapted trait complexes, 151–56
components of, 82, 133
conservative nature of, 81, 176–81, 184,
341, 344–46, 371
convergent, 82–85, 156–60, 175–76, 372–
73
discovering constraint and completing studies, 164–69
divergent, 82–85, 160–64, 175–76, 372–
73
formulating the question, 141–46
key adaptations (or key innovations), 130,
181–82, 344–45, 352
speciation and, 80, 95, 100–104, 125, 130,
178
statistical approach, 362–68
temporal sequence of, 146–51
adaptive radiations, 170–83
ecological preferences, 171–73
life cycle patterns, 173–81
species richness, 181–82, 352–54
additive binary coding, 54–55, 57–58, 208
aggression and ritualization in sticklebacks,
evolution of, 153–54
allocation, spatial and resource, 276, 320,
322, 328
alloparapatric speciation. *See* speciation
allopatric
cospeciation model. *See* coevolution
speciation. *See* speciation
anagenesis, 75–77, 88
analogy, 6
apomorphic, 32, 35–36, 46, 68, 144–45,
371–72
arms race model. *See* coevolution

association
by colonization, 193, 205
by descent, 193, 205
multispecies, 20–23, 189, 252–55, 318–
42, 345–46
asymptotic equilibrium model. *See* community evolution
asymptotic nonequilibrium model. *See* community evolution

Bauplan, 18
BPA, 275–76

character
coding, 33, 52–62
defn., 32
evolution, 74–75, 77–78, 89, 146, 141–42
optimization, 33, 134–41, 365–67
polarization, 33, 46–52, 58–62
qualitative, 72, 368
quantitative, 72, 362–68
transformation series, 32–33, 53–56, 141–
42
cladogenesis, 27, 77, 88
cladogram
area, 197, 206–48, 328
host, 206, 248–74, 328
versus phylogenetic tree, 206
coaccommodation, 200–201
coadaptation, 190, 200–203, 277–342
defn., 190, 200–201, 277
directional, 313
reciprocal, 312
coadapted trait complexes, 151–56
coevolution, 189–200, 277–317
allopatric cospeciation model, 201, 280–
83, 285, 290, 292, 300–307

Taxonomic Index

DATE DUE

NOV 2 7 1993	